Storage and Shelf-Life Assessment of Food Products

Storage and Shelf-Life Assessment of Food Products

Guest Editor

Rui M. S. Cruz

MDPI

Basel • Beijing • Wuhan • Barcelona • Belgrade • Novi Sad • Cluj • Manchester

Guest Editor
Rui M. S. Cruz
Department of Food
Engineering
University of the Algarve
Faro
Portugal

Editorial Office
MDPI AG
Grosspeteranlage 5
4052 Basel, Switzerland

This is a reprint of the Special Issue, published open access by the journal *Foods* (ISSN 2304-8158), freely accessible at: https://www.mdpi.com/journal/foods/special_issues/262OAB70X6.

For citation purposes, cite each article independently as indicated on the article page online and as indicated below:

Lastname, A.A.; Lastname, B.B. Article Title. *Journal Name* **Year**, *Volume Number*, Page Range.

ISBN 978-3-7258-5173-7 (Hbk)
ISBN 978-3-7258-5174-4 (PDF)
https://doi.org/10.3390/books978-3-7258-5174-4

Contents

About the Editor

Rui M. S. Cruz

Rui M. S. Cruz holds a Bachelor's degree in Food Engineering (2001) and a Ph.D. in Biotechnology, with expertise in Food Science and Engineering, from the Portuguese Catholic University (2009). He is currently an Adjunct Professor at the Department of Food Engineering, ISE, at the University of Algarve, and a researcher at MED (Mediterranean Institute for Agriculture, Environment and Development) and CHANGE (Global Change and Sustainability Institute). His interests include applying emerging and sustainable technologies to food preservation, especially in developing new biodegradable packaging systems and innovative food processing technologies. Cruz has collaborated with several food companies on the development of new products, the application of processing techniques, and the characterization and determination of the shelf life of various food items. He has evaluated and participated in numerous research projects, authored over 40 research papers and book chapters, and presented multiple posters and oral presentations at national and international conferences. Additionally, he edited five international books on food processing, food analysis, and food packaging. He has reviewed over 3250 research papers and served as an editor for more than 150 research papers in the field of Food Science & Technology. He is also a member of the editorial boards of several international journals, including the *Journal of Food Processing and Preservation, All Life, Foods, Applied Food Research, Frontiers in Food Science and Technology, Journal of Food Biochemistry, Open Agriculture, Food Control, Discover Food,* and *LWT-Food Science and Technology.*

foods

MDPI

Editorial

Storage and Shelf-Life Assessment of Food Products

Rui M. S. Cruz [1,2]

[1] Department of Food Engineering, Institute of Engineering, Universidade do Algarve, Campus da Penha, 8005-139 Faro, Portugal; rcruz@ualg.pt
[2] MED-Mediterranean Institute for Agriculture, Environment and Development and CHANGE-Global Change and Sustainability Institute, Faculty of Sciences and Technology, Campus de Gambelas, Universidade do Algarve, 8005-139 Faro, Portugal

The assessment of food product storage and shelf life is a critical aspect of ensuring food safety, quality, and sustainability in the food industry. This entails a methodical assessment of numerous parameters that affect food-storage circumstances and the time frame over which the desired characteristics of a food product can be preserved. In today's food industry, there is a growing focus on increasing shelf life while reducing the use of artificial additives and preservatives. Innovations in packaging technology, processing techniques, and natural preservatives have resulted from this focus [1–3]. The use of bioactive compounds and bio-based materials to produce effective and sustainable new packaging materials is a new trend to extend the shelf life of food products and help reduce the environmental impact of plastic materials [4–6]. Moreover, food processing and storage techniques aimed at the production of healthier, safer, and higher-quality food products and, consequently, their shelf life extension are another very important area in the food industry [7–10]. Therefore, the aim of this Special Issue was to present new challenges and new technological approaches related to the storage and shelf life of food products.

Zhao et al. (contribution 1) studied the effects of storage temperature (T1 (−18 °C) and T2 (−25 °C)) on the lipidomics profile in Pacific saury (*Cololabis saira*) for three months. The results showed that free fatty acids increased significantly; the content of triglyceride and phospholipid decreased; and the content of total cholesterol was maintained. Moreover, an increasing trend in the peroxide value, acid value, and content of thiobarbituric acid reactive substances was observed after the freezing process. The results showed that lipid oxidation was significantly higher in the group stored at −18 °C compared to the group stored at −25 °C. This study also identified 4854 lipid molecules in *C. saira*. Triglycerides (TG), phosphatidylcholine (PC), phosphatidylethanolamine (PE), phosphatidylinositol (PI), phosphatidylglycerol (PG), and diglycerides (DG) were among the lipid subclasses. The most prevalent lipid was TG, followed by PC. This work will contribute to the quality control of the frozen storage of pelagic fish.

Grigorakis et al. (contribution 2) assessed the quality of wild and farmed tropical fish species from national aquaculture and fisheries in the Saudi Arabian market and, subsequently, compared their freshness/deterioration. The studied fish species were either farmed, such as barramundi (*Lates calcarifer*), snubnose pompano (*Trachinotus blochii*), and sobaity bream (*Sparidentex hasta*), or caught in the wild, including cobia (*Rachycentron canadum*), coral trout (*Plectropomus leopardus*), giant trevally (*Caranx ignobilis*), milkfish (*Chanos chanos*), and mangrove red snapper (*Lutjanus argentimaculatus*). The results showed that fish shelf lives stored at 4 °C ranged from 8 (coral trout) to 18 days (sobaity bream). The wild-caught species always presented a significantly shorter shelf life than the farmed species. At the time of sensory rejection, the microbial load reached or exceeded a level of

 https://doi.org/10.3390/foods14162795

6 log CFU/g while the K-values, depending on the species, ranged from 30 to 80%. The results obtained in this study contributed to freshness control and quality assurance during commercialization in local Saudi Arabian markets and the rest of the world.

The article by Ahmadi et al. (contribution 3) addressed the quality enhancement of stored chilled farmed rainbow trout (*Oncorhynchus mykiss*) using algal (*Cystoseira myrica* and *Cystoseira trinodis*) extract–ice combinations. The icing system was composed of ice containing different kinds of *C. myrica* and *C. trinodis* extracts during a 16-day chilled storage period. The results showed that all alga-treated fish presented lower pH values, as well as decreased lipid hydrolysis (measured through free fatty acid) and oxidation development (evaluated using the thiobarbituric acid index) compared to control samples. In terms of microbiological analyses (including aerobic, psychrophilic, Enterobacteriaceae, proteolytic, and lipolytic counts), most of the fish samples from the alga-treated batches presented lower values; preservation effect was particularly significant at longer storage times. Water and water–ethanol extracts showed greater inhibitory effects than their corresponding ethanol extracts. Furthermore, a higher total polyphenol content was identified in the water and water–ethanol extracts from both algae compared to those obtained solely with ethanol. This technique is an alternative to replace synthetic preservative compounds with those sourced from natural origins, aimed at producing high-quality seafood and food in general.

Custodio-Mendoza et al. (contribution 4) investigated targeted and untargeted carbonyl compounds and volatile organic compounds (VOCs) in adult formulas stored at three different temperatures (room temperature, 4 °C, and 60 °C) for one month. High-protein formulas presented the highest levels of carbonyl compounds, with 80% of samples showing levels of malondialdehyde, acrolein, and formaldehyde. 2-Heptanone and hexanal were also present in samples stored at higher temperatures. The results recommend that, after opening, storage should be performed in the original packaging, guaranteeing light, moisture, and heat protection. Formulas presenting high levels of protein should be kept refrigerated. These results contribute to the consumer safety and quality control of carbonyl compounds and VOCs in adult formulas.

Using gas chromatography–mass spectrometry (GC-MS) and a variety of bioassays, Verešová et al. (contribution 5) investigated the chemical composition and antimicrobial properties of *Rosa damascena* essential oil (RDEO) to determine its potential for food preservation. *Salmonella enterica* inoculated eggplant was treated with *Rosa damascena* essential oil. According to the GC-MS analysis, phenylethyl alcohol, an antimicrobial with aromatic properties, makes up 70% of RDEO. The minimal inhibitory concentration (MIC) was tested against five *Candida* yeast strains, four Gram-positive, and four Gram-negative bacteria, including biofilm-forming *Salmonella enterica*. MIC showed the strongest effects on Gram-negative species, with MIC50 values as low as 0.250 mg/mL for *S. enterica*. Moreover, an in situ assessment using kiwi and banana showed that the vapor phase of RDEO significantly suppressed microbial growth. RDEO was found to have insecticidal activity against *Megabruchidius dorsalis* (Fahraeus, 1839), an inhibitory effect on *S. enterica* during storage, and an antimicrobial effect on the microbiota of sous vide processed eggplant. Applications for RDEO, a natural antimicrobial agent, include food safety and preservation.

The article by Faria et al. (contribution 6) studied the use of locust bean gum/κ-carrageenan film with the extract of blueberry or beetroot as intelligent films for hake (*Merluccius merluccius*) freshness monitoring. At the end of storage, the BLE films' color changed from pink to blue. According to the hake deterioration profile, these results are associated with pH values (from 6.60 ± 0.04 to 8.02 ± 0.03), total viable count (TVC, from 4.61 ± 0.36 to 8.61 ± 0.21 log CFU/g), and total volatile basic nitrogen content (TVB-N, from 10.21 ± 1.97 to 66.78 ± 4.81 mg/100 g) above the spoilage threshold. This study produced a novel bio-based responsive indicator film that incorporates a natural dye, BLE, which could help reduce food waste and enhance food safety.

Komeroski et al. (contribution 7) evaluated the post-harvest storage and shelf life of arugula microgreens (*Eruca sativa*) in different types of packaging (open air, vacuum sealed, and an under-modified atmosphere (MAP)) through microbiological, physico-chemical, and sensory parameters for 10 days at 5 °C. *Salmonella* spp., *Escherichia coli*, and *Listeria monocytogenes* were not observed in arugula microgreens, regardless of the day of storage and packaging used. On the other hand, the total Enterobacteriaceae was around 7 log CFU/g after 10 days of storage for all packaging systems. The results of the total count of mesophilic, psychrotrophic, and molds and yeasts were between 8 and 9 log CFU/g after 10 days of storage for all packaging systems. Open packaging proved to be promising, with less weight loss and slower chlorophyll degradation. The microgreens stored in the vacuum-sealed packaging showed a decrease in quality from the fifth day onwards for all attributes (sensory analysis). On the other hand, MAP showed good scores similar to the fresh microgreens (better visual quality).

In the article by Dermesonlouoglou et al. (contribution 8), the quality of osmo-dehydrofrozen cherry tomatoes was kinetically discussed during frozen storage. The optimum processing conditions were as follows: OD: T_{OD} = 36 °C, t_{OD} = 72 min, and $C_{glycerol}$ + 61.5% w/w. This indicated water loss WL ≤ 5, color change $\Delta E \leq 8$, and reduced water activity (a_w) at the end of the pretreatment. OD-pretreated frozen cherry tomatoes during frozen storage presented an acceptable color, an a_w decrease from 0.95 to 0.92, higher firmness, lower drip loss, and higher vitamin C/lycopene retention. Based on the loss of sensory quality, OD extended the shelf life of frozen cherry tomatoes by up to 3.5 times.

Chen et al. (contribution 9) studied the changes in the quality of kelp gel edible granules during storage. The results showed that the total bacterial counts were maintained within the national standard range, while the hardness and chewiness increased during storage at 4 °C. Conversely, during storage at 25 °C, the total bacterial count exceeded the national standard on the 20th day. Additionally, the gel strength, chewiness, and hardness all increased before declining. Significant water loss, decreased volume, a shift in color from bright green to dark yellow-brown, and a pronounced smell of decaying algae were observed in the product's spoilage. Throughout the storage period, no coliform bacteria were found in any of the products. The product presented a 16-day shelf life at 25 °C, while this exceeded 20 days when stored at 4 °C. Storage at low temperature contributed to the product's quality, maintaining its overall color.

The final article from de Medeiros et al. (contribution 10) is a review regarding the use of mucilage and bioactive compounds from *Cactaceae* to develop innovative and sustainable edible films and coatings that can preserve fruits and vegetables. Important production techniques, technological and functional characteristics, and the chemical composition were all updated in this review. Furthermore, the impact of these edible films and coatings on the fruits and vegetables, as well as market challenges and prospects, was presented.

Future research on new packaging technology, processing techniques, and natural preservatives will be needed to guarantee the quality and shelf life of food products.

Acknowledgments: As Guest Editor of the Special Issue "Storage and Shelf-Life Assessment of Food Products", I would like to thank all authors for their valuable work, which contributed to the success of this Special Issue. Finally, I would like to dedicate this work to you, "My love 🧡".

Conflicts of Interest: The author declares no conflicts of interest.

List of Contributions

1. Zhao, L.; Wang, S.; Liu, Q.; Cao, R.; Zhang, Y.; Su, D.; Yu, Y. Storage Stability and Lipidomic Analysis Reveal the Effect of Frozen Storage Temperature on Pacific Saury (*Cololabis saira*). *Foods* **2025**, *14*, 756. https://doi.org/10.3390/foods14050756
2. Grigorakis, K.; Kogiannou, D.; Kotsiri, M.; Kleidas, I.; de Mello, P.H.; Habiballah, S.; Alshaikhi, A.; Alhafedh, Y.S.; Mohamed, A.H.W. Freshness and Spoilage Patterns of Wild and Farmed Tropical Fish Species with Major Commercial Importance Originating from Saudi Arabian Waters. *Foods* **2025**, *14*, 690. https://doi.org/10.3390/foods14040690
3. Ahmadi, S.; Pourashouri, P.; Shabanpour, B.; Aubourg, S.P. Enhancement of the Storage Potential of Farmed Rainbow Trout (*Oncorhynchus mykiss*) by Using Algal (*Cystoseira myrica* and *Cystoseira trinodis*) Extract–Ice Combinations. *Foods* **2025**, *14*, 371. https://doi.org/10.3390/foods14030371
4. Custodio-Mendoza, J.A.; Rangel Silva, A.; Kurek, M.A.; Almeida, P.J.; Santos, J.R.; Rodrigues, J.A.; Carro, A.M. Assessment of Lipid Peroxidation Products in Adult Formulas: GC-MS Determination of Carbonyl and Volatile Compounds Under Different Storage Conditions. *Foods* **2024**, *13*, 3752. https://doi.org/10.3390/foods13233752
5. Verešová, A.; Vukic, M.D.; Vukovic, N.L.; Terentjeva, M.; Ban, Z.; Li, L.; Bianchi, A.; Kollár, J.; Ben Saad, R.; Ben Hsouna, A.; et al. Chemical Composition, Biological Activity, and Application of *Rosa damascena* Essential Oil as an Antimicrobial Agent in Minimally Processed Eggplant Inoculated with *Salmonella enterica*. *Foods* **2024**, *13*, 3579. https://doi.org/10.3390/foods13223579
6. Faria, C.S.V.; Vieira, J.M.; Vicente, A.A.; Martins, J.T. Locust Bean Gum/κ-Carrageenan Film Containing Blueberry or Beetroot Extracts as Intelligent Films to Monitoring Hake (*Merluccius merluccius*) Freshness. *Foods* **2024**, *13*, 3088. https://doi.org/10.3390/foods13193088
7. Komeroski, M.R.; Beninca, T.; Portal, K.A.; Malheiros, P.S.; Klug, T.V.; Flores, S.H.; Rios, A.O. Postharvest Quality of Arugula (*Eruca sativa*) Microgreens Determined by Microbiological, Physico-Chemical, and Sensory Parameters. *Foods* **2024**, *13*, 3020. https://doi.org/10.3390/foods13193020
8. Dermesonlouoglou, E.; Pittas, L.; Taoukis, P.; Giannakourou, M. Osmodehydrofreezing of Tomatoes: Optimization of Osmotic Dehydration and Shelf Life Modeling. *Foods* **2024**, *13*, 2689. https://doi.org/10.3390/foods13172689
9. Chen, T.; Li, Y.; Wang, Y.; Chen, J.; Fan, L.; Liu, Z. Study on Quality Changes of Kelp Gel Edible Granules during Storage. *Foods* **2024**, *13*, 2267. https://doi.org/10.3390/foods13142267
10. de Medeiros, V.P.B.; de Oliveira, K.Á.R.; Queiroga, T.S.; de Souza, E.L. Development and Application of Mucilage and Bioactive Compounds from *Cactaceae* to Formulate Novel and Sustainable Edible Films and Coatings to Preserve Fruits and Vegetables—A Review. *Foods* **2024**, *13*, 3613. https://doi.org/10.3390/foods13223613

References

1. Michel, M.; Eldridge, A.L.; Hartmann, C.; Klassen, P.; Ingram, J.; Meijer, G.W. Benefits and Challenges of Food Processing in the Context of Food Systems, Value Chains and Sustainable Development Goals. *Trends Food Sci. Technol.* **2024**, *153*, 104703. [CrossRef]
2. Lisboa, H.M.; Pasquali, M.B.; dos Anjos, A.I.; Sarinho, A.M.; de Melo, E.D.; Andrade, R.; Batista, L.; Lima, J.; Diniz, Y.; Barros, A. Innovative and Sustainable Food Preservation Techniques: Enhancing Food Quality, Safety, and Environmental Sustainability. *Sustainability* **2024**, *16*, 8223. [CrossRef]
3. Nunes, C.; Silva, M.; Farinha, D.; Sales, H.; Pontes, R.; Nunes, J. Edible Coatings and Future Trends in Active Food Packaging–Fruits' and Traditional Sausages' Shelf Life Increasing. *Foods* **2023**, *12*, 3308. [CrossRef] [PubMed]
4. Shi, X.; Cui, L.; Xu, C.; Wu, S. Next-Generation Bioplastics for Food Packaging: Sustainable Materials and Applications. *Materials* **2025**, *18*, 2919. [CrossRef] [PubMed]
5. Gururani, P.; Bhatnagar, P.; Dogra, P.; Chandra Joshi, H.; Chauhan, P.K.; Vlaskin, M.S.; Chandra Joshi, N.; Kurbatova, A.; Irina, A.; Kumar, V. Bio-Based Food Packaging Materials: A Sustainable and Holistic Approach for Cleaner Environment—A Review. *Curr. Res. Green Sustain. Chem.* **2023**, *7*, 100384. [CrossRef]
6. Arshad, M.T.; Hassan, S.; Shehzadi, R.; Sani, M.A.; Ikram, A.; Maqsood, S.; Ahmad, A.; Hussain, M.F.; Abdullah, Z.; Gnedeka, K.T. Emerging Trends in Sustainable Packaging of Food Products: An Updated Review. *J. Nat. Fibers* **2025**, *22*, 2505608. [CrossRef]
7. Yuan, Y.; Wei, X.; Mao, Y.; Zheng, Y.; He, N.; Guo, Y.; Wu, M.; Dumpler, J.; Li, B.; Chen, X.; et al. Innovative Food Processing Technologies Promoting Efficient Utilization of Nutrients in Staple Food Crops. *Engineering* **2025**, *50*, 229–244. [CrossRef]
8. Gazda, P.; Glibowski, P. Advanced Technologies in Food Processing—Development Perspective. *Appl. Sci.* **2024**, *14*, 3617. [CrossRef]

9. Can Karaca, A.; Cravotto, G.; Zhao, H.; Amin, T.; Rastrelli, L. Editorial: Advances in Food Processing and Analysis: Product Quality and Green Revolution. *Front. Nutr.* **2025**, *12*, 1559027. [CrossRef] [PubMed]
10. Pandey, V.K.; Dar, A.H.; Rohilla, S.; Mahanta, C.L.; Shams, R.; Khan, S.A.; Singh, R. Recent Insights on the Role of Various Food Processing Operations towards the Development of Sustainable Food Systems. *Circ. Econ. Sustain.* **2023**, *3*, 1491–1514. [CrossRef] [PubMed]

foods

MDPI

Article

Storage Stability and Lipidomic Analysis Reveal the Effect of Frozen Storage Temperature on Pacific Saury (*Cololabis saira*)

Ling Zhao [1,2], Shanyu Wang [2], Qi Liu [2], Rong Cao [2], Yating Zhang [1], Dong Su [1] and Yueqin Yu [1,*]

[1] State Key Laboratory of Eco-Chemical Engineering, College of Chemistry and Molecular Engineering, Qingdao University of Science and Technology, Qingdao 266042, China; zhaoling@ysfri.ac.cn (L.Z.); stella95730@163.com (Y.Z.); ziyousweet@163.com (D.S.)

[2] Yellow Sea Fisheries Research Institute, Chinese Academy of Fishery Sciences, Qingdao 266071, China; wsy8587@foxmail.com (S.W.); liuqi@ysfri.ac.cn (Q.L.); caorong@ysfri.ac.cn (R.C.)

* Correspondence: qustyu@163.com

Abstract: Objectives: This study aimed to assess the effects of storage temperature on the lipidomics profile change in Pacific saury (*Cololabis saira*). Methods: In this paper, *C. saira* underwent frozen storage at two different temperatures, T1 (−18 °C) and T2 (−25 °C), for a duration of three months. Chemical and lipidomic methods were used to determine the changes in lipids during the storage process. Results: Results showed that the content of triglyceride and phospholipid decreased significantly ($p < 0.05$), and free fatty acid increased significantly ($p < 0.05$), while the content of total cholesterol remained relatively constant across different storage temperatures. Additionally, an increasing trend in AV, POV, and TBARS contents was observed after the freezing process, with lipid oxidation being significantly higher in the −18 °C group compared to the −25 °C group ($p < 0.05$). A comprehensive analysis identified 4854 lipid molecules in the muscles of *C. saira*, categorized into 46 lipid subclasses, predominantly including triglycerides (TG), phosphatidylcholine (PC), phosphatidylethanolamine (PE), phosphatidylinositol (PI), phosphatidylglycerol (PG), and diglycerides (DG). Among them, TG was the most abundant lipid, followed by PC. Using orthogonal partial least squares discriminant analysis (OPLS-DA) with a variable importance in projection (VIP) score > 1 and p value < 0.05 as criteria, 338, 271, and 103 highly significantly differentiated lipids were detected in the comparison groups CK vs. T1, CK vs. T2, and T1 vs. T2, respectively. The results indicated that storage at −18 °C had a more pronounced effect than storage at −25 °C. During the freezing process, TG expression was significantly down-regulated, and TG(18:4_14:0_20:5), TG(20:5_13:0_22:6), TG(22:6_14:1_22:6), and TG(18:4_13:0_22:6) were the most predominant individuals. The CK group was initially present in *C. saira* before storage. Differential lipid molecules in the CK vs. T1 and CK vs. T2 groups were screened using a fold change (FC) > 2 or FC < 0.5. In the CK vs. T2 group, 102 highly significant differential lipid molecules were identified, with 55 being down-regulated across seven subclasses. In contrast, the CK vs. T1 group revealed 254 highly significant differential lipid molecules, with 85 down-regulated across 13 subclasses. The results showed that more PCs and PEs were down-regulated, with a higher differential abundance of PE and PC in the −25 °C group compared to the −18 °C group. The differential metabolites were primarily enriched in 17 metabolic pathways, with glycerophospholipid metabolism being the most prominent, followed by sphingolipid metabolism during the frozen storage. Conclusions: Overall, −25 °C storage in production was more favorable for maintaining the lipid stability of *C. saira*. This work could provide useful information for aquatic product processing and lipidomics.

Keywords: *Cololabis saira*; oxidation; frozen storage temperature; lipidomics

1. Introduction

Cololabis saira is a significant pelagic fish, which is widely distributed in the subtropical and temperate waters of the Pacific Ocean along the coasts of Asia and America [1]. In 2022, China's commercial harvest of *C. saira* reached 35,477 tons [2]. *C. saira* is rich in protein, lipids, and other essential nutrients, and is particularly abundant in omega-3 polyunsaturated fatty acids [3–5]. The origin of *C. saira* is far from inland China, and it needs to undergo a long period of freezing and transport. Currently, sales of *C. saira* primarily consists of primary processed products such as frozen items, with fewer deep-processed products available, including convenient ready-to-eat pieces, canned, and boneless *C. saira*.

Freezing is one of the most widely used methods for preserving marine foods [6]. Currently, the standard storage temperature for most aquatic products in China is set at −18 °C. However, this temperature has not been optimal for lipid-rich aquatic products. For instance, the ideal freezing temperature for Atlantic salmon ranges between −45 °C and −60 °C [7]. Similarly, tuna requires storage and transportation at a minimum of −55 °C [8]. These specific temperature requirements are essential to slow down the deterioration in quality.

Although it partially inhibits most of the processes leading to quality deterioration, it does not completely prevent lipid oxidation and unfavorable changes in lipid quality and sensory properties [9]. A series of changes have occurred in fish meat during storage and processing, such as protein denaturation and lipid oxidation [10,11]. Many studies on *C. saira* have focused on lipid and flavor changes during processing [10,12,13] and the effects of natural antioxidants on lipid oxidation [4,12], but little work has been conducted on lipid oxidation of *C. saira* at different freezing temperatures; that is, apart from Tanaka et al. [14], who have investigated the quality changes of *C. saira* under different temperature storage conditions. However, comprehensive lipidomic profile differences due to the storage process had not been analyzed until now. Temperatures of −18 °C and −25 °C are commonly used in actual production. In this study, an untargeted lipidomic strategy was used to investigate the effect of storage temperature on lipid composition and lipid molecules during a −18 °C and −25 °C freezing storage process over three months. We screened the different lipids during storage at different temperatures and explored their metabolic pathways by detecting the dynamic changes in lipid profiles. This study aims to enhance the understanding of lipid profile changes during different frozen storage conditions and provide data support for the actual production of *C. saira*. Additionally, it offers a theoretical basis for the quality control of pelagic fishery catches.

2. Materials and Methods

2.1. Materials

A quantity of 180 kg of *Cololabis saira* was purchased from Penglai Jinglu Fisheries Co., Ltd. (Penglai, China) for this study. It was harvested from the North Pacific Ocean and initially frozen at −35 °C on the fishing vessel. The transportation from the North Pacific Ocean to China typically needed two or three months, during which the temperature was maintained at −18 °C. Upon arrival at Yantai port, *C. saira* were promptly transferred to the laboratory for further processing. The initial *C. saira* was labeled as CK, and the following fish were placed at −18 °C (labeled as T1) and −25 °C (labeled as T2) for frozen storage, respectively. In total, 60 kilogram fish were placed in storage for each group.

2.2. Reagents

Chloroform, methanol, isopropanol, sodium chloride, trichloroacetic acid (TCA), 95% ethanol, sodium hydroxide, sodium thiosulfate, acetic acid, starch indicator, potassium iodide, Triton X100, petroleum ether, thiobarbituric acid (TBA), toluene, copper acetate pyridine, ammonium formate, concentrated nitric acid, perchloric acid, and anhydrous sodium sulfate were analytically pure (Sinopharm Chemical Reagent Co., Ltd., Shanghai, China). MS-grade methanol, MS-grade acetonitrile, and HPLC-grade 2-propanol were purchased from Thermo Fisher Scientific Inc. (Waltham, MA, USA). HPLC-grade formic acid and HPLC-grade ammonium formate were purchased from Sigma Chemical Co. (St. Louis, MO, USA).

2.3. *Lipid Composition and Lipid Oxidation Assay*

The lipid composition of *C. saira* was analyzed at monthly intervals over a three-month period. Referring to the method of Wang et al. [15], 0.1 to 0.2 g of the total lipid sample was added to 20 mL digestive solution for wet digestion, the volume ratio of which (concentrated nitric acid and perchloric acid) was 4:1. Digestion continued until the digestive solution was essentially colorless. The digestion was fixed to 50 mL, and the phospholipid (PL) content in the total lipids was determined by the molybdenum blue colorimetric method. In total, 100 μL total lipid sample was diluted to the appropriate concentration. The diluted solution was made of isopropanol and TritonX100; its volume mass ratio was 9:1. Triglyceride (TG) and total cholesterol (TC) content were determined by following the kit instructions. Amounts of 5 mL of toluene and 1 mL of copper reagent (pH 6.1) were added to 0.10 g of the total lipid sample, and shaken for 2 min. Centrifugation was performed at 1000× *g* for 5 min, and the supernatant was taken to determine the Optical Density (OD) value at 715 nm. The standard curve was plotted with oleic acid to calculate the free fatty acid (FFA) content in the *C. saira* samples. Lipid oxidation of *C. saira* samples were assessed by measuring the acid value (AV), peroxide value (POV), and thiobarbituric acid reactive substances (TBARSs). AV was determined according to a thermal ethanol method of the Chinese standard GB 5009.229-2016 [16]. POV was determined by a titrimetric method of the Chinese standard GB 5009.227-2016 [17]. TBARSs were determined according to the spectrophotometric method of the Chinese standard GB 5009.181-2016 [18]. The detailed procedures of the assay methods can be found in the Supplementary Materials.

2.4. *Determination of Lipid Profile*

2.4.1. Lipid Extraction

The lipidomic analysis was conducted at the end of the storage period (the third month), and the tissue of *C. saira* was rapidly frozen using liquid nitrogen and subsequently stored at −80 °C. Lipids of *C. saira* were extracted according to the Methyl tert-Butyl Ether (MTBE) method [19]. Briefly, a 200 μL volume of water was added to the sample, and then the sample was vortexed for 5 s. Subsequently, 240 μL of precooling methanol was added and the mixture was vortexed for 30 s. After that, 800 μL MTBE was added and the mixture was ultrasonicated for 20 min at 4 °C followed by sitting still for 30 min at room temperature. The solution was centrifuged at 14,000× *g* for 15 min at 10 °C and the upper organic solvent layer was obtained and dried under nitrogen.

2.4.2. LC-MS/MS Method for Lipid Analysis

Reverse-phase chromatography was selected for LC separation using a CSH C18 column (1.7 μm, 2.1 mm × 100 mm, Waters Corporation, Milford, MA, USA). The lipid extracts were re-dissolved in 200 μL 90% isopropanol/acetonitrile, centrifuged at 14,000× *g* for 15 min, and, finally, 3 μL of sample was injected. Solvent A was acetonitrile–water (6:4, *v*/*v*) with 0.1% formic acid and 0.1 mM ammonium formate and solvent B was acetonitrile–isopropanol (1:9, *v*/*v*) with 0.1% formic acid and 0.1 mM ammonium formate. The initial mobile phase was 30% solvent B at a flow rate of 300 μL/min. It was held for 2 min, and then linearly increased to 100% solvent B in 23 min, followed by equilibrating at 5% solvent B for 10 min.

Mass spectra were acquired by Thermo Fisher Q-Exactive Plus in positive and negative modes. ESI parameters were optimized and preset for all measurements as follows: source temperature, 300 °C; capillary temp, 350 °C; the ion spray voltage for positive and negative ions was set at 3000 V; the S-Lens RF Level was set at 50%; and the scan range of the instruments was set at m/z 200–1800.

2.4.3. Identification by Lipid Search

Lipid Search software version 4.2 (Thermo Scientific™) was used to identify the lipid molecules and lipid species based on MS/MS calculations [20]. It contains more than 30 lipid classes and more than 1,500,000 ion fragments in the database. Adducts of +H and $+NH_4$ were selected for positive mode searches, and −H and $+CH_3COO$ were selected for negative mode searches since ammonium acetate was used in the mobile phases. Both mass tolerances for the precursor and fragment were set to 5 ppm, with a product ion threshold of 5%.

2.5. Statistical Analysis

The experimental data for lipid composition and oxidation were analyzed using SPSS 22.0 software, and significant differences were analyzed with Duncan's multiple range test at $p < 0.05$. All lipidomic measurements were conducted in eight replicates. Regarding the data extracted from Lipid Search, lipids with a relative standard deviation (RSD) greater than 30% and missing values greater than 50% in each group were removed. After normalization and integration using the Perato scaling method, the processed data were imported into SIMPCA-P 16.0 (Umetrics, Umea, Sweden) for principal component analysis (PCA).

3. Results

3.1. Lipid Composition Change and Lipid Oxidation Analysis of C. saira *at Different Storage Temperatures*

Figure 1A illustrates the lipid composition of *C. saira* at various storage temperatures, providing insights into the evolution of lipids during the frozen storage process. Similarly to most marine fish, the lipid composition of *C. saira* was predominantly comprised of TG and PL, with smaller quantities of FFA and very low levels of TC. During storage, the contents of TG and PL decreased significantly ($p < 0.05$), while FFA levels increased significantly ($p < 0.05$). However, the TC content remained relatively constant at different storage temperatures. Notably, the changes in lipid composition were more pronounced in the T1 group (−18 °C) compared to the T2 group (−25 °C), suggesting that storage temperature significantly influences lipid composition. Consequently, a storage temperature of −25 °C was more suitable for the long-term frozen storage of *C. saira*.

To assess the levels of primary and secondary oxidation products during the lipid oxidation process, AV, POV, and TBARS were utilized as key indicators [21]. Figure 1 illustrates the lipid oxidation of *C. saira* at various frozen storage temperatures. It is evident that lipid oxidation was typically unavoidable during frozen storage, leading to significant increases in the AV, POV, and TBARS values of *C. saira*.

Figure 1B illustrates the changes in the AV of *C. saira* at various storage temperatures, revealing a gradual increase in AV with the extension of the storage period. Notably, the AV in the T1 group increased significantly more than in the T2 group ($p < 0.05$). This observation aligned with the results from the lipid composition analysis, suggesting a potential link to the hydrolysis of TC and PL. Furthermore, the rate of lipid hydrolysis was influenced by several factors, including the moisture content and its state within the samples, as well as the activity of endogenous lipase and other variables [22,23]. POV was an index used to determine the content of primary oxidation products, such as hydroperoxides, which can reflect the degree of the primary oxidation of lipids [24]. Aquatic products, such as *C. saira*, are rich in unsaturated fatty acids, which were prone to oxidative decomposition during storage, leading to the formation of malondialdehyde. The TBARS assay was utilized to evaluate lipid secondary oxidation by measuring the malondialdehyde content produced during this process [25]. Lipid oxidation was generally unavoidable during transportation or storage, resulting in significant increases in both POV and TBARS values in *C. saira*. Figure 1C,D illustrate the changes in POV and TBARSs, respectively, at different storage temperatures. The POV and TBARS values of the T1 group increased significantly compared to the T2 group ($p < 0.05$).

Overall, an increasing trend in AV, POV, and TBARS contents was observed during the freezing process. These results indicated that storage at −25 °C is more conducive to the preservation of *C. saira*.

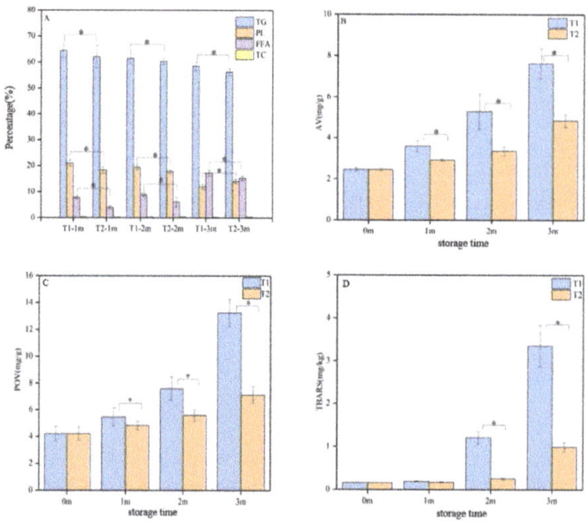

Figure 1. Lipid composition (**A**) and lipid oxidation (**B–D**) analysis of *C. saira* at different storage temperatures; asterisks indicate statistical significance (* $p < 0.05$).

3.2. Lipidomics Analysis of *C. saira* at Different Storage Temperatures

During the storage process, lipid oxidation can be influenced by various factors, including temperature, duration, and packaging. When lipid peroxidation occurs, the appearance of *C. saira* may become tawny, and it may develop an unpleasant odor or a hard texture, which negatively impacts consumer sensory acceptability. Therefore, understanding the fundamental processes and mechanisms of lipid oxidation is crucial for ensuring the storage stability of *C. saira*.

Table 1 presented the statistical results obtained from both positive and negative ion modes. In the muscles of *Cololabis saira*, a total of 4854 lipid molecules were identified and categorized into 46 lipid subclasses. These subclasses primarily included triglycerides (TG), phosphatidylcholine (PC), phosphatidylethanolamine (PE), phosphatidylinositol (PI), phosphatidylglycerol (PG), and diglycerides (DG). Notably, TG was the most abundant lipid in *C. saira*, serving as the primary storage form of fatty acids in organisms, as well as an essential energy source and carrier [26]. PC was the second most prevalent lipid, recognized as a major component of fish phospholipids, and it plays a critical role in maintaining cell membrane permeability and structural integrity [27].

Table 1. Number of lipid subclasses and compounds.

Lipid Subclass	Number of Compounds
Triglycerides, TG	1182
Phosphatidylcholine, PC	430
Phosphatidylethanolamine, PE	409
Phosphatidylinositol, PI	330
Phosphatidylglycerol, PG	327
Diglyceride, DG	316
Phosphatidylserine, PS	224
Cardiolipin, CL	205
Ceramide, Cer	127
Sulfatide, ST	108
Hexosylceramide, Hex1Cer	97
Sphingomyelin, SM	75
Dihexosylceramide, Hex2Cer	74
Lysophosphatidylcholine, LPC	58
Simple Glc series, Hex3Cer	51
N-acetylhexosyl ceramide, CerG2GNAc1	45
Phytosphingosine, phSM	44
Lysophosphatidylethanolamine, LPE	39
Phosphatidylinositol phosphate, PIP	29
Phosphatidylinositol bisphosphate, PIP2	29
Wax exters, WE	26
Aacyl carnitine, AcCa	22
Sphingosine bases, SPH	22
Monoglyceride, MG	21
Phosphatidic acid, PA	20
Ganglioside, monosialo trihexosyl ceramide, GM3	16
Lysophosphatidylglycerol, LPG	14
Ceramides phosphate, CerP	13
Fatty acid, FA	13
Lysophosphatidylinositol, LPI	10
Lysophosphatidylserine, LPS	9
Zymosterol ester, ZyE	8
(O-acyl)-1-hydroxy fatty acid, OAHFA	4
Cholesteryl esters, ChE	3
Coenzyme, Co	3
Phosphatidylinositol triphosphate, PIP3	3
Ganglioside, disialo tetrahexosyl ceramide, GD1a	2
Ganglioside, disialo dihexosyl ceramide, GD2	2
monosialotetrahexosyl ganglioside, GM1	2
Ganglioside, trisialo trihexosyl, GT3	2
Sphin-gosine phosphate, SPHP	2
Dihexosyl N-acetylhexosyl ceramide, CerG3GNAc1	1
Disialoganglioside, GD3	1
Lysophosphatidic acid, LPA	1
Sitosteryl ester, SiE	1
Stigmasterol ester, StE	1

3.3. Multivariate Statistical Analysis of C. saira at Different Storage Temperatures

In order to elucidate the changes in the lipid profile of *C. saira* under different storage temperatures and identify potential differences in lipid types, a principal component analysis (PCA) model was established, as illustrated in Figure 2. The figure demonstrated the distribution of the lipid profiles in the three groups of *C. saira*, from which it could be seen that the distribution of the same group of *C. saira* was concentrated in a region. The samples of TI and T2 groups were dispersed in different regions and far away from each other, which indicated that there were significant differences in the lipid profiles of the T1 and T2 group of *C. saira*, and further confirmed that storage temperature has a substantial effect on the lipid profiles of *C. saira*.

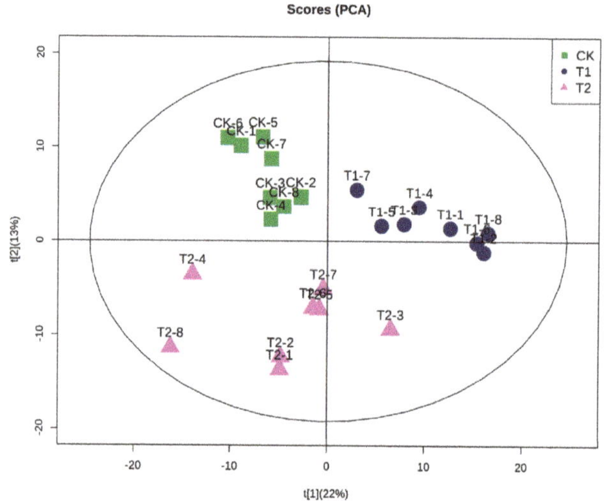

Figure 2. Principal component analysis.

3.4. Differentially Abundant Lipid Analysis

In lipidomics studies, a high Variable Importance in Projection (VIP) score combined with a low *p*-value has indicated significant differences in target lipid samples [28]. By employing Orthogonal Partial Least Squares Discriminant Analysis (OPLS-DA) with VIP > 1 and *p*-value < 0.05 as the criteria, we identified highly significantly differentiated lipid molecules (Figure 3). Specifically, 338, 271, and 103 highly significantly differentiated lipids were detected in the comparison groups CK vs. T1, CK vs. T2, and T1 vs. T2, respectively. Figure 3 presented volcano plots of the three comparison groups, illustrating the changes in lipid compounds. In these plots, each point represents a lipid compound, with larger values along the horizontal and vertical axes indicating more significant alterations in the selected lipid compounds. The figure reveals that the relative contents of most lipid metabolites were either significantly up-regulated or down-regulated across the three sample groups. These findings suggested that the most pronounced differences in the lipid composition of *C. saira* occurred during T1 storage. The results indicated that the T1 group storage had a more substantial impact on lipid quality compared to the T2 group.

The lipid molecules that were either up-regulated or down-regulated in the CK vs. T1 and CK vs. T2 groups were identified using a fold change (FC) criterion of >2 or <0.5. The analysis revealed that the down-regulation of triglycerides (TG) was the main reason for the changes in the lipid profiles of *C. saira* during storage, and thus the lipid compounds that were down-regulated lipid molecules in the different comparison groups were listed in Tables 2 and 3, separately. In the CK vs. T2 group, 102 highly significant differential lipid molecules were detected, with 55 being down-regulated. These down-regulated molecules belonged to seven subclasses, including 40 species of TG, 4 species of sphingomyelins (SM), 3 species of sterols (ST), 3 species of sphingolipids (SPH), 3 species of PI, 1 species of PG, and 1 species of phosphatidylserine (PS). In comparison, the CK vs. T1 group exhibited 254 highly significant differential lipid molecules, with 85 being down-regulated across 13 subclasses. These included 33 species of TG, 17 species of PE, 11 species of phosphatidylcholines (PC), 5 species of PI, 5 species of ST, 4 species of PG, 2 species of SPH, 2 species of SM, 2 species of cardiolipins (CL), 1 species of LPE, 1 species of Dihexosylceramide (Hex2Cer), 1 species of DG, and 1 species of Phosphatidylserine (PS). This suggested a more complex lipid metabolism in this reservoir. These lipid compounds could also serve as differentiating factors. The findings indicated that the lipid profiles of *C. saira* were significantly influenced by different storage temperatures, aligning with previous studies on lipid content and oxidation.

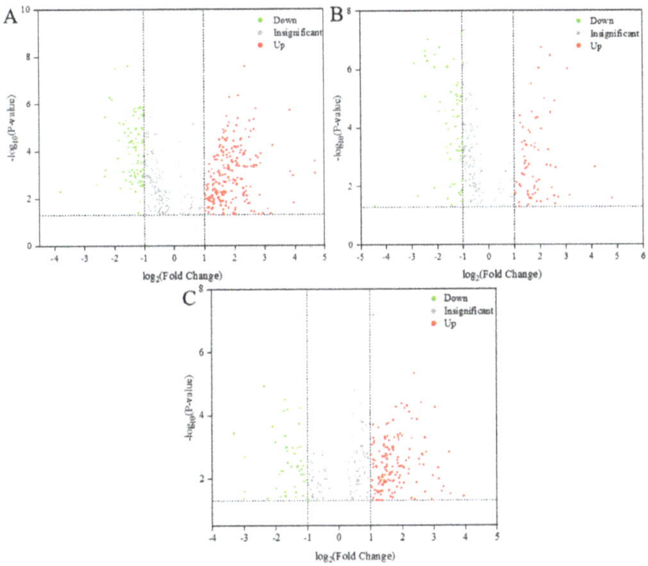

Figure 3. Volcano plot of differential lipid molecules in CK versus T1 sample (**A**), CK versus T2 sample (**B**), T1 versus T2 sample (**C**).

Table 2. Changes in significantly different lipid molecules in CK vs. T1 group during storage process.

Number	Lipid Ion	Class	Fold Change	*p*-Value	VIP
1	PE(18:1p_22:6)+H	PE	0.499098	6.44×10^{-5}	1.042086
2	PC(39:6)+H	PC	0.496688	8.43×10^{-5}	1.38837
3	TG(16:0_14:4_22:6)+NH_4	TG	0.492149	2.84×10^{-6}	1.881346
4	TG(20:5_13:0_22:6)+NH_4	TG	0.488128	3.24×10^{-6}	3.105524
5	TG(22:6_14:1_22:6)+NH_4	TG	0.487257	0.000175	3.042199
6	PE(16:0_22:6)−H	PE	0.486444	1.94×10^{-6}	2.852529
7	PE(35:2e)+Na	PE	0.48616	0.000594	1.108355
8	SM(d20:0_24:5)+H	SM	0.480833	0.00303	1.631861
9	CL(78:11)−2H	CL	0.480744	1.49×10^{-6}	2.871019
10	TG(22:6_14:3_22:6)+NH_4	TG	0.47991	0.002727	1.572998
11	PS(43:1e)+H	PS	0.478902	3.56×10^{-5}	2.906576
12	PC(42:9)+H	PC	0.47835	0.001059	1.089234
13	TG(18:4_14:0_20:5)+NH_4	TG	0.477786	0.001276	9.197348
14	PC(22:3_14:1)+H	PC	0.47614	1.19×10^{-5}	1.162553
15	TG(18:4_14:0_18:4)+NH_4	TG	0.475079	4.68×10^{-5}	6.977423

Table 2. *Cont.*

Number	Lipid Ion	Class	Fold Change	*p*-Value	VIP
16	PI(16:0_22:6)−H	PI	0.469722	0.000323	1.639228
17	PC(22:1_11:2)+Na	PC	0.467923	1.05×10^{-5}	4.199848
18	PG(43:3e)+NH_4	PG	0.466328	3.02×10^{-6}	3.178235
19	PE(10:0e_6:0)+H	PE	0.463883	0.002008	1.500992
20	ST(d17:1_25:0)+NH_4	ST	0.457943	0.000698	1.70078
21	PI(40:0e)+H	PI	0.457439	0.000708	1.704323
22	PC(17:1_22:6)+HCOO	PC	0.455686	1.43×10^{-6}	1.153681
23	PE(16:0_18:2)+Na	PE	0.454466	0.003023	1.162741
24	TG(18:4_12:0_18:4)+NH_4	TG	0.444249	0.001779	2.19355
25	TG(18:4_18:0_22:6)+H	TG	0.441096	4.30×10^{-6}	1.808455
26	PE(18:1p_20:5)+H	PE	0.440797	2.07×10^{-5}	1.528535
27	TG(18:4_18:4_22:6)+H	TG	0.439262	1.60×10^{-6}	1.637022
28	PG(32:2e)+NH_4	PG	0.436992	1.91×10^{-5}	1.147128
29	TG(18:4_18:3_20:5)+NH_4	TG	0.435734	0.000153	4.625362
30	LPE(16:0)−H	LPE	0.435246	0.002236	2.086989
31	TG(22:6_12:2_22:6)+NH_4	TG	0.431817	0.000265	2.296603
32	PG(18:1e_23:1)+NH_4	PG	0.427813	4.44×10^{-5}	1.735106
33	TG(18:4_14:1_18:4)+NH_4	TG	0.425433	0.000918	1.645381
34	PC(20:5_22:6)+HCOO	PC	0.424829	0.000338	1.648933
35	PC(44:11)+H	PC	0.417744	0.000639	1.998421
36	PC(22:6_22:6)+HCOO	PC	0.414333	1.13×10^{-5}	2.111974
37	TG(18:4_13:0_22:6)+NH_4	TG	0.411841	2.24×10^{-6}	3.32173
38	TG(18:4_18:4_20:5)+NH_4	TG	0.411485	6.69×10^{-5}	5.206043
39	TG(18:4_18:3_18:4)+NH_4	TG	0.411481	9.19×10^{-5}	2.389019
40	TG(18:4_13:0_20:5)+NH_4	TG	0.408598	1.28×10^{-5}	1.884269
41	PE(18:2e_22:6)−H	PE	0.4084	2.98×10^{-5}	1.0287
42	PE(18:1_22:6)−H	PE	0.404044	0.005235	1.597052
43	CL(82:13)−2H	CL	0.403207	0.005028	1.601206
44	TG(20:4_22:6_23:1)+NH_4	TG	0.402977	0.001303	1.084609
45	ST(m42:0 + O)+NH_4	ST	0.401593	7.04×10^{-5}	4.169806
46	ST(d15:0_25:0)+NH_4	ST	0.398087	0.001072	1.264458
47	PI(33:1)+NH_4	PI	0.397697	1.66×10^{-6}	1.082869
48	DG(21:5e)+NH_4	DG	0.39523	0.000806	1.177416
49	PG(31:2e)+NH_4	PG	0.390305	3.11×10^{-6}	1.162939
50	TG(16:0_14:1_22:6)+H	TG	0.389657	4.88×10^{-5}	1.350519
51	TG(18:4_12:0_20:5)+NH_4	TG	0.387985	2.06×10^{-6}	2.156707
52	TG(22:4_12:4_22:4)+NH_4	TG	0.3853	7.13×10^{-5}	2.344739
53	PC(20:5_22:6)+H	PC	0.383516	0.00035	1.10243
54	PC(17:1_22:6)+H	PC	0.375831	0.002524	2.368831
55	PC(37:5)+H	PC	0.372954	3.74×10^{-5}	2.477881
56	TG(18:4_14:3_22:6)+NH_4	TG	0.372685	0.000631	1.676399
57	PI(42:1e)+H	PI	0.364115	1.67×10^{-5}	1.105667
58	SPH(m18:1)+H	SPH	0.363043	4.33×10^{-5}	1.437811
59	PE(16:0_20:5)−H	PE	0.362863	0.001141	1.06783
60	PE(39:2)+Na	PE	0.359881	0.003976	1.106558
61	PE(29:1_11:3)+Na	PE	0.356409	0.00195	2.475733
62	TG(18:4_18:4_18:4)+NH_4	TG	0.350723	4.43×10^{-5}	3.849431
63	PE(20:1_22:6)−H	PE	0.348642	0.000763	1.674806
64	TG(18:4_14:2_20:5)+NH_4	TG	0.348242	0.000135	1.611142
65	TG(11:0_18:4_22:6)+NH_4	TG	0.344356	2.29×10^{-5}	1.024028
66	PE(16:1e_22:6)−H	PE	0.343037	1.16×10^{-5}	1.600446
67	TG(18:4_14:3_20:5)+NH_4	TG	0.342678	7.13×10^{-5}	1.382108
68	TG(22:1_10:4_22:6)+H	TG	0.339259	2.49×10^{-8}	1.694093

Table 2. *Cont.*

Number	Lipid Ion	Class	Fold Change	p-Value	VIP
69	ST(d17:1_24:0)+NH_4	ST	0.324952	5.26×10^{-6}	1.074646
70	TG(20:5_12:1_22:6)+H	TG	0.324425	0.001746	1.569025
71	SPH(d20:0)+H-H_2O	SPH	0.314818	1.68×10^{-5}	1.334997
72	ST(d48:3)+Na	ST	0.306939	7.41×10^{-6}	1.166141
73	PE(18:2_22:6)+Na	PE	0.305299	0.000763	1.272167
74	TG(20:2_14:4_22:5)+NH_4	TG	0.285599	0.003817	1.646126
75	PE(22:6_22:6)−H	PE	0.280681	6.49×10^{-6}	2.91534
76	PE(20:3_22:6)+Na	PE	0.27609	3.58×10^{-5}	3.615534
77	PE(20:5_22:6)−H	PE	0.271036	0.000203	1.163258
78	TG(18:4_20:5_22:6)+H	TG	0.255424	3.27×10^{-8}	1.147957
79	TG(20:5_14:2_22:6)+H	TG	0.236534	6.83×10^{-7}	1.309602
80	TG(18:1_11:3_21:0)+Na	TG	0.224134	5.49×10^{-7}	1.208512
81	TG(18:4_18:4_20:5)+H	TG	0.202142	0.000627	2.141881
82	TG(22:6_11:2_21:0)+H	TG	0.201635	3.72×10^{-6}	2.215741
83	Hex2Cer(m39:3)+H-H_2O	Hex2Cer	0.199907	0.001191	1.061588
84	SM(d20:0_22:0)+H	SM	0.166036	0.002705	8.710905
85	PI(25:1_11:2)+NH_4	PI	0.071681	0.005298	7.23351

Table 3. Changes in significantly different lipid molecules in CK vs. T2 group during storage process.

Number	LipidIon	Class	Fold Change	p-Value	VIP
1	TG(20:5_22:1_22:6)+NH_4	TG	0.49851	5.68×10^{-7}	5.215054
2	TG(20:5_22:6_22:6)+NH_4	TG	0.491599	0.000221	2.567459
3	TG(18:1_20:5_22:6)+NH_4	TG	0.490677	6.05×10^{-8}	4.438805
4	TG(20:0_22:3_23:0)+Na	TG	0.489055	7.87×10^{-6}	4.994488
5	TG(24:1_22:1_24:1)+NH_4	TG	0.486977	0.006783	1.564398
6	TG(22:6_14:1_22:6)+NH_4	TG	0.486156	0.000355	2.537809
7	TG(18:4_16:0_20:5)+NH_4	TG	0.482545	4.57×10^{-6}	9.422562
8	TG(18:4_18:4_20:5)+NH_4	TG	0.475521	0.000612	3.973146
9	TG(20:5_14:1_22:6)+NH_4	TG	0.471436	8.49×10^{-6}	1.658027
10	TG(22:4_12:4_22:4)+NH_4	TG	0.470116	0.000904	1.769338
11	TG(26:1_22:1_22:6)+NH_4	TG	0.467412	0.001272	1.034124
12	TG(20:5_18:2_22:6)+NH_4	TG	0.460452	1.21×10^{-5}	3.218636
13	TG(24:1_22:1_22:1)+NH_4	TG	0.456162	0.000451	4.562797
14	ST(m42:0 + O)+NH4	ST	0.451775	0.000622	3.284288
15	TG(18:4_16:1_20:5)+NH_4	TG	0.449565	3.97×10^{-5}	4.416179
16	TG(20:5_20:5_22:6)+NH_4	TG	0.440813	2.87×10^{-6}	5.177796
17	TG(22:5_20:4_22:6)+NH_4	TG	0.433843	0.000243	1.074286
18	TG(22:1_22:1_22:1)+NH_4	TG	0.432794	8.21×10^{-6}	9.731679
19	ST(d46:2)+NH_4	ST	0.432073	8.33×10^{-6}	5.279914
20	SM(d18:2_24:1)+H	SM	0.430173	4.26×10^{-5}	1.308388
21	TG(18:4_20:5_22:6)+NH_4	TG	0.429418	1.56×10^{-5}	6.177515
22	TG(14:0_20:5_22:6)+NH_4	TG	0.429232	1.12×10^{-5}	2.991447
23	TG(20:1_11:3_22:6)+NH_4	TG	0.426299	3.61×10^{-6}	1.616654
24	TG(18:4_14:1_18:4)+NH_4	TG	0.424032	0.000601	1.434173
25	TG(18:4_18:3_22:6)+NH_4	TG	0.41912	7.34×10^{-6}	1.093412
26	TG(16:0_14:1_22:6)+H	TG	0.406657	9.14×10^{-5}	1.130789
27	TG(20:5_12:1_22:6)+H	TG	0.395436	0.008554	1.019339
28	TG(16:0_14:4_22:6)+NH_4	TG	0.395024	4.52×10^{-7}	1.802223
29	TG(18:4_14:3_16:0)+NH_4	TG	0.389088	4.19×10^{-5}	1.988512

Table 3. *Cont.*

Number	LipidIon	Class	Fold Change	p-Value	VIP
30	PG(43:3e)+NH4	PG	0.380387	2.56×10^{-5}	2.920494
31	PI(37:0_9:0)+H	PI	0.371176	0.008509	1.172954
32	SM(d20:0_24:3)+H	SM	0.36418	5.99×10^{-6}	2.831523
33	SM(d20:0_22:2)+H	SM	0.35894	0.000375	2.3627
34	TG(22:6_11:2_21:0)+H	TG	0.355089	0.001193	1.531593
35	TG(18:4_12:0_18:4)+NH_4	TG	0.338135	0.000114	2.230751
36	TG(16:1_14:0_20:5)+NH_4	TG	0.331458	8.32×10^{-7}	7.620491
37	TG(12:0_14:0_22:6)+NH_4	TG	0.328791	0.000737	3.230222
38	TG(18:4_12:0_20:5)+NH_4	TG	0.327852	7.66×10^{-7}	1.988383
39	TG(18:4_12:0_14:0)+NH_4	TG	0.31954	0.000463	1.849902
40	TG(18:4_14:0_20:5)+NH_4	TG	0.318295	7.37×10^{-5}	10.2416
41	TG(18:4_14:3_20:5)+NH_4	TG	0.311052	2.61×10^{-5}	1.263194
42	TG(18:4_14:2_20:5)+NH_4	TG	0.270653	1.31×10^{-5}	1.556192
43	TG(28:1_22:3_22:6)+NH_4	TG	0.269109	1.76×10^{-7}	1.040046
44	SM(d20:0_24:5)+H	SM	0.259478	2.83×10^{-5}	1.932346
45	TG(18:4_13:0_22:6)+NH_4	TG	0.237456	3.20×10^{-7}	3.431831
46	PI(42:1e)+H	PI	0.233915	8.54×10^{-7}	1.113729
47	PS(43:1e)+H	PS	0.232173	8.28×10^{-7}	3.321953
48	SPH(m18:1)+H	SPH	0.195451	9.37×10^{-8}	1.459295
49	ST(d17:1_25:0)+NH_4	ST	0.19223	5.03×10^{-7}	2.081955
50	PI(40:0e)+H	PI	0.191481	5.31×10^{-7}	2.086004
51	ST(d15:0_25:0)+NH_4	ST	0.18241	8.61×10^{-6}	1.444491
52	SPH(d20:0)+H-H_2O	SPH	0.181566	3.37×10^{-7}	1.301676
53	TG(22:6_12:2_22:6)+NH_4	TG	0.180922	2.34×10^{-7}	2.711327
54	TG(18:4_14:3_22:6)+NH_4	TG	0.176824	8.09×10^{-6}	1.854093
55	TG(22:6_14:3_22:6)+NH_4	TG	0.134366	6.24×10^{-7}	2.144745

During the freezing storage process, the expression of TG was significantly down-regulated (Figure 4), primarily decomposing into FFA, DG, and monoglycerides (MG). These metabolites served as energy sources and substrates for the proliferation of cryophilic microorganisms, facilitating microbial growth [26]. TG(18:4_14:0_20:5), TG(20:5_13:0_22:6), TG(22:6_14:1_22:6), and TG(18:4_13:0_22:6) were the most predominant species. The freezing storage process also led to an increase in AV, POV, and TBARSs, indicating a comparative deterioration of the lipid profiles in *C. saira*. This reduction in TG levels can be attributed to oxidative degradation, which is triggered by hydrolysis due to the enzymatic action of lipases [28]. This process might convert TG into substances such as DG, while simultaneously releasing fatty acids [29]. DG, an intermediate product of lipid metabolism, is closely associated with the biosynthesis and catabolism of TG [30]. Notably, the T1 and T2 groups exhibited two up-regulated DG lipid species. PE and PC played crucial roles in regulating energy, lipid metabolism, and lipoprotein secretion [31]. The expression of functional ingredients such as PE, PC, and PI was also significantly down-regulated, suggesting their conversion to lysophosphatidylethanolamine (LPE), lysophosphatidylcholine (LPC), and lysophosphatidylinositol (LPI), respectively, through pathways such as partial hydrolysis or lipid oxidation [27]. These findings underscored the critical role of lipid oxidation in lipid transformations during the freezing storage process of *C. saira*, in accordance with the results of previous studies [22,32].

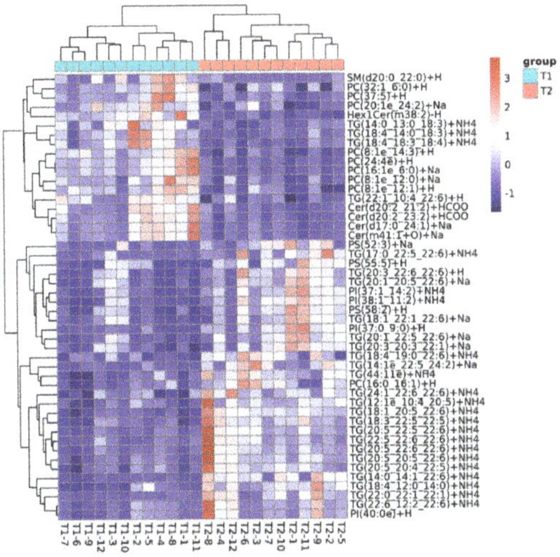

Figure 4. The correlation analysis between the lipid metabolites and different frozen storage temperature.

Comprehensive analysis of the differential lipids at freezing temperatures of −18 °C and −25 °C revealed significant findings regarding lipid regulation and oxidation in muscle tissue. The study found that more phosphatidylcholines (PCs) and phosphatidylethanolamines (PEs) exhibited down-regulation, likely due to oxidation reactions in muscle tissue. These reactions may cause hydrogen rearrangement on the PC/PE chains and lead to the breakage of the C-C layer of α-bonds [33]. Notably, the differential abundance of PE and PC was higher in the −25 °C group compared to the −18 °C group, indicating that a freezing temperature of −25 °C could more effectively inhibit lipid oxidation in the muscle of *C. saira*. These results underscored the significant impact of different freezing storage temperatures on the lipid profile, with −25 °C being more favorable for the frozen storage of *C. saira*. This finding aligned with previous results concerning lipid composition and oxidation.

3.5. Lipid Metabolic Pathway Analysis of *C. saira* During Storage Process

The analysis of metabolic pathways provided insights into how frozen storage temperatures affect lipid distribution in samples, as indicated in previous studies [6]. Utilizing the KEGG database, the pathway annotation analysis of lipid differential metabolites in *C. saira* at varying storage temperatures was depicted in Figure 5. Comparative analysis revealed that these differential metabolites were predominantly enriched in 17 metabolic pathways of the T1 (−18 °C) vs. T2 (−25 °C) group. Among these, glycerophospholipid metabolism emerged as the primary pathway enriched during the frozen storage process. Key lipid components such as phosphatidylcholine (PC), phosphatidylethanolamine (PE), phosphatidylinositol (PI), and phosphatidylserine (PS) were annotated within this pathway. These lipids were crucial as they constitute major components of cell membranes, thereby maintaining cellular physiological functions and facilitating the transport of triglycerides [34]. Additionally, sphingolipid metabolism was identified as another significant pathway enriched during frozen storage, with ceramide (Cer) and sphingomyelin (SM) being the two significantly different lipid metabolites annotated in this pathway. Further investigation into the lipid oxidation pathway is necessary, in conjunction with metabolomics, to uncover the potential key lipid molecules involved in the storage process of *C. saira*.

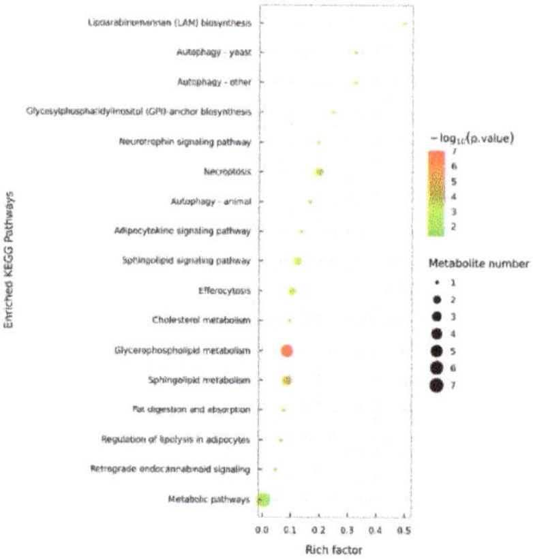

Figure 5. KEGG pathway analysis of highly significant differential lipids in T1 vs. T2 group.

4. Conclusions

In conclusion, the findings indicated that storage temperature had a significant influence on the lipid stability of *C. saira* during the frozen storage process. The lipid stability of *C. saira* stored at −18 °C was much worse than that at −25 °C. Thus, −25 °C storage was more conducive to maintain the storage stability of *C. saira* in production. The study will provide valuable insights into production quality control of pelagic fishery catches.

Supplementary Materials: The following supporting information can be downloaded at https://www.mdpi.com/article/10.3390/foods14050756/s1, Determination of lipid oxidation.

Author Contributions: Writing—original draft, visualization, investigation, L.Z.; data curation, S.W.; conceptualization, project administration, Q.L.; methodology, D.S. and Y.Z.; funding acquisition, R.C.; supervision, Y.Y. All authors have read and agreed to the published version of the manuscript.

Funding: This research was funded by the National Key Research and Development Plan (No. 2020YFD0901203) and Central Public-interest Scientific Institution Basal Research Fund, CAFS (No. 2023TD72).

Institutional Review Board Statement: Not applicable.

Informed Consent Statement: Not applicable.

Data Availability Statement: The original contributions presented in the study are included in the article/Supplementary Materials, further inquiries can be directed to the corresponding author.

Conflicts of Interest: The authors declare no conflicts of interest.

References

1. Wang, X. International fisheries dynamics. *Fish. Inf. Strategy* **2023**, *38*, 317–324.
2. NPFC. NPFC-2023-AR-Annual Summary Footprint-Pacific Saury. 2023. Available online: https://www.npfc.int/summary-footprint-pacific-saury-fisheries (accessed on 11 July 2024).
3. Zhang, J.; Tao, N.; Wang, M.; Shi, W.; Ye, B.; Wang, X.; Zhu, Q.; Hua, C. Characterization of phospholipids from Pacific saury (*Cololabis saira*) viscera and their neuroprotective activity. *Food Biosci.* **2018**, *24*, 120–126. [CrossRef]
4. Liu, H.M.; Yang, G.Q.; Zhao, Q.; Li, H.Y.; Niu, L.H.; Wu, H.Y.; Yu, H. Antioxidant effects of Stevia rebaudiana leaf and stem extracts on lipid oxidation in salted Pacific Saury (*Cololabis saira*) during processing. *Eur. J. Lipid Sci. Technol.* **2022**, *124*, 210022. [CrossRef]
5. Tao, X.Y.; Yin, M.Y.; Liu, L.; Song, R.Z.; Wang, X.D.; Tao, N.P.; Wang, X.C. UPLC-ESI-MS/MS strategy to analyze fatty acids composition and lipid profiles of Pacific saury (*Cololabis saira*). *Food Chem. X* **2024**, *23*, 101682. [CrossRef] [PubMed]

6. Suárez-Medina, M.D.; Sáez-Casado, M.; Martínez-Moya, T.; Rincón-Cervera, M.Á. The Effect of Low Temperature Storage on the Lipid Quality of Fish, Either Alone or Combined with Alternative Preservation Technologies. *Foods* **2024**, *13*, 1097. [CrossRef] [PubMed]
7. Indergård, E.; Tolstorebrov, I.; Larsen, H.; Eikevik, T.M. The influence of long-term storage, temperature and type of packaging materials on the quality characteristics of frozen farmed Atlantic Salmon (*Salmo salar*). *Int. J. Refrig.* **2014**, *41*, 27–36. [CrossRef]
8. Lan, W.Q.; Liu, L.; Xiao, L.; Mei, J.; Xie, J. Effect of Temperature Fluctuation on the Quality of Big-Eye Tuna (*Thunnus obesus*) during Low Temperature Circulation. *Food Sci.* **2021**, *42*, 205–212.
9. Zhao, L.; Wang, L.; Cao, R.; Liu, Q.; Su, D.; Zhang, Y.T.; Yu, Y.Q. The role of ultraviolet radiation in the flavor formation during drying processing of Pacific saury (*Cololabis saira*). *J. Sci. Food Agric.* **2024**, *104*, 8099–8108. [CrossRef] [PubMed]
10. Xu, S.Y.; Chen, Y.; Gong, Y.S. Improvement of Theaflavins on Glucose and Lipid Metabolism in Diabetes Mellitus. *Foods* **2024**, *13*, 1763. [CrossRef]
11. Luo, X.; Sun, W.C.; Luo, Y.H. Research progress in detection and function of sphingomyelin in food. *Food Res. Dev.* **2020**, *41*, 211–218.
12. Luo, H.B.; Wang, W.H.; Chen, W.; Tang, H.Q.; Jiang, L.; Yu, Z.F. Effect of incorporation of natural chemicals in water ice-glazing on freshness and shelf-life of Pacific saury (*Cololabis saira*) during −18 °C frozen storage. *J. Sci. Food Agric.* **2018**, *9*, 3309–3314. [CrossRef] [PubMed]
13. Ding, K.H.; Wang, Y.F.; Luan, D.L. Effects of high-temperature short-time processing on nutrition quality of Pacific saury (*Cololabis saira*) using extracted fatty acids as the indicator. *Food Sci. Nutr.* **2022**, *1*, 157–167. [CrossRef] [PubMed]
14. Tanaka, R.; Naiki, K.; Tsuji, K.; Nomata, H.; Sugiura, Y.; Matsushita, T.; Kimura, I. Effect of antioxidative treatment on lipid oxidation in skinless fillet of Pacific Saury (*Cololabis saira*) in frozen storage. *J. Food Process. Preserv.* **2013**, *37*, 325–334. [CrossRef]
15. Wang, S.Y.; Jian, C.; Hu, M.Y.; Zhao, L.; Sun, H.H.; Liu, Q.; Cao, R.; Xue, Y. Lipid changes and volatile compounds formation in different processing stages of dry-cured Spanish mackerel. *Food Qual. Saf.* **2024**, *8*, fyae026. [CrossRef]
16. *GB 5009.181-2016*; National Standards for Food Safety Determination of Malondialdehyde in Food. The National Health and Family Planning Commission of the People's Republic of China: Beijing, China, 2016.
17. *GB 5009.227-2016*; National Standards for Food Safety Determination of Peroxide Value in Food. The National Health and Family Planning Commission of the People's Republic of China: Beijing, China, 2016.
18. *GB 5009.229-2016*; National Standards for Food Safety Determination of Acid Value in Food. The National Health and Family Planning Commission of the People's Republic of China: Beijing, China, 2016.
19. Shi, C.P.; Guo, H.; Wu, T.T.; Tao, N.P.; Wang, X.C.; Zhong, J. Effect of three types of thermal processing methods on the lipidomics profile of tilapia fillets by UPLC-Q-Extractive Orbitrap mass spectrometry. *Food Chem.* **2019**, *298*, 125029. [CrossRef] [PubMed]
20. Liu, Z.Q.; Zhao, M.T.; Wang, X.W.; Li, C.; Liu, Z.Y.; Shen, X.R.; Zhou, D.Y. Investigation of oyster *Crassostrea gigas* lipid profile from three sea areas of China based on non-targeted lipidomics for their geographic region traceability. *Food Chem.* **2022**, *386*, 132748. [CrossRef]
21. Zhang, Q.; Chen, X.C.; Ding, Y.T.; Ke, Z.G.; Zhou, X.X.; Zhang, J.Y. Diversity and succession of the microbial community and its correlation with lipid oxidation in dry-cured black carp (*Mylopharyngodon piceus*) during storage. *Food Microbiol.* **2021**, *98*, 103686. [CrossRef]
22. Fang, C.D.; Chen, H.S.; Yan, H.B.; Shui, S.S.; Benjakul, S.; Zhang, B. Investigation of the changes in the lipid profiles in hairtail (*Trichiurus haumela*) muscle during frozen storage using chemical and LC/MS-based lipidomics analysis. *Food Chem.* **2022**, *390*, 133140. [CrossRef] [PubMed]
23. Russo, A.; Caputo, S.; Pantusa, M.; Perri, E.; Sindona, G.; Sportelli, L. Amino acids as modulators of lipoxygenase oxidation mechanism. The identification and structural characterization of spin adducts intermediates by electron spin resonance and tandem mass spectrometry. *Food Chem.* **2010**, *119*, 533–538. [CrossRef]
24. Wang, Y.Y.; Cai, Z.C.; Sang, X.H.; Deng, W.T.; Zeng, L.X.; Wang, J.M.; Zhang, J.H. LC-MS-based lipidomics analyses of alterations in lipid profiles of Asian sea bass (*Lates calcarifer*) induced by plasma-activated water treatment. *Food Res. Int.* **2024**, *177*, 113866. [CrossRef] [PubMed]
25. Soyer, A.; Özalp, B.; Dalmıs, Ü.; Bilgin, V. Effects of freezing temperature and duration of frozen storage on lipid and protein oxidation in chicken meat. *Food Chem.* **2010**, *120*, 1025–1103. [CrossRef]
26. Hu, L.L.; Ren, S.J.; Shen, Q.; Ye, X.Q.; Chen, J.C.; Ling, J.G. Protein oxidation and proteolysis during roasting and in vitro digestion of fish (*Acipenser gueldenstaedtii*). *J. Sci. Food Agric.* **2018**, *98*, 5344–5351. [CrossRef] [PubMed]
27. Chen, J.; Kong, Q.; Sun, Z.T.; Liu, J.Y. Freshness analysis based on lipidomics for farmed Atlantic salmon (*Salmo salar* L.) stored at different times. *Food Chem.* **2022**, *373*, 131564. [CrossRef]
28. Wang, L.; Zang, M.W.; Cheng, X.Y.; Wang, S.W.; Zhao, X.; Zhao, B.; Li, D. Evaluation of changes in the lipid profiles of dried shrimps (*Penaeus vannamei*) during accelerated storage based on chemical and lipidomics analysis. *LWT—Food Sci. Technol.* **2024**, *191*, 115564. [CrossRef]

29. Guo, X.; Shi, D.; Liu, C.J.; Huang, Y.L.; Wang, Q.L.; Wang, J.Y.; Pei, L.Y.; Lu, S.L. UPLC-MS-MS-based lipidomics for the evaluation of changes in lipids during dry-cured mutton ham processing. *Food Chem.* **2022**, *377*, 131977. [CrossRef]
30. Fu, Y.H.; Cao, S.Y.; Yang, L.; Li, Z.L. Flavor formation based on lipid in meat and meat products: A review. *J. Food Biochem* **2022**, *46*, e14439. [CrossRef] [PubMed]
31. He, C.; Sun, Z.X.; Qu, X.C.; Cao, J.; Shen, X.R.; Li, C. A comprehensive study of lipid profiles of round scad (*Decapterus maruadsi*) based on lipidomic with UPLCQ-Exactive Orbitrap-MS. *Food Res. Int.* **2020**, *133*, 109138. [CrossRef] [PubMed]
32. Yu, M.M.; Gang, K.Q.; Li, C.; Wang, J.H.; Liu, Y.X.; Zhou, D.Y.; Zhu, B.W. Change of lipids in whelks (*Neptunea arthritica cumingi* Crosse and *Neverita didyma*) during cold storage. *Food Res. Int.* **2020**, *136*, 109330. [CrossRef]
33. Wang, X.S.; Zhang, H.W.; Song, Y.; Cong, P.X.; Li, Z.J.; Xu, J.; Xue, C.H. Comparative Lipid Profile Analysis of Four Fish Species by Ultraperformance Liquid Chromatography Coupled with Quadrupole Time-of-Flight Mass Spectrometry. *J. Agric. Food Chem.* **2019**, *67*, 9423–9431. [CrossRef]
34. Cui, Z.; Liu, C.; RAO, W.X.; Chen, P.; Lei, K.K.; Mai, K.S.; Zhang, W.B. Dietary phospholipids improve growth performance and change the lipid composition and volatile flavor compound profiles in the muscle of abalone *Haliotis discus hannai* by affecting the glycerophospholipid metabolism. *Aquac. Rep.* **2023**, *30*, 101567. [CrossRef]

foods

MDPI

Article

Freshness and Spoilage Patterns of Wild and Farmed Tropical Fish Species with Major Commercial Importance Originating from Saudi Arabian Waters

Kriton Grigorakis [1,*], Dimitra Kogiannou [1], Mado Kotsiri [1], Ioannis Kleidas [1], Paulo H. de Mello [2], Salaheldeen Habiballah [2], Ali Alshaikhi [3], Youssef S. Alhafedh [3] and Asaad H. W. Mohamed [2]

1 Hellenic Center for Marine Research (HCMR), Institute of Marine Biology, Biotechnology & Aquaculture, 19013 Anavyssos, Greece; dkogiannou@hcmr.gr (D.K.); mkotsiri@hcmr.gr (M.K.); i.kleidas@hcmr.gr (I.K.)
2 Kaust Beacon Development, King Abdullah University of Science and Technology, Thuwal, Jeddah 23955, Saudi Arabia; paulo.demello@kaust.edu.sa (P.H.d.M.); salaheldeen.habiballah@kaust.edu.sa (S.H.); asaad.mohamed@kaust.edu.sa (A.H.W.M.)
3 Ministry of Environment, Water and Agriculture, King AbdulAziz Rd., Riyadh 11195, Saudi Arabia; ali.alshaikhi@mewa.gov.sa (A.A.); yhafedh@nfdp.gov.sa (Y.S.A.)
* Correspondence: kgrigo@hcmr.gr

Abstract: Ice-stored farmed barramundi (*Lates calcarifer*), snubnose pompano (*Trachinotus blochii*) and sobaity bream (*Sparidentex hasta*), as well as wild-caught cobia (*Rachycentron canadum*), coral trout (*Plectropomus leopardus*), giant trevally (*Caranx ignobilis*), milkfish (*Chanos chanos*) and mangrove red snapper (*Lutjanus argentimaculatus*), were compared for their freshness/spoilage using sensory, chemical and microbiological methods. Quality Index Method schemes were developed to determine alterations in the sensory freshness. The shelf lives ranged from 8 (coral trout) to 18 days (sobaity bream). The farmed species always exhibited a significantly longer shelf life than the wild-caught species. The adenosine triphosphate (ATP) breakdown followed different patterns in the studied species. The K-values at the time of sensory rejection ranged from 30 to 80% depending on the species, while the microbial load reached or exceeded a level of 6 log cfu/g. Although the shelf life duration was dependent on the origin of the fish (wild or farmed), the ATP breakdown scheme, as well as the K-values and microbial loads at the time of rejection, were species-dependent and independent of the origin.

Keywords: QIM; K-value; fish quality; microbiological changes; seafood; sensory; wild; farmed; chilling

1. Introduction

Freshness is considered one of the four pillars of fish quality, along with safety, traceability and authentication [1]. Fish is a food item that is characterized by its high perishability; this, in practice, is translated to a rejection rate of 10–50% of all products due to post-slaughter spoilage. Therefore, freshness and spoilage are of capital importance in the commercial chain of fish. Three main postmortem processes appear to happen, namely enzymatic autolysis, oxidation and microbial degradation, and these are responsible for the transformation of certain substances in fish tissues that result in spoilage. In fresh ice-stored fish, autolysis and microbial degradation prevail, while oxidation is more pronounced in frozen storage [2,3].

Different species exhibit different spoilage patterns [2,4] and speeds; therefore, individual monitoring of freshness reduction/spoilage is required to understand these patterns.

The study of freshness and its reduction requires tools such as sensory, chemical and microbiological evaluations [2,4,5]. Among the various methods for sensory freshness evaluation, the Quality Index Method (QIM) seems to be the preferred tool, as it is species-specific and highly analytic, thus providing a numerical score that is more representative of the spoilage stage [5,6]. While significant attention has been paid to the freshness and spoilage of fish species that are popular in the European, Japanese and US markets, scarce data are available for popular tropical species that are highly important in other Asian and African fish markets.

Barramundi (*Lates calcarifer*) is a species mainly originating from the Indian and Pacific Oceans; it is also widely farmed, with a global production of 154,280.7 (FishstatJ, 2024 v4.04.00, https://www.fao.org/fishery/en/statistics/software/fishstatj, accessed on 20 December 2024) metric tons in 2022 and market sizes ranging from 500 g to 2 kg. Cobia (*Rachycentron canadum*) is a globally distributed warm-water species, with a global harvesting volume of 19,147.73 (2022) and global production of 35,613.45 metric tons in 2022 (FishstatJ, 2024). China and Taiwan are the main producers, with market sizes of 5–10 kg and net exports of its fresh form reaching USD 30 million in 2020 [7]. Coral trout species (*Plectropomus* sp.) are sold in the local Saudi fish market, with a global production of about 250 metric tons; however, their value is ranked highest among the fish that are traded in the local Red Sea markets, exceeding 25 USD/kg [8]. The giant trevally (*Caranx ignobilis*) is an important species with Indo-Pacific distribution, a global harvesting volume of 109,000 metric tons in 2017, and farming attempts taking place in recent years in the Philippines (https://www.fao.org/fishery/aquagris/en/aqgr/phl__nxi?status=pending, accessed on 20 December 2024). The milkfish (*Chanos chanos*), an herbivorous species that is commercialized in market sizes of 250–500 g, accounted for a global aquaculture production exceeding 1,196,017.29 metric tons in 2022 (FishstatJ, 2024). It is highly valued in Asian markets and has a long and robust farming tradition that started in Indonesia, the Taiwan Province of China and the Philippines four to six centuries ago, but a significant number of wild fishery catches also occur [7]. The mangrove red snapper (*Lutjanus argentimaculatus*) is a widespread Indo-Pacific species with significant market importance in the region but is never found in significant quantities, with annual production reaching approx. 15,000 tons from fisheries and 5308 (2022) from aquaculture, according to the species factsheet of the FAO (https://www.fao.org/fishery/en/aqspecies/3134/en, accessed on 20 December 2024; FishstatJ, 2024). The snubnose pompano (*Trachinotus blochii*) is a widely farmed species in China, India, Indonesia, the Philippines, Taiwan, Thailand and Vietnam, with significant importance in these markets. There is high demand for this species, including in the US market, with a dockside price of USD 8 in international markets, as reported in the respective FAO factsheets (https://www.fao.org/fishery/en/culturedspecies/trachinotus_spp/en, accessed on 20 December 2024). The sobaity bream (*Sparidentex hasta*) is a species with significant importance in fish farming in the Arabian Peninsula, a marketable size of 400 g and very high market potential. The latest estimation of its aquaculture production was 241 metric tons in 2022 (FishstatJ, 2024).

Despite the capital economic importance of these species, which is outlined herein, there is a significant lack of knowledge regarding their freshness and spoilage patterns. Such knowledge is important in ensuring their quality during commercialization.

Within this context, the aim of this study was to systematically assess the freshness and spoilage of the aforementioned commercially important fish originating from national aquaculture and fisheries in the Saudi Arabian market and, subsequently, to compare their freshness/deterioration. For this purpose, the chemical, microbial and sensory spoilage of eight different species were analytically studied, and the individual characteristics of their spoilage patterns were assessed.

2. Materials and Methods

2.1. Fish

In total, eight locally (Saudi Arabia) marketed fish species were studied, selected on the basis of their market importance, availability and lack of data on their freshness and spoilage patterns. The characteristics of the studied fish, including their average weights, appear in Table 1.

Table 1. The studied fish species, their origin and/or fishing methods and their average weights.

Common Name	Scientific Name	Origin/Fishing Method	Weight of the Studied Fish (Average in g)
Barramundi (Asian seabass)	*Lates calcarifer*	Farmed	1500
Cobia	*Rachycentron canadum*	Wild/hand line	6500
Coral trout	*Plectropomus leopardus*	Wild/hand line	920
Giant trevally	*Caranx ignobilis*	Wild/hand line	2560
Milkfish	*Chanos chanos*	Wild/surface gillnet	400
Mangrove red snapper	*Lutjanus argentimaculatus*	Wild/hand line	3200
Snubnose pompano	*Trachinotus blochii*	Farmed	1200
Sobaity bream	*Sparidentex hasta*	Farmed	1000

All the analyzed fish were whole, unprocessed or fresh and provided from the Central Fish Market in Jeddah. They were freshly landed or slaughtered (farms) and remained in Styrofoam boxes with ice thereafter. The farmed fish were slaughtered, packed with ice and transferred to the market within a day, thus ensuring the maintenance of absolute freshness until market distribution. The wild-caught fish were iced on the vessel, although no accurate estimation of the slaughter time was provided. For each species, one batch of 15 fish was selected in order to have adequate samples for studying the sensory freshness and developing the QIM schemes; subsequently, a second batch with an equal number of individuals of each species was used for scoring in the developed QIM, analyzing their chemical freshness and measuring the microbial spoilage.

The fish were transferred to King Abdullah University of Science and Technology (KAUST) facilities for sampling and a sensory freshness evaluation during the first 8 days of storage. To assess the late spoilage stages, the sampled fish were subjected to identical analyses at the Hellenic Centre for Marine Research (HCMR) facilities in Greece. Storage always occurred under refrigeration in cold rooms (4 °C). The permanent presence of ice ensured a temperature of 0 °C around the fish while at the same time reducing the bacterial load and retaining moisture in the fish skin. Therefore, periodic ice replacement and continuous ice presence were ensured throughout the experimentation.

For all species, the sensory freshness, microbial status and chemical freshness were evaluated at specific time intervals from the time 0 (the time of obtainment) to a time when the fish had reached/passed the acceptability limit (the shelf life). The acceptability limit was detected by means of human senses, i.e., the perceived sensory freshness.

2.2. QIM

The QIM is a sensory freshness evaluation tool that differs for each species. Depending on the individual sensory changes that a species undergoes during freshness reduction/spoilage, a system of demerit scores is developed based on the changes in specific features, as perceived organoleptically.

With the exception of cobia, where QIM schemes are already available [9], the QIM schemes for the rest of the species had to be developed from scratch. For the development of a QIM, changes in the external appearance/visual quality, external smells and texture were analytically monitored. For each feature, these changes were evaluated from the start, namely the first day after slaughtering for farmed fish or the first day after purchase for wild ones, until the point when a fish passed its shelf life and could not be commercialized (i.e., beyond the sensory acceptability limit). A total of 3–5 specimens were used at each sampling point to evaluate freshness and determine their characteristics and the changes that they underwent.

For cobia, the already existing scheme was generally followed, with minor alterations after the preliminary screening procedures.

Three assessors were used, each of whom scored 3 individual fish (for each species) at every sampling point, based on the respective previously developed QIM schemes. The total scores that they received were averaged to establish the freshness score at each time interval. Standard deviations and the two-tail Pearson correlation with days of ice storage were used to check the validity of the created schemes.

2.3. Chemical Freshness: K-Value and ATP Breakdown Products

The adenosine triphosphate (ATP) breakdown products were analyzed using high-performance liquid chromatography (HPLC), and K-value determination was performed to evaluate the chemical freshness and determine the shelf life in chemical terms. The determination of ATP breakdown products, which include ATP, adenosine diphosphate (ADP), adenosine monophosphate (AMP), inosine monophosphate (IMP), inosine (Ino) and hypoxanthine (Hx), took place after extraction from dorsal white muscle tissue. A quantity of 5 g was homogenized on ice with 25 mL of perchloric acid (PCA; $HClO_4$, 0.6 M) for 2 min. Subsequently, centrifugation at 4 °C and 7000× *g* for 5 min took place, and 10 mL supernatant was obtained. The pH was adjusted to 6.7–6.9, with KOH 1 M initially and 0.1 M subsequently for the fine adjustment.

The HPLC determination of the ATP breakdown products took place according to the previously published methodology in [10]. The method for the separation and detection of ATP metabolites was originally developed and validated by Ryder [11]. The apparatus that was used combined a Waters 600 Pump, a 600 Pump system Controller, a Waters 717 Plus Autosampler set at a 10 °C injection temperature, a Waters 2487 UV detector set at 254 nm and Empower Chromatography Software v5.00.00.00 (Waters, Milford, MA, USA). The mobile phase consisted of 0.4 M KH_2PO_4 and 0.06 M K_2HPO_4 in HPLC water. The flow was set at 1.5 mL/min. An aliquot of 5 µL was injected. The chromatographic column was a C18, using 5 µm, 100 RP and 4 × 250 mm (Phenomenex, Torrance, CA, USA). The total running time per sample was 20 min.

Subsequently, the identification of individual ATP metabolites took place, and we determined the K-value as follows:

$$K\ value\% = 100 \times \frac{Ino + Hx}{ATP + ADP + AMP + IMP + Ino + Hx}$$

2.4. Microbial Spoilage

Microbial spoilage was evaluated following previously published methodologies [12,13], with minor modifications. Briefly, 10 g from the dorsal muscle of each fish was placed into stomacher bags with 90 mL of sterile MRD (maximum recovery diluent: 8.5 g/L of NaCl and 1.0 g/L of bacteriological peptone) and homogenized for 1.5 min using a stomacher. Then, volumes of 0.1 mL from the serial dilutions in the MRD were spread onto the surface of water plates to count the agar in order to enumerate the total viable counts (TVCs) and incubated at 27 °C for 48 h. Samples of 1 mL of serial dilution in the MRD were used for the pour plate technique for the enumeration of (a) H_2S -producing bacteria (presumably *Shewanella* sp.) on iron agar by counting only black colonies after incubation at 27 °C for 48 h, and (b) Enterobacteriaceae (*E. coli*) on Violet Red Bile Glucose agar, which was incubated at 37 °C for 24 h. The results were expressed as the mean log cfu/g ± standard deviation of 3 replicates (3 fish per species).

2.5. Statistical Analysis

For all fish species and for every measured factor, the average and standard deviations were calculated. Differences in the means in the K-values and the ATP breakdown products were statistically tested by performing an analysis of variance (ANOVA), followed by Tukey's significant difference test (IBM SPSS Statistics Version 26 software). The levels of significance were set at $p = 0.05$ (5%).

3. Results and Discussion

The QIM is considered the most widely used and reliable method for sensory freshness evaluation, not only for ice-stored fish but also for a variety of both fresh and frozen seafood and other muscular foods [4,5,14–17]. To accomplish the aim of this study, new QIM schemes were designed for most of the tested fish species. For cobia, the already developed scheme [9] was slightly modified based on the observations of the present study. The analytic QIM schemes for each species can be found in Supplementary Tables S1–S8 (Supplementary-material QIM schemes). Representative photos of the fish in the initial (fresh) and final (at the acceptability limit) stages are provided in Supplementary Figures S1–S8.

Following the evaluations using the developed QIM schemes, each of the studied species was scored at various intervals during ice storage (Figure 1). For snubnose pompano, the total demerit score at the acceptability limit was 10; for sobaity bream, it was 11; for coral trout and barramundi, it was 12; for cobia and mangrove red snapper, it was 13; and for milkfish and giant trevally, it was 14.

In general, the spoilage schemes of the different fish species have commonalities, namely, the resolution of rigor mortis and softening of muscle tissue, the loss of skin color intensity, the development of undesirable odors in the gills and skin, and the loss of eye convexity and shine [2]. Interestingly, the present study revealed individualities in the tested species. The coral trout initially exhibited soft muscles, with muscle hardening occurring toward the later stages of spoilage. The milkfish exhibited a remarkable black discoloration in the belly area and a mass loss of scales. The mangrove red snapper exhibited excessive discoloration patches in the red skin as spoilage progressed. Freitas, Vaz-Pires and Câmara [15] indicated that the main disadvantage of QIM is its species specificity, which leads to the necessity of individually developing different schemes. The aforementioned study also highlighted the limitations relating to the variability of the specimens that are assessed and the assessors' judgments and proposed a validation scheme for QIM. However, it was stated that these procedures add to the complexity of the QIM establishment process. The low standard deviations of all QIM scores (Figure 1) and the QIM correlations with days of ice storage were >0.98 in all cases for each species, indicating the proper function of the herein-developed schemes.

Based on the QIM sensory evaluation, the acceptability limit for each species was evaluated, and the awarded demerit scores appear in Figure 1. Among the studied species, coral trout and milkfish were the most perishable ones, with shelf lives of 8 and 9 days, respectively, while sobaity bream was found to be the most durable, with an 18-day shelf life. These were followed by the remaining farmed species (barramundi and snubnose pompano), which had a shelf life of 16 days of ice storage. The shelf life of the aforementioned farmed species was quite similar to that of gilthead seabream, which is in the range of 15–17 days [10,18,19], and red seabream.

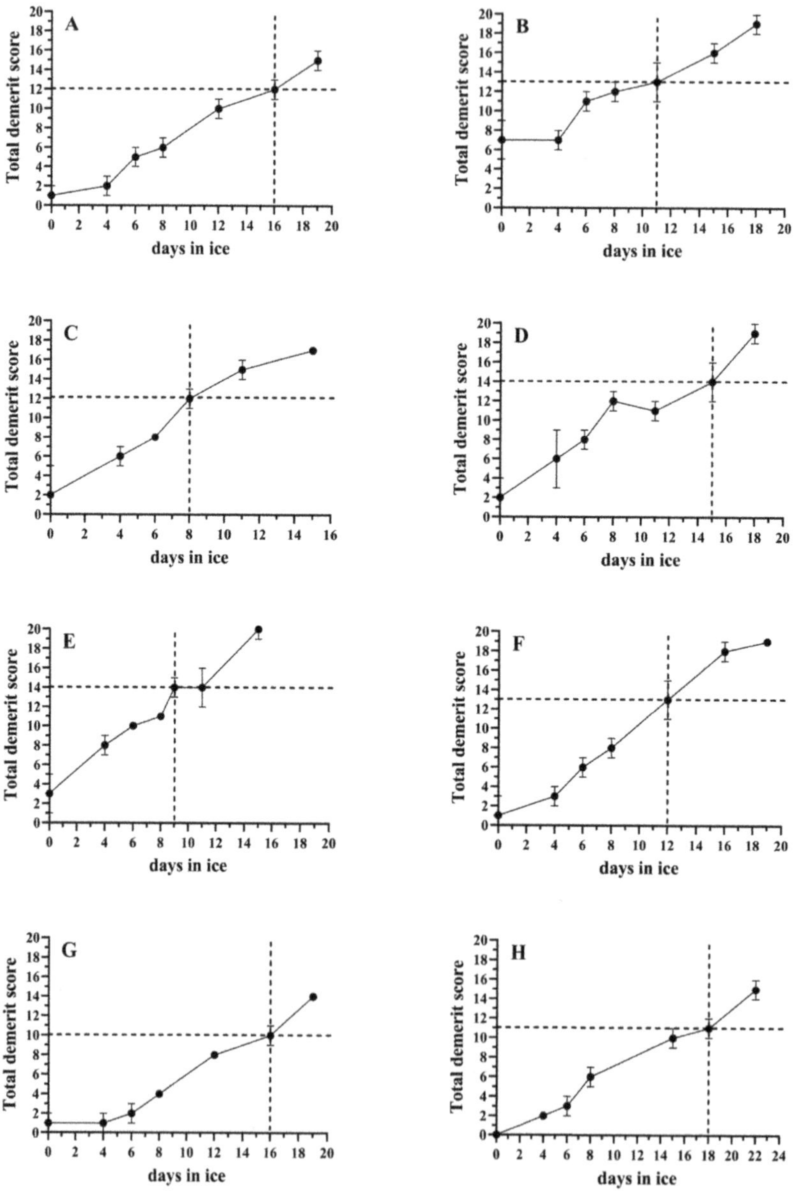

Figure 1. QIM sensory freshness evaluations of the studied species during ice storage. (**A**) Barramundi, (**B**) cobia, (**C**) coral trout, (**D**) giant trevally, (**E**) milkfish, (**F**) mangrove red snapper, (**G**) snubnose pompano and (**H**) sobaity bream. The vertical dotted line in the diagrams marks the acceptability limit (in days), while the dotted horizontal line indicates the QIM score that the fish received at this point. The bars indicate the standard deviations (n = 3).

The significantly longer shelf life of the farmed species compared with that of their wild counterparts can be explained by the use of controlled slaughtering conditions and an undisrupted cold chain. Additionally, the lower amount of stress that occurs during slaughtering compared with the stress that wild fish undergo during the fishing process may also be a factor. The longer shelf life of farmed fish compared with their wild counterparts of the same species has been previously reported and is in accordance with the present observations [20]. The shelf life of cobia that was determined herein was significantly shorter than that mentioned in [9] for the same species, which was 15 days of ice storage. A reasonable explanation for this is that the individuals that were studied herein were captured from the wild, unlike those in the aforementioned study, which were farmed. In a study referring to a grouper species of the same genus (*Plectropomus* sp.) as the coral trout, a shelf life of 18 days of ice storage was indicated when the fish were slaughtered and immediately stored in ice [21]. However, in the same study, a mere two-hour delay in icing, i.e., storage at an ambient tropical temperature, resulted in a shelf life of 8 days. This duration matches the herein-found shelf life for coral trout.

A basic autolytic procedure that occurs after fish death is the anaerobic metabolism of glycogen (due to the absence of respiratory function), which produces ATP. This, unlike in living fish, cannot be re-generated to provide energy to the cell. The ATP is inevitably further metabolized into other substances [22,23]. The initial pathway steps from ATP to IMP are very quick—almost immediate—and IMP is the first intermediate product that accumulates and is further gradually metabolized [24].

It has been shown that the ATP breakdown is dictated by degradation pathways, including endogenous enzymes, as well as microbial mechanisms of action. The enzymes that are produced by the spoilage microorganisms are the ones that are mainly responsible for the degradation of IMP and related products during the mid- to late stage of the storage of fish [23]. The two final products, i.e., Ino and Hx, gradually increase as fish spoilage proceeds. The measurement of these individual products' concentrations, as well as their ratio, namely, the K-value, provides useful information on the freshness/spoilage stage of fish. It has been found the K-value is the most reliable among the chemical indices and that it correlates better with the sensory freshness of fish [19,25].

The K-values of the eight studied species are presented in Figure 2. The red dotted line marks the shelf life duration and the K-values of the species at the end of their shelf life. For most of the species, K-values of around 40–50% characterize fish that are at their acceptability limit. However, there are exceptions, such as in the giant trevally, which had a K-value of around 80% at the acceptability limit, or the barramundi and coral trout, which exhibited lower K-values of around 30% at the end of their shelf lives. Similarly to our results, a K-value of 40% has been previously mentioned to generally describe spoiled fish [26].

The changes in the concentrations of the three basic individual ATP metabolites, namely the IMP, Ino, and Hx, for the studied species are presented in Figure 3.

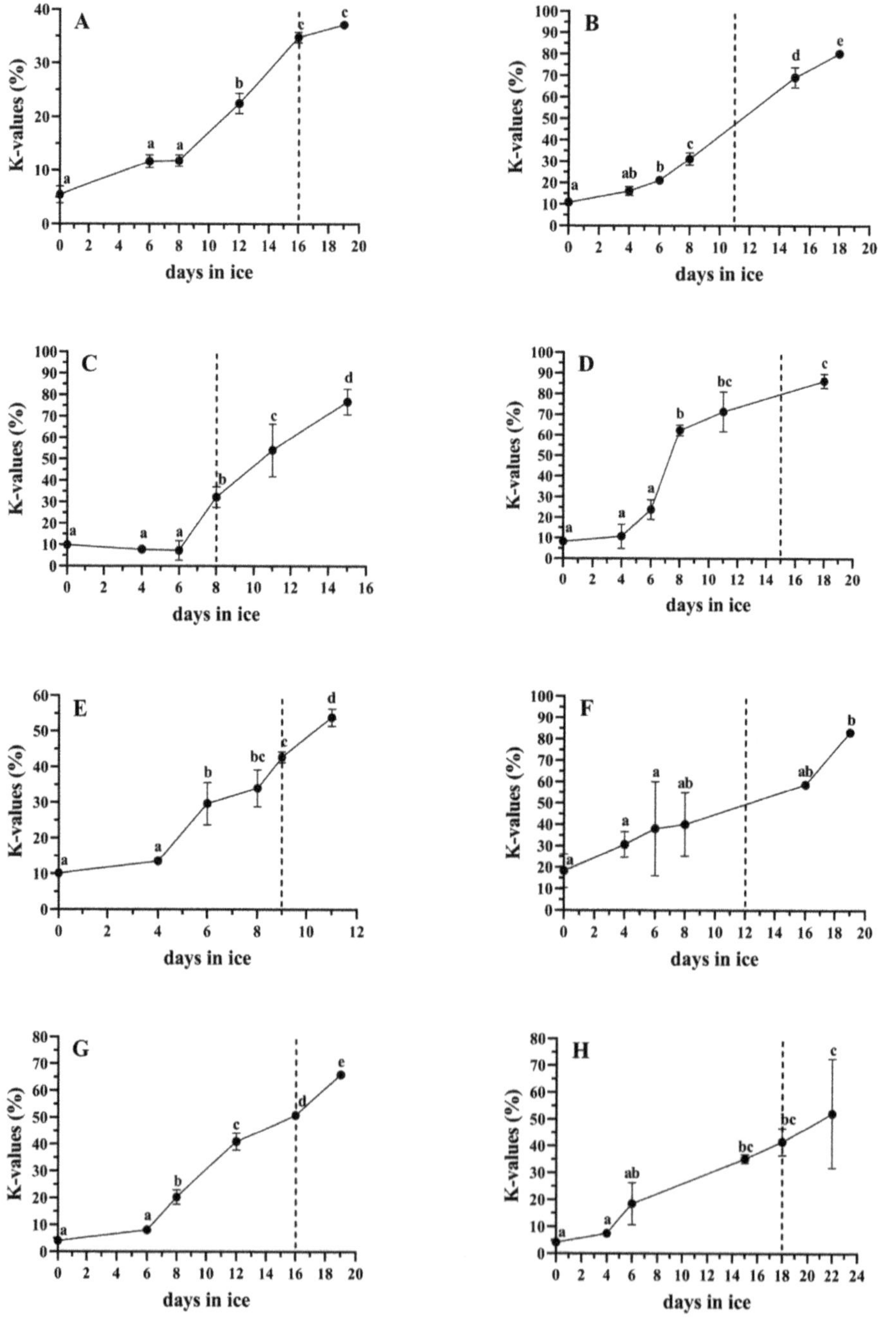

Figure 2. K value changes during ice storage of (**A**) barramundi, (**B**) cobia, (**C**) coral trout, (**D**) giant trevally, (**E**) milkfish, (**F**) mangrove red snapper, (**G**) snubnose pompano and (**H**) sobaity bream. The dotted line represents the day that the acceptability limit was reached (end of shelf life). Different letters represent statistically significant differences ($p < 0.05$). The bars represent the standard deviations (n = 3).

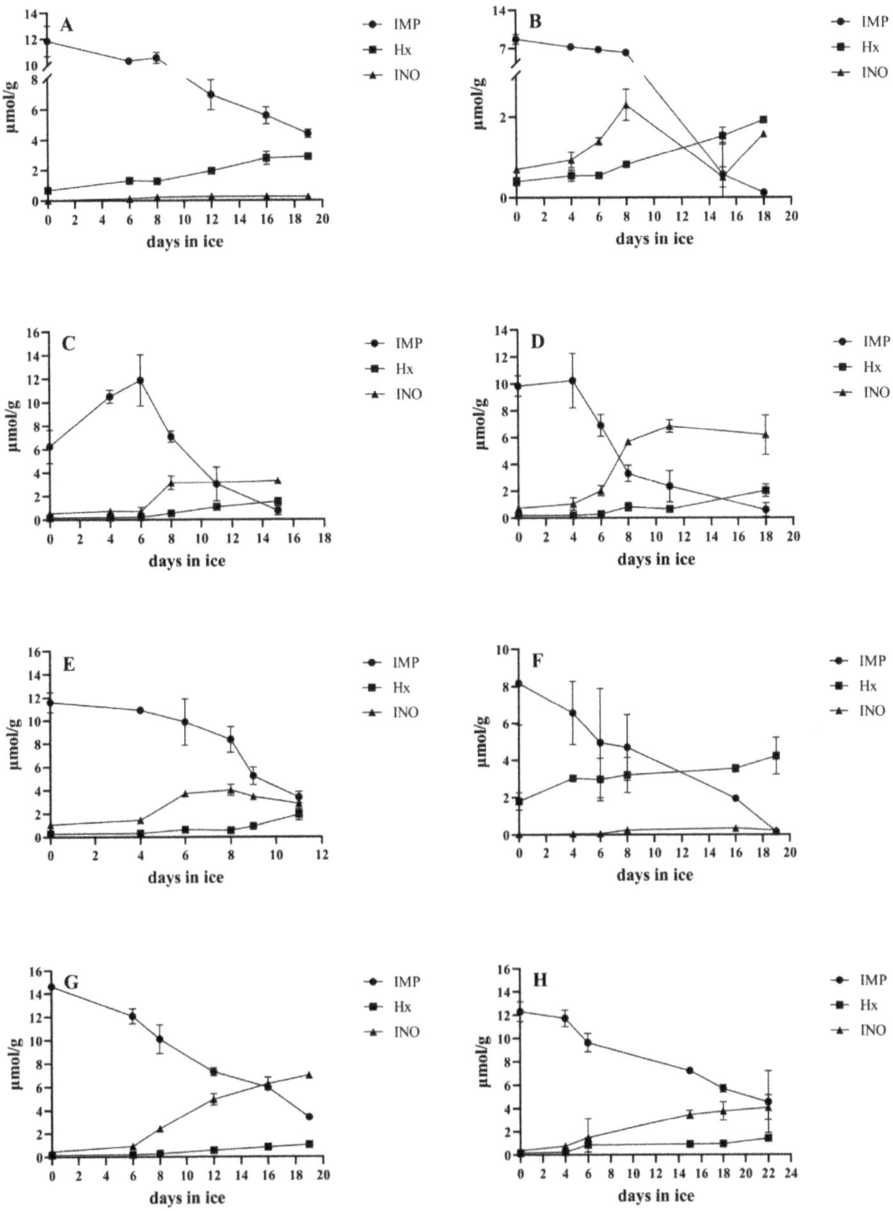

Figure 3. IMP, Ino and Hx concentrations of the studied species during ice storage. (**A**) Barramundi, (**B**) cobia, (**C**) coral trout, (**D**) giant trevally, (**E**) milkfish, (**F**) mangrove red snapper, (**G**) snubnose pompano and (**H**) sobaity bream. The bars represent the standard deviations (n = 3).

Although the speed of ATP breakdown and the concentrations of ATP metabolites seem to be affected by various factors, including the season, aquaculture pattern, storage temperature, stunning method, and harvest treatments like bleeding [23,27], the pattern of ATP metabolite changes is strongly species-specific [24]. In our study, we found that snubnose pompano and sobaity bream had some commonalities in their ATP breakdown patterns. These included slow gradual IMP breakdown and subsequent Ino formation, while the Hx levels remained low with a small and delayed increase. The final concentrations of Hx in their muscle never surpassed 1–1.5 μmoL/g. The Ino levels increased significantly ($p < 0.05$) after the 8th day of ice storage in snubnose pompano and on the 14th day in sobaity bream. These two species can be characterized as non-Hx forming species. A similar ATP breakdown pattern was described for gilthead seabream [19]. In contrast, the mangrove red snapper was found to be a characteristic of Hx-forming species, with Hx increasing at high levels from the beginning and further increasing with spoilage. Barramundi also seemed to be a Hx-forming species but with a more gradual Hx increase. These two species both have high levels of Hx but also very low and fluctuating levels of Ino.

In the giant trevally, Ino gradually increased, peaked and then plateaued around day 10 (with a significant change, $p < 0.05$), followed by a very slow decrease to the level of Hx thereafter. The Hx only increased significantly ($p < 0.05$) on day 18, i.e., at the last sampling point. Thus, the giant trevally can be categorized between the two categories, i.e., it forms Hx at significant levels, but only after the fish has passed its shelf life, while Ino is present at high levels even during the late stage of spoilage.

Cobia showed a gradual decrease in IMP down to negligible levels, fluctuating Ino levels and a slow/gradual increase in Hx, reaching intermediate concentrations. Similarly, milkfish showed a pattern that fell between the two basic categories, i.e., it gradually formed Hx at significant concentrations, but Ino also remained at high levels. The shelf life of the species was indicated by the significant ($p < 0.05$) increase in Ino on day 8, while Hx was mainly formed long after the fish had passed the acceptability limit. The coral trout can also be grouped into the intermediate category (between Hx-forming and non-forming species), but more toward the non-forming ones; the rapid increase ($p < 0.05$) in Ino on day 8 also indicated the end of its shelf life. Hx was formed slowly and increased gradually, but it remained at relatively low levels even after the fish had surpassed its shelf life. The existence of different kinetics in IMP breakdown and the variability in the formation of Ino or Hx as final products in different fish species have been previously reviewed [24], but none of the herein-studied species were included, and therefore, this is an important new finding.

The microbial load in the muscle tissue of the studied species under ice storage is graphically presented in Figure 4. Although there is no legal safety barrier related to microbial presence in fish, a microbial load of log 6 to log 7 is generally considered the microbial threshold for fish spoilage, indicating unsatisfactory fish in good manufacturing practice [28,29]. There were cases, such as the snubnose pompano, giant trevally and barramundi, where this threshold was reached before the sensory determination of the end of shelf life. In particular, it was reached on day 12 in snubnose pompano (sensory acceptability limit: day 16), in giant trevally on day 11 (sensory acceptability limit: day 15) and in barramundi on day 12 (sensory acceptability limit: day 16). In a previous study, where a similar pompano species (*Pompano falcatus*) was stored under refrigeration, a shelf life of 18 days coincided with a total microbial load of log 6 [30]. *E. coli* is not normally part of fish's microbiota and is considered a contaminant [31,32]. The enterobacteria load that was measured in a shelf life study of ice-stored cobia [9] was much higher after 15 days of storage than our *E. coli* measurements in cobia, but this may well be due to contamination of the stored fish, since the rest of the microbial population was similar to our findings.

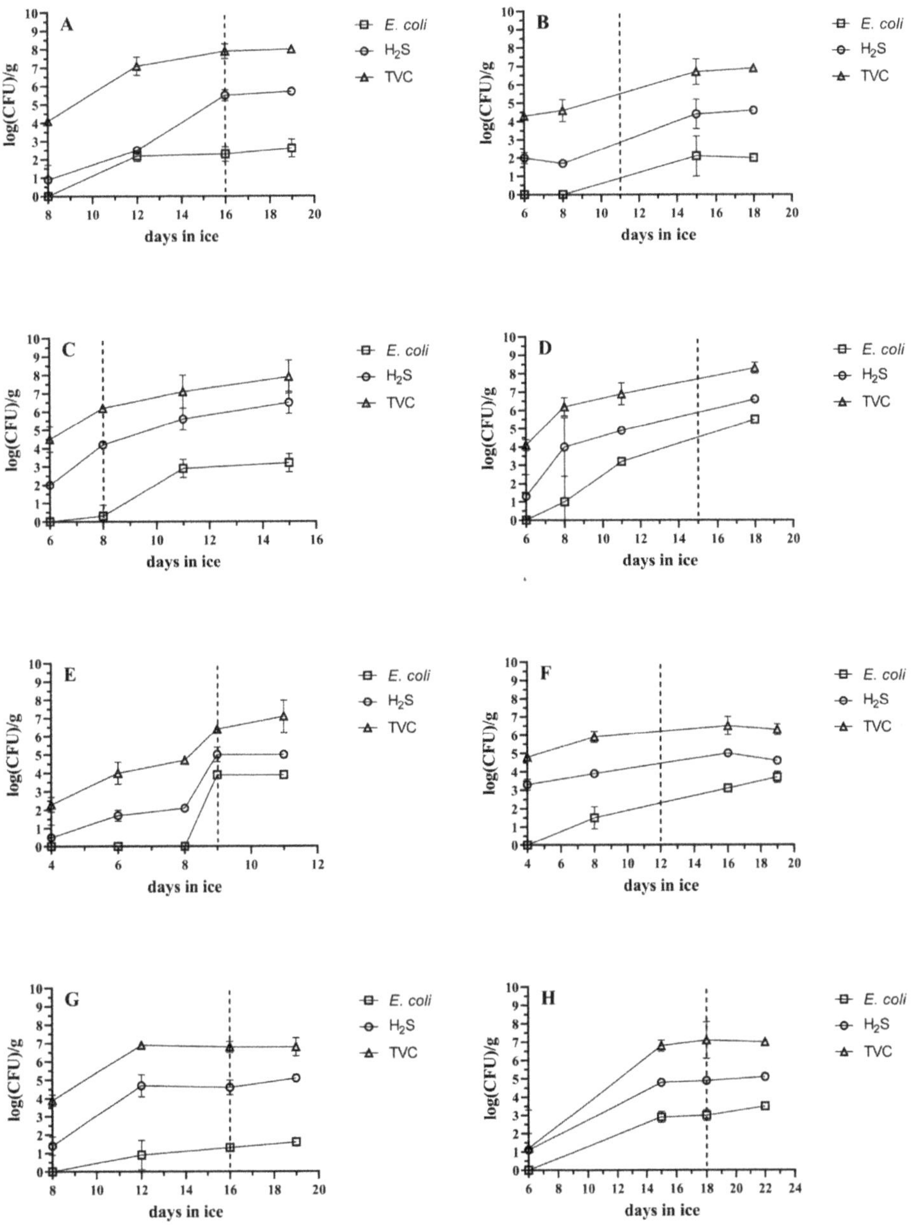

Figure 4. Microbial spoilage of (**A**) barramundi, (**B**) cobia, (**C**) coral trout, (**D**) giant trevally, (**E**) milkfish, (**F**) mangrove red snapper, (**G**) snubnose pompano and (**H**) sobaity bream during ice storage. TVC (△), H2S-producing bacteria (○) and *E. coli* (□). The dotted line represents the day that the acceptability limit was reached (end of shelf life). The bars represent the standard deviations (n = 3).

4. Conclusions

The spoilage patterns of eight important wild and farmed Saudi Arabia-produced fish species were studied. Since there are scarce results on the freshness and shelf lives of the studied species when they are stored in ice, these results may serve as a valuable tool for freshness control and quality assurance during commercialization in local Saudi Arabian markets and the rest of the world. Our study showed that wild-caught species (cobia, coral trout, giant trevally, milkfish and mangrove red snapper) had shorter shelf lives than farmed ones (barramundi and sobaity bream), based on our evaluations using the species-specific QIM schemes. The ATP breakdown product concentrations and the K-value schemes were strictly dependent on the species, and the values at the point of rejection did not relate to whether the fish were a wild-caught or farmed species. Some species, such as the mangrove red snapper and barramundi, accumulated hypoxanthine as a final ATP product, while others, such as sobaity bream, formed inosine. The total microbial loads were within a range of log 6–8 at the end of the fish's shelf life and were also independent of whether the fish was of farmed or wild origin.

Supplementary Materials: The following supporting information can be downloaded at https://www.mdpi.com/article/10.3390/foods14040690/s1. Tables S1–S8: Supplementary-meterial QIM schemes; Figure S1: baramundi; Figure S2: cobia; Figure S3: coral trout; Figure S4: giant trevaly; Figure S5: milkfish; Figure S6: mangrove red snapper; Figure S7: snubnose pompano; Figure S8: sobaity bream.

Author Contributions: K.G.: Supervision, investigation, methodology and writing—original draft. D.K.: Investigation, methodology, data analysis and writing—review and editing. M.K.: Investigation and methodology. I.K.: Investigation. P.H.d.M.: Resources and project administration. S.H.: Resources and project administration. A.A.: Conceptualization and funding resources. Y.S.A.: Conceptualization and funding resources. A.H.W.M.: Conceptualization and funding acquisition. All authors have read and agreed to the published version of the manuscript.

Funding: This research was funded by the Ministry of Environment, Water, and Agriculture, Riyadh, Saudi Arabia (Applied Research Support for Enhancing Fisheries Production, Initiative No. 368).

Data Availability Statement: The original contributions presented in this study are included in the article/Supplementary Material. Further inquiries can be directed to the corresponding author.

Acknowledgments: The authors are grateful for the award to KAUST Beacon Development, King Abdullah University of Science and Technology, Thuwal, Jeddah, Saudi Arabia.

Conflicts of Interest: The authors declare no conflicts of interest.

References

1. Freitas, J.; Vaz-Pires, P.; Câmara, J.S. From aquaculture production to consumption: Freshness, safety, traceability and authentication, the four pillars of quality. *Aquaculture* **2020**, *518*, 734857. [CrossRef]
2. Huss, H.H. Quality and quality changes in fresh fish. In *FAO Fisheries Technical Paper*; Food and Agriculture Organization (FAO): Rome, Italy, 1995.
3. Tan, K.; Lim, L.; Peng, Y.; Cheong, K.L. Effects of food processing on the lipid nutritional quality of commercially important fish and shellfish. *Food Chem. X* **2023**, *20*, 101034. [CrossRef]
4. Duarte, A.M.; Silva, F.; Pinto, F.R.; Barroso, S.; Gil, M.M. Quality assessment of chilled and frozen fish—Mini review. *Foods* **2020**, *9*, 1739. [CrossRef]
5. Walayat, N.; Tang, W.; Wang, X.; Yi, M.; Guo, L.; Ding, Y.; Liu, J.; Ahmad, I.; Ranjha, M.M. Quality evaluation of frozen and chilled fish: A review. *eFood* **2023**, *4*, e67. [CrossRef]
6. Hyldig, G.; Martinsdóttir, E.; Sveinsdóttir, K.; Schelvis, R.; Bremner, A. Quality index methods. In *Sensory Analysis of Foods of Animal Origin*; CRC Press: Boca Raton, FL, USA, 2010; pp. 267–284.
7. FAO. *Fishery and Aquaculture Statistics—Yearbook 2020*; FAO: Rome, Italy, 2023.
8. Shellem, C.T.; Ellis, J.I.; Coker, D.J.; Berumen, M.L. Red Sea fish market assessments indicate high species diversity and potential overexploitation. *Fish. Res.* **2021**, *239*, 105922. [CrossRef]
9. Fogaça, F.H.d.S.; Gonzaga Junior, M.A.; Vieira, S.G.A.; Araujo, T.D.S.; Farias, E.A.; Ferreira-Bravo, I.A.; Silva, T.F.A.; Calvet, R.M.; Pereira, A.L.M.; Prentice-Hernández, C. Appraising the shelf life of farmed cobia, Rachycentron canadum, by application of a Quality Index Method. *J. World Aquac. Soc.* **2017**, *48*, 70–82. [CrossRef]

10. Alasalvar, C.; Taylor, K.; Öksüz, A.; Garthwaite, T.; Alexis, M.; Grigorakis, K. Freshness assessment of cultured sea bream (*Sparus aurata*) by chemical, physical and sensory methods. *Food Chem.* **2001**, *72*, 33–40. [CrossRef]
11. Ryder, J.M. Determination of adenosine triphosphate and its breakdown products in fish muscle by high-performance liquid chromatography. *J. Agric. Food Chem.* **1985**, *33*, 678–680. [CrossRef]
12. Katsouli, M.; Semenoglou, I.; Kotsiri, M.; Gogou, E.; Tsironi, T.; Taoukis, P. Active and Intelligent Packaging for Enhancing Modified Atmospheres and Monitoring Quality and Shelf Life of Packed Gilthead Seabream Fillets at Isothermal and Variable Temperature Conditions. *Foods* **2022**, *11*, 2245. [CrossRef]
13. Tsironi, T.; Ntzimani, A.; Gogou, E.; Tsevdou, M.; Semenoglou, I.; Dermesonlouoglou, E.; Taoukis, P. Modeling the Effect of Active Modified Atmosphere Packaging on the Microbial Stability and Shelf Life of Gutted Sea Bass. *Appl. Sci.* **2019**, *9*, 5019. [CrossRef]
14. Bernardo, Y.A.; Rosario, D.K.; Delgado, I.F.; Conte-Junior, C.A. Fish Quality Index Method: Principles, weaknesses, validation, and alternatives—A review. *Compr. Rev. Food Sci. Food Saf.* **2020**, *19*, 2657–2676. [CrossRef] [PubMed]
15. Freitas, J.; Vaz-Pires, P.; Câmara, J.S. Quality Index Method for fish quality control: Understanding the applications, the appointed limits and the upcoming trends. *Trends Food Sci. Technol.* **2021**, *111*, 333–345. [CrossRef]
16. Rajic, S.; Simunovic, S.; Djordjevic, V.; Raseta, M.; Tomasevic, I.; Djekic, I. Quality multiverse of beef and pork meat in a single score. *Foods* **2022**, *11*, 1154. [CrossRef] [PubMed]
17. Souza, J.T.; da Silva Oliveira, M.É.; Gonçalves, A.A.; Ozogul, Y. Seafood Quality Index Method (QIM)–Are All Studies Going in the Same Direction? *J. Aquat. Food Prod. Technol.* **2024**, *33*, 612–636. [CrossRef]
18. Huidobro, A.; Pastor, A.; Tejada, M. Quality index method developed for raw gilthead seabream (*Sparus aurata*). *J. Food Sci.* **2000**, *65*, 1202–1205. [CrossRef]
19. Grigorakis, K.; Taylor, K.; Alexis, M. Seasonal patterns of spoilage of ice-stored cultured gilthead sea bream (*Sparus aurata*). *Food Chem.* **2003**, *81*, 263–268. [CrossRef]
20. Šimat, V.; Bogdanović, T.; Krželj, M.; Soldo, A.; Maršić-Lučić, J. Differences in chemical, physical and sensory properties during shelf life assessment of wild and farmed gilthead sea bream (*Sparus aurata*, L.). *J. Appl. Ichthyol.* **2012**, *28*, 95–101. [CrossRef]
21. Surti, T.; Taylor, A.; Ma'Ruf, F. The effect of storage at tropical ambient temperature on the quality and shelf life of grouper (*Plectropomus maculatus*). *Int. J. Food Sci. Technol.* **2001**, *36*, 517–522. [CrossRef]
22. Huss, H.H. *Fresh Fish—Quality and Quality Changes: A Training Manual Prepared for the FAO/DANIDA Training Programme on Fish Technology and Quality Control*; Food & Agriculture Organization: Rome, Italy, 1988.
23. Chen, B.; Yan, Q.; Li, D.; Xie, J. Degradation mechanism and development of detection technologies of ATP-related compounds in aquatic products: Recent advances and remaining challenges. *Crit. Rev. Food Sci. Nutr.* **2025**, *65*, 101–122. [CrossRef] [PubMed]
24. Howgate, P. A review of the kinetics of degradation of inosine monophosphate in some species of fish during chilled storage. *Int. J. Food Sci. Technol.* **2006**, *41*, 341–353. [CrossRef]
25. Grigorakis, K.; Alexis, M.; Gialamas, I.; Nikolopoulou, D. Sensory, microbiological, and chemical spoilage of cultured common sea bass (*Dicentrarchus labrax*) stored in ice: A seasonal differentiation. *Eur. Food Res. Technol.* **2004**, *219*, 584–587. [CrossRef]
26. Mallouchos, A.; Mikrou, T.; Gardeli, C. Gas chromatography–mass spectrometry-based metabolite profiling for the assessment of freshness in gilthead sea bream (*Sparus aurata*). *Foods* **2020**, *9*, 464. [CrossRef]
27. Ahimbisibwe, J.B.; Inoue, K.; Shibata, T.; Aoki, T. Effect of bleeding on the quality of amberjack Seriola dumerili and red sea bream Pagrus major muscle tissues during iced storage. *Fish. Sci.* **2010**, *76*, 389–394. [CrossRef]
28. ICMSF. International Commission on Microbiological Specifications for Foods. In *Microorganisms in Foods 8*, 1st ed.; Swanson, K.M., Ed.; Springer: New York, NY, USA, 2011; pp. 33–40.
29. Alberghini, G.; Ben Mhenni, N.; Di Leva, V.; Forzano, R.; Miotti Scapin, R.; Pappalardo, P.M.; Giacometti, F.; Giaccone, V. A challenge test on *Pseudomonas* spp. as spoiling microorganism in fish fillets. *Heliyon* **2024**, *10*, e32170. [CrossRef] [PubMed]
30. Erikson, U.; Truong, H.T.; Le, D.V.; Pham, P.D.; Svennevig, N.; Phan, V.T. Harvesting procedures, welfare and shelf life of ungutted and gutted shortfin pompano (*Trachinotus falcatus*) stored in ice. *Aquaculture* **2019**, *498*, 236–245. [CrossRef]
31. Costa, R.A. Escherichia coli in seafood: A brief overview. *Adv. Biosci. Biotechnol.* **2013**, *4*, 450–454. [CrossRef]
32. Devane, M.L.; Moriarty, E.; Weaver, L.; Cookson, A.; Gilpin, B. Fecal indicator bacteria from environmental sources; strategies for identification to improve water quality monitoring. *Water Res.* **2020**, *185*, 116204. [CrossRef] [PubMed]

foods

MDPI

Article

Enhancement of the Storage Potential of Farmed Rainbow Trout (*Oncorhynchus mykiss*) by Using Algal (*Cystoseira myrica* and *Cystoseira trinodis*) Extract–Ice Combinations

Shima Ahmadi [1], Parastoo Pourashouri [1], Bahareh Shabanpour [1] and Santiago P. Aubourg [2,*]

[1] Department of Seafood Processing, Faculty of Fisheries and Environmental Sciences, Gorgan University of Agricultural Sciences and Natural Resources, Shahid Beheshti St., Gorgan 4913815739, Golestan, Iran; sh.ahmadi171@yahoo.com (S.A.); pourashouri.p@gau.ac.ir (P.P.); shabanpour@gau.ac.ir (B.S.)

[2] Department of Food Technology, Marine Research Institute (CSIC), c/Eduardo Cabello 6, 36208 Vigo, Spain

* Correspondence: saubourg@iim.csic.es

Abstract: An attempt to apply extracts of the brown algae *Cystoseira myrica* and *Cystoseira trinodis* for the quality enhancement of fish was carried out. Aqueous, ethanolic, and aqueous–ethanolic (1:1, *v*/*v*) extracts of both algae were included, respectively, in the icing system employed for the chilled storage of farmed rainbow trout (*Oncorhynchus mykiss*). Chemical and microbiological quality indices were determined for a 0–16-day storage period. At the end of the experiment, all alga-treated fish revealed lower ($p < 0.05$) pH values and lower ($p < 0.05$) lipid hydrolysis (free fatty acid assessment) and oxidation (thiobarbituric acid index) development when compared to Control samples. Regarding microbial activity development (aerobe, psychrophilic, *Enterobacteriaceae*, proteolytic, and lipolytic counts), lower average values were detected in most cases in fish corresponding to alga-treated batches; preservative effects were found more important at advanced storage times. In general, water and water–ethanol extracts led to higher ($p < 0.05$) inhibitory effects than their counterpart ethanol extracts. Higher ($p < 0.05$) total polyphenol values were detected in water and water–ethanol extracts of both algae than in their counterpart extracts obtained only with ethanol. A novel, simple, and practical strategy for the quality enhancement and commercialization of chilled farmed rainbow trout is proposed by employing different extracts obtained from both *Cystoseira* species.

Keywords: *Cystoseira* sp.; extracts; farmed rainbow trout; chilling; ice medium; microbial development; lipid damage; quality; polyphenol content

1. Introduction

Rainbow trout (*Oncorhynchus mykiss*) is a highly valuable species for aquaculture due to its ability to be raised and high growing capabilities [1,2]. With a production volume of approximately 740,000 tons in 2020, it is considered the most widely farmed salmonid species in freshwater [3]. Currently, this species is farmed all over the world and shows the ability to live within a wide range of water temperatures, i.e., from 0 to 23 °C. According to its high nutritional quality, rainbow trout farming can contribute to addressing the increasing demand for farmed fish production to help feed the ever-growing human population [2,4]. However, the shelf-life time of this species is short due to being highly susceptible to chemical deterioration and microbial spoilage. Although several preservative strategies (packaging and the addition of essential oils or plant extracts) have been successfully applied for maintaining the quality of rainbow trout as a fresh product [5,6], further research is still needed in order to offer the fish trade and consumers simple and practical strategies for the quality retention of this seafood.

Ensuring the quality and safety of seafood has become a paramount concern worldwide for consumers, producers, and regulatory agencies. Temperature control plays a crucial role in preventing enzymatic and microbiological reactions that can lead to spoilage in seafood [7,8]. Among the various preservation techniques employed for seafood, the use of ice is a widely accepted method [9]. Although this method significantly extends the average shelf-life time of fish and shellfish, biochemical and microbial processes can still lead to a noticeable decrease in the sensory quality of chilled fish [10,11]. To address this concern, the industries involved in fish preservation are nowadays exploring more effective natural approaches to maintain the high quality of the product and ensure the consumers' safety [12,13]. The development of ice with preservative properties has gained remarkable interest in recent years due to its potential to reinforce food safety and promote human health [14,15]. Recent studies have focused on enhancing the storage properties of marine species by employing ice containing different kinds of preservative compounds. These studies have investigated the efficacy of various natural preservatives, such as organic acids (ascorbic, lactic, and citric acids) [16,17] and different kinds of plant extracts like betel leaf (*Piper bettle*) (4-methoxy-isophthalic acid, phenol, and propanoic acid) [18], rosemary (water-soluble compounds) (*Rosmarinus officinalis*) [19], pistachio (*Pistachia vera*) (phenolic compounds) [20], or *Garcinia* sp. (aqueous phenolic compounds) [21].

Macroalgae have attracted great attention as a natural source of compounds with antioxidant, antimicrobial, anti-inflammatory, and anti-cancer properties [7,22,23]. These beneficial properties can be explained on the basis of the presence of preservative metabolites, such as polyphenols, terpenes, phlorotannins, and fucoxanthin. Brown algae, in particular, have proved promising antioxidant activity due to the presence of various polysaccharides including fucoidan, laminarin, and alginic acid [24–26]. Among brown algae, the *Cystoseira* genus, belonging to the *Sargassaceae* family, comprises approximately 40 species and is found along the Eastern Atlantic and Mediterranean coastlines [27]. Previous studies have focused on the extraction and purification of bioactive compounds from species belonging to this genus. According to the different kinds of in vitro assays, water-soluble compounds (i.e., fucoidan and sodium alginate compounds) obtained from *Cystoseira compressa* showed remarkable antioxidant properties [28]. A valuable presence of phlorotannin compounds and a notable antioxidant activity (DPPH assay) were detected in crude extracts from *Cystoseira trinodis* collected from the Indian coast [29]. Debromolaurinterol and fucosterol, obtained from different kinds of extracts (petroleum ether, dichloromethane, or methanol) from *Cystoseira myrica* and *C. trinodis*, were found to be responsible for the antioxidant and antimicrobial properties in different in vitro assays [30]. A brassinosteroid-related metabolite with anticarcinogenic properties was extracted and isolated from ethanolic extracts of *C. myrica* [31]. Regarding seafood systems, the presence of an aqueous ethanol extract of *C. compressa* in the icing medium led to an inhibitory effect on the lipid damage and microbial activity development of chilled horse mackerel (*Trachurus trachurus*) [32]. Additionally, the addition of an aqueous extract of *Cystoseira stricta* to the glazing medium led to a remarkable decrease in the lipid oxidation development in frozen Atlantic Chub mackerel (*Scomber colias*) [33].

In agreement with the above-mentioned considerations, the present study addressed the quality enhancement of chilled farmed rainbow trout (*O. mykiss*). For this, ice containing different kinds of *C. myrica* and *C. trinodis* extracts was employed as the icing system during the chilled storage of this fish species. The evolution of the chemical changes and microbial development were determined in the fish muscle throughout a 16-day storage period. The validity of this novel, simple, and practical strategy was verified via a comparison to a fish Control batch stored under traditional ice, i.e., ice prepared only with water.

2. Materials and Methods

The chemical reagents and solvents employed were of reagent grade (Merck, Darmstadt, Germany); otherwise, the supplier is expressed.

2.1. Extraction of Seaweed

The brown algae *C. myrica* and *C. trinodis* were collected from the coast of Qeshm Island (Iran) in the Persian Gulf and were identified in the Faculty of Science, Guilan University (Rasht, Iran). A voucher specimen was deposited at the herbarium of Guilan University. Once in our laboratory, the algae were thorough washed, dried, and finely powdered in the shade [31].

Both powdered algae were then subjected to extraction using absolute ethanol, distilled water, or a water–ethanol (1:1, v/v) mixture (15 g dried alga·240 mL^{-1} in all cases), respectively. For this, the suspensions were homogenized (Ultraturrax, IKA, Taufkirchen, Germany) for 30 s, immersed in a magnetic stirrer (Vortex, Scientific Industries, Bohemia, NY, USA) for 12 h, and sonicated (J. P. Selecta, S. A., Barcelona, Spain) for 5 min at 40 KHz (1 s on and 1 s off) in a container including water and ice. Then, the samples were centrifuged (Beckman Coulter ALLEGRA X12R, Brea, CL, USA) at 3500× g (10 min, 4 °C), and the resulting supernatants were isolated and subjected to lyophilization (FD8515-C60, Ilshin Biobase Europe, Ede, The Netherlands).

2.2. Icing Systems Preparation

To prepare the icing systems, 4.2 mg of each lyophilized extract was mixed with distilled water (2 × 12 mL). The mixture was stirred for 30 s and centrifuged at 3500× g for 10 min at 4 °C. Then, the supernatant (20 mL) was taken and diluted to 6 L with distilled water (0.7 mg lyophilized extract·L^{-1} water solution). This solution was stored in a polyethylene bag at −18 °C and employed as the icing system for fish storage. This process was carried out for each kind of extract for both algae. Additionally, traditional or common ice was obtained from distilled water (6 L), packaged, and stored at −18 °C. This ice was employed as the Control condition.

Prior to being used, all icing systems were ground in order to obtain flaked ice. Then, flaked ice was added to the fish in order to carry out the different storage experiments.

2.3. Preliminary Trials

The concentration of algal extracts in the icing media used in this study was chosen according to previous preliminary trials. As a first step, the three kinds of lyophilized extracts of both algae were assessed for their antibacterial activities (MIC and MBC assays) on different kinds of bacteria; as a result, increased microbial inhibition (Gram-negative and Gram-positive bacteria) was detected by increasing the extract concentration. Additionally, icing systems containing different concentrations of all kinds of lyophilized extracts (0.1–1 mg·L^{-1} solution range) of both algae were prepared and employed for a storage trial of rainbow trout. During this storage period, possible modification of the natural external color and odor of the fish due to the presence of the algal extract was analyzed. This evaluation aimed to identify the appropriate concentration of the extract that did not compromise the natural color and odor of the fish and maintained a valuable antimicrobial activity. As a result, a concentration of 0.7 mg lyophilized extract·L^{-1} solution showed to be the most concentrated one that did not influence the mentioned sensory descriptors. Consequently, this concentration was considered for carrying out the present study.

2.4. Raw Fish, Chilled Storage, and Sampling

A total of 105 rainbow trout (*O. mykiss*) (130 ± 25 g and 22 ± 2 cm) were obtained from a breeding pond (Gorgan, Golestan Province, Iran). The fish were transferred to the laboratory on ice and divided into 7 groups with 15 fish in each. The different fish groups were introduced in separated boxes and directly surrounded by different types of ice. The different batches were as follows: ice without algal extract (Control batch), ice containing aqueous extract from *C. myrica* (MW batch) and *C. trinodis* (TW batch), ice containing ethanolic extract from *C. myrica* (ME batch) and *C. trinodis* (TE batch), and ice containing aqueous–ethanolic extract from *C. myrica* (MWE batch) and *C. trinodis* (TWE batch).

All fish samples were stored in isolated boxes and kept in a refrigerated room (4 ± 1 °C). The fish-to-ice ratio was maintained at 1:1 (*w*/*w*). Throughout the storage period, the water resulting from ice melting was drained and appropriate amounts of new ice were replenished. The analysis of chilled fish was performed after 0, 4, 8, 12, and 16 days of chilled storage.

2.5. Determination of Total Phenolic Content (TPC) of Lyophilized Extracts

The TPC of lyophilized extracts from both algae was quantified using the Folin–Ciocalteu colorimetric method [34]. For this, an aliquot of the sample was added to a sodium carbonate solution and to the Folin–Ciocalteu reagent. The mixture was stored in the dark at room temperature for a 30 min period; then, measurement at 720 nm absorbance was carried out. For quantitative purposes, gallic acid was used as a standard to construct the calibration curve. The results were calculated as mg gallic acid·g^{-1} dried algal extract.

2.6. Determination of pH and Free Fatty Acid (FFA) Values

The pH value of rainbow trout muscle was determined (390 pH ISE Meter, Beckman Coulter Inc., Fullerton, CA, USA) via dilution of the fish muscle in distilled water in a 1:10 (*w*/*v*) ratio [35].

The content of FFAs was measured following the method proposed by Rukunudin et al. [36]. For this, 5 g of fish fillet was cut into small pieces and homogenized with 30 mL of chloroform (Ultraturrax, IKA, Taufkirchen, Germany). Then, the fillet particles were separated from the fat-containing fraction via centrifugation (5000× *g*, 5 min, 4 °C). Finally, the lipid fraction was titrated with ethanolic potassium hydroxide by employing a 1% phenolphthalein solution as an indicator. The results are expressed as g FFAs·100 g^{-1} lipids.

2.7. Measurement of Lipid Oxidation Development

Lipids from rainbow trout muscle were extracted by using the Bligh and Dyer [37] procedure; this method employs a chloroform/methanol (1/1, *v*/*v*) mixture with a solvent–tissue ratio of 4:1 (*v*/*w*). Quantification was carried out in agreement with Herbes and Allen [38]. The lipid content is expressed as g·100 g^{-1} fish muscle.

The peroxide value (PV) was assessed according to previous research [39]. For this, samples of minced fish were mixed with an acetic acid–chloroform (3:2, *v*/*v*) mixture and filtered by employing a Whatman No. 1 filter paper. To the filtered solution, a saturated potassium iodide solution and a starch solution were added. Titration was carried out against a standard solution of sodium thiosulfate. The PV was calculated using the following equation and expressed as peroxide milliequivalents per kg of sample: $PV = \{(S \times N)/W\} \times 1000$, where "S" is the volume of titration (mL), "N" is the normality of the sodium thiosulfate solution, and "W" is the sample weight (g).

The thiobarbituric acid index (TBA-i) was assessed in agreement with Buege and Aust [40]. The method is based on the reaction between a trichloroacetic acid extract of the fish muscle and a thiobarbituric acid (TBA) solution. The content of TBA reactive substances (TBARSs) was calculated by employing a standard curve obtained with 1,1,3,3-tetraethoxy-propane (TEP). This curve was prepared by employing an aqueous solution of TEP (0.24 moles TEP·L^{-1}); for this, a range of 0.1×10^{-8} to 4×10^{-8} moles of malondialdehyde was used. The value of this index was calculated as mg malondialdehyde·kg^{-1} muscle.

2.8. Assessment of Microbial Activity

The bacterial content of the chilled rainbow trout muscle was determined following previous methodologies [41–43]. Thus, 10 g muscle portions were thoroughly homogenized with 90 mL of a sodium chloride solution. Then, the homogenized suspension was subjected to successive dilutions, and various culture media were used for surface culturing. For total viable bacteria counts, plate count agar (PCA) culture media were used, with the plates being incubated at 30 °C for 48 h. Similarly, PCA culture medium was utilized for psychrophilic bacteria counts, the plates being incubated at 4 °C for 5 to 10 days. To assess *Enterobacteriaceae* counts, Violet Red Bile Agar was used for surface culturing, and the samples were incubated at 37 °C for 24 h. For the identification of bacteria with proteolytic and lipolytic phenotypes, casein–agar and tributyrin–agar media were employed, respectively. The incubations of both media were performed at 30 °C for 48 h. In all bacterial groups, counts were calculated as log CFU·g^{-1} muscle.

2.9. Statistical Analysis

Statistical analysis of chemical and microbial values was carried out by employing SAS® software version 9.4. (SAS Campus Drive, Cary, NC, USA). A completely randomized design was employed to analyze the data using a split plot test over time. Subsequently, a mean comparison was performed using the Tukey test at a 95% confidence level. The effect of algal extract in the icing medium was studied. Each processing condition was performed by carrying out three replicates (n = 3).

3. Results and Discussion

3.1. Determination of the TPC of Dried Algal Extracts

The TPCs of crude algal extracts are presented in Figure 1. The range of TPCs was 30.72–55.87 mg gallic acid·g^{-1} dry algal extract. Among the different extracting systems, the highest average contents were found in the aqueous extract (55.87 ± 1.72 mg gallic acid·g^{-1}) and in the aqueous–ethanolic extract (53.76 ± 3.83 mg gallic acid·g^{-1}) of *C. trinodis*, followed by the aqueous extract (52.60 ± 2.31 mg gallic acid·g^{-1}) of *C. myrica*. Such algal extracts showed higher ($p < 0.05$) TPCs than the remaining algal extracts. Notably, ethanol extracts exhibited the lowest ($p < 0.05$) TPCs in both species.

Seaweeds contain a great diversity of water-soluble compounds such as proteins, peptides, sulfated polysaccharides, glycosides, salts, and low-molecular-weight organic acids [24–26]. Compared to green and red algae, brown ones exhibit higher levels of phenolic compounds with a remarkable presence of phlorotannin molecules [44,45]. The number of phenolic compounds in algae can fluctuate because of factors such as environmental conditions, season, or habitat [34,46]. Additionally, the choice of the extracting solvent can significantly influence the chemical composition of the extracts, as highlighted by Sathya et al. [29] and Begum et al. [30]. Compared to the present results, higher TPC values (40.8 ± 8.3 mg gallic acid·g^{-1} dry alga) were detected in the brown alga *Bifurcaria bifurcata* [47] but lower (11.0 ± 1.0 mg gallic acid·g^{-1} dried alga) in the brown alga *Undaria pinnatifida* [48].

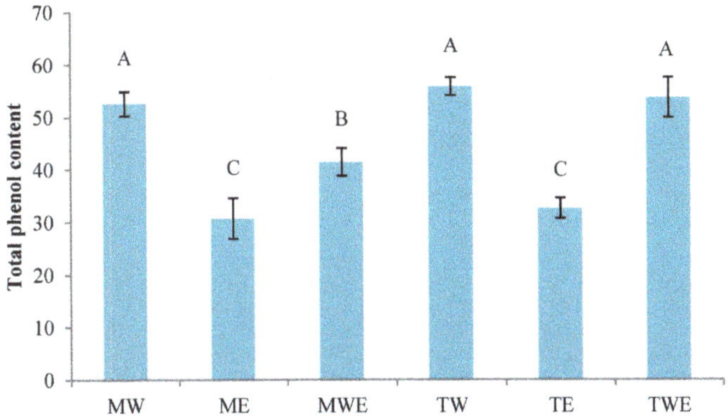

Figure 1. Determination of total phenolic compound contents (mg gallic acid·g^{-1} dried algal extract) in different algal extracts. Mean values of three (n = 3) replicates; standard deviations are indicated with bars. Mean values with different letters (A–C) indicate significant ($p < 0.05$) differences. Algal extract abbreviations: MW (water extract of *C. myrica*), ME (ethanolic extract of *C. myrica*), MWE (water–ethanolic extract of *C. myrica*), TW (water extract of *C. trinodis*), TE (ethanolic extract of *C. trinodis*), and TWE (water–ethanolic extract of *C. trinodis*).

A remarkable correlation between antioxidant capacity and polyphenol content was proved during the extraction using different solvents (water, ethanol, methanol, and water–methanol, 1:1, v/v) of the brown alga *Stypocaulon scoparium* [46]; thus, the aqueous extract showed the highest antioxidant activity (DPPH and RSA assays) and the highest phenolic compound content, with gallic acid being the most abundant. The authors indicated that water, being more polar than ethanol and methanol, facilitated the extraction of a greater amount of hydrophilic phenolic compounds [46]. This finding is in agreement with the results obtained in the current study, which indicates that extracts including water (i.e., aqueous and aqueous ethanol extractions) yielded better outcomes in terms of phenolic compounds. Water-soluble compounds (proteins, peptides, sulfated polysaccharides, salts, glycosides, and low-molecular-weight organic acids) have shown to be present in seaweed and be responsible for the preservation properties [49,50]. Notably, aqueous extracts of brown algae have been reported to contain phenolic preservative compounds like chlorogenic acid, vanillic acid, and caffeic acid, which were not observed in ethanol extracts [51]. Furthermore, previous studies have reported that water can lead to higher extraction yields as compared to other extracting systems [26,51,52].

3.2. Determination of the pH Value

The pH evolution of rainbow trout muscle stored in different kinds of ice is included in Figure 2. At day 0, pH values were found in the 6.24–6.27 range in all treatments ($p > 0.05$). However, during the storage period, a remarkable increase in all batches was detected. Notably, samples corresponding to the Control condition revealed higher pH values ($p < 0.05$) than the treated fish throughout the 8–16-day period. At the end of storage, batches corresponding to ice with the *C. trinodis* extract exhibited lower pH values ($p < 0.05$) than their counterparts corresponding to *C. myrica*-treated and Control samples.

The pH value plays a decisive role in influencing the activities of microorganisms and enzymes, thereby affecting the shelf-life time [9]. Researchers have attributed the increase in the pH value during the storage period to the activity of internal enzymes within the carcass, as well as to the activity of decay bacteria. These microorganisms produce volatile nitrogen compounds, including ammonia, trimethylamine, and other amine compounds, which subsequently contribute to the pH increase [53,54].

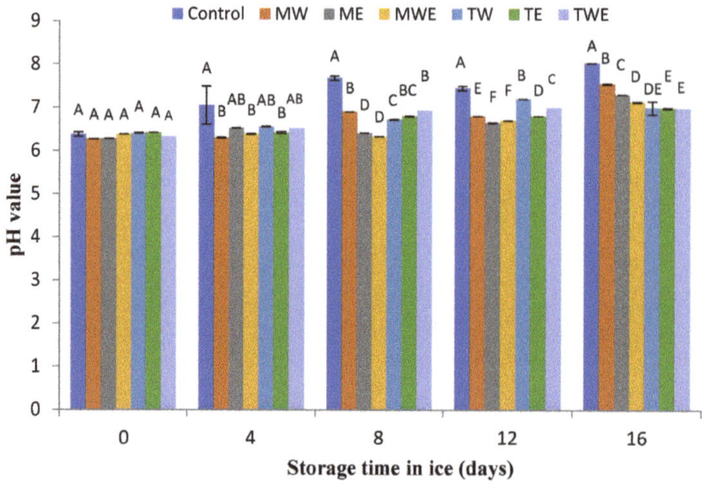

Figure 2. Assessment of the pH value in rainbow trout subjected to different icing systems. Mean values of three (n = 3) replicates; standard deviations are indicated with bars. Mean values with different letters (A–E) indicate significant ($p < 0.05$) differences. Abbreviations of icing systems are as expressed in Figure 1.

According to the present results, previous research accounts for the inhibition of a pH increase during the chilled storage by including brown algal extracts in the icing medium. This is the case of a combined ethanol–water extract of *C. stricta* during the chilled storage of horse mackerel (*T. trachurus*) for 11 days [32] and an ethanolic extract of *Bifurcaria bifurcata* during the megrim (*Lepidorhombus whiffiagonis*) storage for 14 days [47]. Regarding red algae, Arulkumar et al. [55] observed a pH decrease by including a methanolic extract of *Gracilaria verrucosa* in the icing medium employed during the chilled storage of Indian mackerel (*Rastrelliger kanagurta*) for 15 days.

3.3. Assessment of the FFA Content

The data illustrated in Figure 3 show the changes in the FFA content of fish stored in different kinds of ice for 16 days. At day 0, fillets from all kinds of samples were characterized by a ca. 0.75 FFA level. Then, a progressive increase in all samples was observed with the storage time. However, during the 12–16-day period, the FFA content of fish samples corresponding to all alga-treated batches showed lower mean values than their corresponding Control samples; differences were found to be significant ($p < 0.05$) in all cases at the end of the experiment. At this time, the ice treatment with aqueous extracts of *C. trinodis* and aqueous–ethanolic extracts of both algae displayed the lowest average values (4.51, 4.79 and 4.70 g FFAs·100 g^{-1} lipids, respectively).

The current findings suggest the protective properties of extracts obtained from the studied *Cystoseira* algae during the storage period, thus preventing the hydrolytic decay of fatty acids in the fish tissue. This result can be justified by the presence of preservative compounds in the seaweed extracts.

During the chilled storage of fish species, FFA formation has been reported to originate from the activities of fish endogenous enzymes and microorganisms [8,55–57]. Before the end of the microbial lag phase is reached, FFAs are signaled to be mostly produced by the activities of endogenous enzymes like phospholipases and lipases [58,59]. Then, an increase of processes related to bacterial catabolism would imply a significant increase in the microbial activity [9,56]. Despite the fact that the formation of FFAs itself does not lead to any notable nutritional loss, the evolution of the FFA content is considered of great interest due to remarkable implications related to quality loss. Thus, increased FFA levels can facilitate the presence of off-tastes and off-odors, detrimental protein changes, and, ultimately, negative effects on the overall quality of the product [9,56,60]. Furthermore, FFA presence has a remarkable effect on lipid oxidation development; thus, FFAs have been reported to present a lower stability to oxidation development than do high-molecular-weight lipid classes (i.e., triacylglycerol and phospholipid molecules) because of reduced steric hindrance to the initial development of oxidative reactions [61,62].

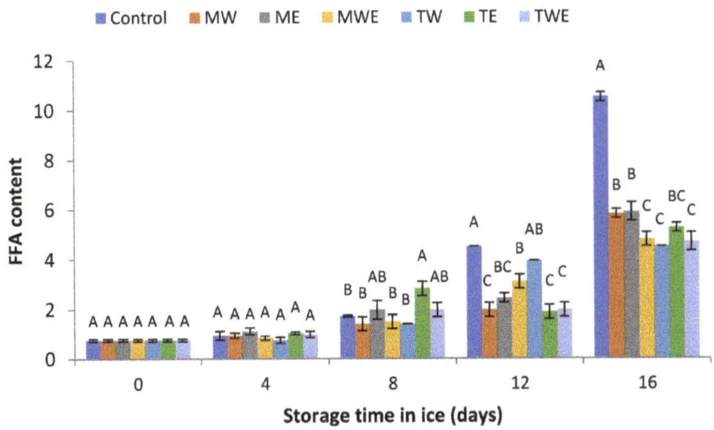

Figure 3. Assessment of the free fatty acid value (g·100 g^{-1} lipids) in rainbow trout subjected to different icing systems. Mean values of three (n = 3) replicates; standard deviations are represented with bars. Mean values with different letters (A–C) indicate significant ($p < 0.05$) differences. Icing systems are as expressed in Figure 1.

Regarding FFA formation, previous research related to algal species within the present algal genus have been reported for their valuable presence of different kinds of preservative secondary metabolites, such as fatty acids, terpenoids, triacylglycerols, phlorotannins, phenolic compounds, steroids, and polysaccharides [27]. In a seafood system, the addition of a combined ethanol–water extract of *C. stricta* to the icing system employed during horse mackerel (*T. trachurus*) storage for 11 days led to an inhibitory effect on lipid hydrolysis development [32]. Furthermore, the presence of an aqueous extract of the alga *C. stricta* to the glazing medium used during the frozen storage (−18 °C for 9 months) of mackerel (*S. colias*) [33] led to a lower FFA value of fish muscle throughout the frozen storage period.

Regarding brown algae in general, the addition of different kinds of extracts to the icing system used for the chilled storage of marine species has shown to inhibit the evolution of lipid hydrolysis development. This is the case of an ethanolic extract of *B. bifurcata* during a 14-day chilled storage of megrim (*L. whiffiagonis*) [47], a comparative study of aqueous and ethanolic extracts of *B. bifurcata* during a 13-day chilled storage of hake (*Merluccius merluccius*) [63], and an ethanolic extract of *U. pinnatifida* during a 9-day chilled storage of megrim (*L. whiffiagonis*) [48].

3.4. Determination of Lipid Oxidation Development

The evolution of lipid oxidation in chilled fish was determined via the PV (primary oxidation) and the TBA-i (secondary oxidation) (Figures 4 and 5, respectively).

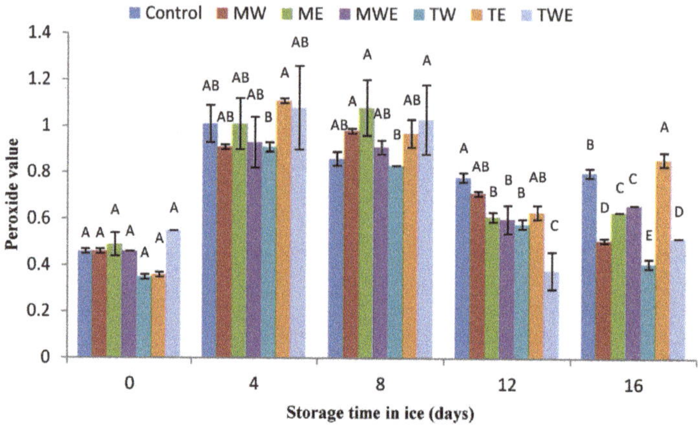

Figure 4. Determination of the peroxide value (meq. active oxygen·kg^{-1} lipids) in rainbow trout subjected to different icing systems. Mean values of three ($n = 3$) replicates; standard deviations are denoted with bars. Mean values with different letters (A–E) indicate significant ($p < 0.05$) differences. Icing systems are as expressed in Figure 1.

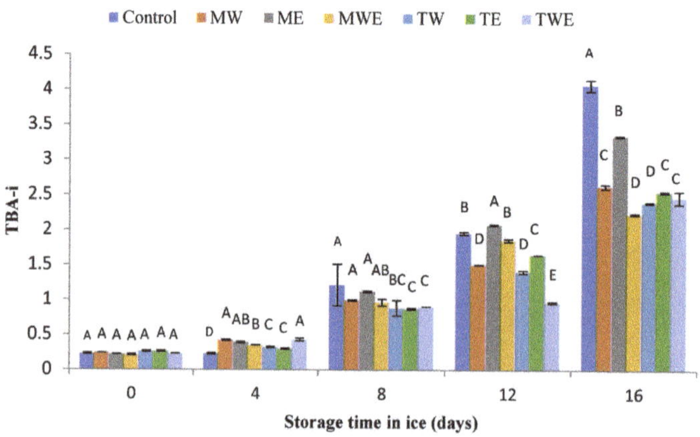

Figure 5. Assessment of the thiobarbituric acid index (TBA-i; mg malondialdehyde·kg^{-1} muscle) in rainbow trout subjected to different icing systems. Average values of three ($n = 3$) replicates; standard deviations are indicated with bars. Average values with different letters (A–E) indicate significant ($p < 0.05$) differences. Icing systems are as expressed in Figure 1.

During the ice storage, the PV exhibited an increase after 4 days in all batches (Figure 4). After this time, no subsequent increase was detected in any of the batches under study. Indeed, a general value decrease was detected for the 12–16-day period. Throughout the whole experiment, PVs were, in all cases, below a 1.2 meq·kg^{-1} lipids score, which can be considered as a low level [64]. Some differences among samples corresponding to the different batches could be observed, especially in the 12–16-day period; however, a definite trend about the effect on the PV of the algal extract presence in the icing medium could not be concluded.

The results of the TBA-i are presented in Figure 5. Low values were observed in all batches for the 0–4-day period. Then, a general increase was observed after 8 days of storage, which was followed in all batches by a progressive increase till the end of the experiment. During the 8–16-day period, lower average values were observed in most alga-treated samples than in Control ones; in the case of *C. trinodis*-treated fish, the differences proved to be significant ($p < 0.05$) in all cases. Regarding *C. myrica*-treated samples, batches corresponding to the water extracts showed lower ($p < 0.05$) TBARS values than their corresponding Control samples in the period corresponding to 12–16 days.

The oxidation mechanism of the lipid fraction is described as a multistep process in which various molecules are originated sequentially. Thus, the molecules produced during the initial steps (i.e., primary compounds like peroxides) are relatively unstable and would produce lower-molecular-weight molecules (i.e., secondary compounds like carbonyls) [9,64]. In the subsequent steps of the development of the lipid oxidation mechanism, both kinds of electrophilic compounds (peroxides and carbonyls) would react with compounds including nucleophilic groups like -NH_2 or -SH (namely proteins, peptides, free amino acids, aminated phospholipids, etc.) in the fish muscle; such reactions would lead to important losses in the fish product regarding acceptability (sensory acceptance decrease) and nutritional value (detrimental changes of the protein fraction) [65–67].

In the present research, the general low peroxide level indicates that a remarkable breakdown of peroxides has occurred (Figure 4). However, the presence of preservative compounds in alga-treated fish led to a lower ($p < 0.05$) level of TBARSs than that detected in the Control fish (Figure 5). The relative reduction in oxidation development in treated rainbow trout could be related to polyphenol presence in the extracts, which may contribute to a reduction in oxidation development on the basis of their antioxidant properties. These compounds are electron rich and are prone to enter into efficient electron-donation reactions and produce phenoxyl radical species as intermediates in the presence of oxidizing agents [68,69]. Among polyphenols, phlorotannins from *C. trinodis* have been shown to stabilize phenoxyl radicals through hydrogen bonds with an adjacent hydroxyl group [29].

Previous research indicates that species within the *Cystoseira* genus include a wide range of secondary metabolites, i.e., terpenoids, steroids, phlorotannins, phenolic compounds, and polysaccharides (merediterpenes and toluquinols), which exhibit potent antioxidant properties [27,70]. Thus, a combined ethanolic-aqueous extract of *C. stricta* was present in the icing system employed for the 11-day chilled storage of horse mackerel (*T. trachurus*) [32]; remarkably, decreased lipid oxidation (fluorescent compound formation) development was detected. Different kinds of extracts (petroleum ether, dichloromethane, or methanol) from *C. myrica* and *C. trinodis* showed a remarkable antioxidant activity (DPPH assay) [30]; this effect was shown to vary according to the extracting solvent employed. Two polysaccharides (i.e., a fucoidan and a sodium alginate) obtained from *C. compressa* showed remarkable antioxidant properties (ferrous ion chelation, ferric ion reduction, and DPPH radical scavenging) [28]. An aqueous extract of the alga *C. stricta* was added to the glazing system used for the frozen (−18 °C for 9 months) storage of Chub mackerel (*S. colias*) [33]; as a result, decreased lipid oxidation (assessment of peroxide, TBA, and fluorescence indices) development and a notable retention of polyunsaturated fatty acid and alpha-tocopherol values was observed.

Previous studies have also shown the antioxidant properties of brown algal extracts in general when employed for the quality retention of chilled marine species. Thus, ethanolic extracts of *B. bifurcata* inhibited the formation of lipid oxidation compounds (i.e., fluorescent compound formation) during the 14-day chilled storage of megrim (*L. whiffiagonis*) [47]. The addition of aqueous, ethanolic, or ethanolic–aqueous extracts of *Fucus spiralis* to the icing media used for the chilled storage of hake (*M. merluccius*) inhibited lipid oxidation development (TBA and fluorescence indices) [71]. An ethanolic extract of *U. pinnatifida* showed a remarkable antioxidant capacity (DPPH assay) and inhibited fluorescent compound formation in chilled megrim (*L. whiffiagonis*) when included in the icing medium during a 9-day storage period [48]. The addition of aqueous or ethanolic extracts of *B. bifurcata* to the icing system used during the 13-day chilled storage of hake (*M.*

merluccius) was comparatively studied [63]; neither extract provided a different effect, but both provoked the decreased formation of secondary lipid oxidation molecules (TBARSs) when compared to the Control condition.

3.5. Determination of Microbial Activity

The determination of the microbial activity evolution in chilled fish muscle was carried out via the assessment of total aerobe, psychrophilic, *Enterobacteriaceae*, proteolytic, and lipolytic counts (Figures 6–10, respectively).

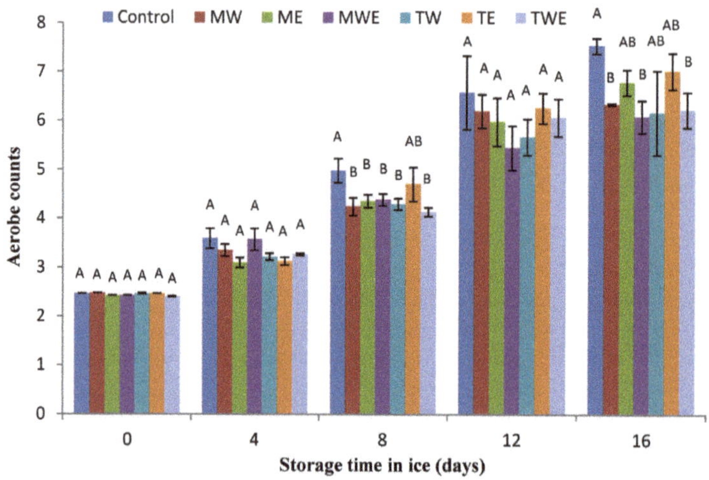

Figure 6. Determination of aerobe counts (log CFU·g^{-1} muscle) in rainbow trout subjected to different icing systems. Average values of three (n = 3) replicates; standard deviations are indicated with bars. Average values with different letters (A,B) indicate significant ($p < 0.05$) differences. Icing systems are as expressed in Figure 1.

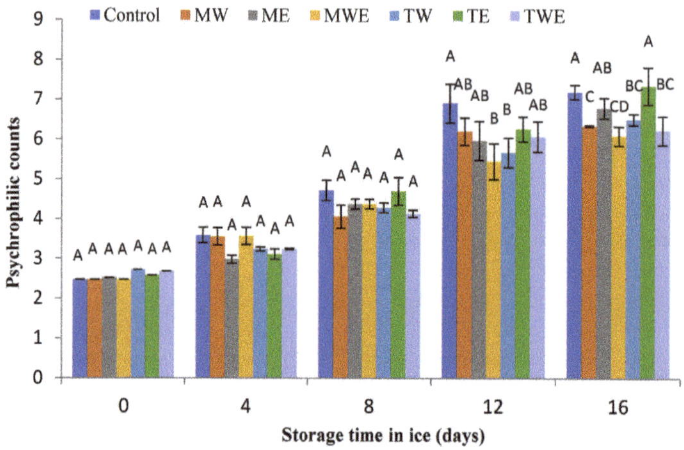

Figure 7. Determination of psychrophilic counts (log CFU·g^{-1} muscle) in rainbow trout subjected to different icing systems. Average values of three (n = 3) replicates; standard deviations are indicated with bars. Average values with different letters (A–D) indicate significant ($p < 0.05$) differences. Icing systems are as expressed in Figure 1.

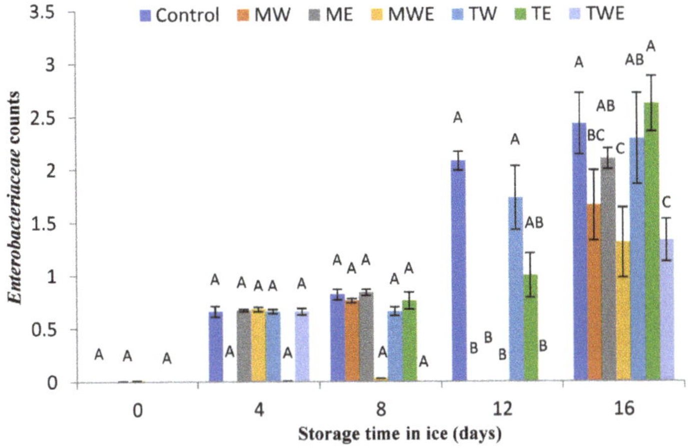

Figure 8. Determination of *Enterobacteriaceae* counts (log CFU·g^{-1} muscle) in rainbow trout subjected to different icing systems. Average values of three (n = 3) replicates; standard deviations are indicated with bars. Average values with different letters (A–C) indicate significant ($p < 0.05$) differences. Icing systems are as expressed in Figure 1.

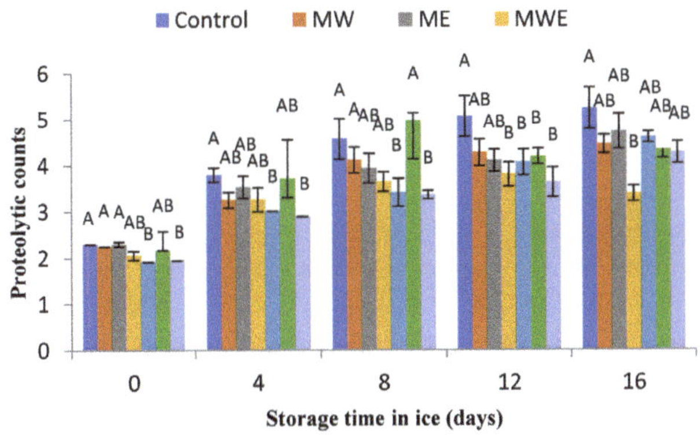

Figure 9. Determination of proteolytic counts (log CFU·g^{-1} muscle) in rainbow trout subjected to different icing systems. Average values of three (n = 3) replicates; standard deviations are indicated with bars. Average values with different letters (A,B) indicate significant ($p < 0.05$) differences. Icing systems are as expressed in Figure 1.

Mesophilic bacterial counting is recognized as a worthwhile tool for evaluating the quality and pollution of fishery products during or after processing [41,42]. In the present study, the initial total bacterial load for rainbow trout muscle ranged from 2.47 to 2.84 CFU·g^{-1} (Figure 6). An overall increase in the total bacterial count across all treatments was detected during storage. No differences ($p > 0.05$) were observed among batches during the 0–4-day period. However, lower average values were detected in all alga-treated fish than in Control samples after 8 days; such differences were found to be significant ($p < 0.05$) in most cases. At day 16, aqueous–ethanolic extracts of both algae led to lower ($p < 0.05$) aerobe counts than in the Control fish.

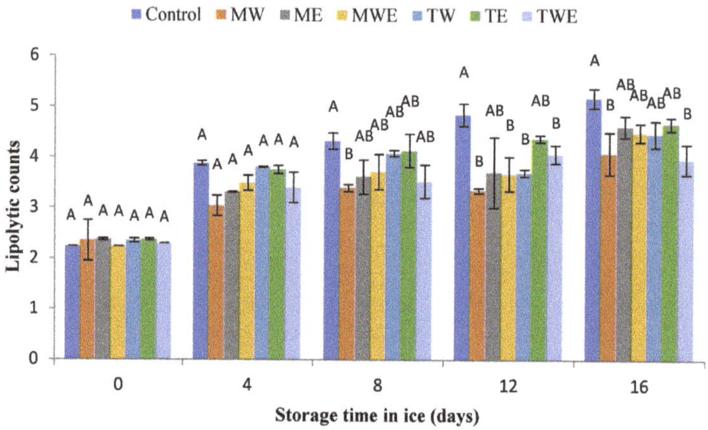

Figure 10. Determination of lipolytic counts (log CFU·g^{-1} muscle) in rainbow trout subjected to different icing systems. Average values of three (n = 3) replicates; standard deviations are indicated with bars. Average values with different letters (A,B) indicate significant ($p < 0.05$) differences. Icing systems are as expressed in Figure 1.

Psychrophilic bacterial counts play a relevant role in the deterioration of fresh fish and shellfish [72]. Figure 7 displays the changes obtained in psychrophilic bacteria counts for the Control and alga-treated fish during the present 16-day storage in ice. A progressive count increase in all batches was detected with the chilling time, so that a similar growth pattern was observed to that of aerobic bacteria. The initial count of psychrophilic bacteria ranged from 2.47 to 2.58 log CFU·g^{-1}. No significant differences ($p > 0.05$) were detected among batches for the 0–8-day period. Then, most alga-treated samples revealed lower average values than those corresponding to the Control condition; the differences were found to be significant ($p < 0.05$) in fish corresponding to the aqueous extract of *C. myrica* (day 16), the aqueous extract of *C. trinodis* (days 12 and 16), the aqueous ethanol extract of *C. myrica* (days 12 and 16), and the aqueous–ethanolic extract of *C. trinodis* (day 16).

During ice storage, psychrophilic bacteria may still maintain certain physiological activities and tolerate cold conditions [41,42]. This activity can be explained on the basis of the development of cold shock proteins, the maintenance of membrane fluidity, and the presence of intracellular enzymes and compatible solutes. It has been shown that common foodborne pathogenic bacteria, such as *Listeria monocytogenes* and *Yersinia enterocolitica*, can survive and grow in cold environments [73]; this fact may signify a potential risk to the microbial safety of ice during fish and shellfish preservation and consumption.

Figure 8 displays the changes in *Enterobacteriaceae* bacteria counts in rainbow trout samples during storage in ice. Up to 8 days of storage, the *Enterobacteriaceae* counts in the Control treatment and groups containing extracts of both algae were not different ($p > 0.05$). However, for the 12–16-day period, the counts in Control fish showed higher average values than most alga-treated samples; the differences were significant ($p < 0.05$) in fish subjected to aqueous extracts of *C. myrica* (days 12 and 16) and in aqueous ethanol extracts of *C. myrica* (days 12 and 16) and *C. trinodis* (days 12 and 16).

In the present study, the count evolution of microorganisms with a proteolytic phenotype is depicted in Figure 9. The results of this bacterial group showed similarities with the above-mentioned aerobe and psychrophilic count evolutions, which showed progressive increases with the storage time. Compared to the Control samples, lower values ($p < 0.05$) were detected in the 0–12-day period in fish corresponding to batches including water and water–ethanol extracts of *C. trinodis*. Additionally, the presence of aqueous–ethanolic extracts of *C. myrica* in the icing system led to lower levels ($p < 0.05$) than in Control samples in the 12–16-day period.

Proteolytic bacteria are specific bacteria that have been signaled to provide unfavorable modifications by contributing to protein breakdown and subsequent textural and nutritional losses of the fish muscle [47,71]. In the current chilled storage, a growing decrease in rainbow trout muscle was concluded for proteolytic bacteria. Therefore, an important preservative effect of the current fish against events related to protein hydrolysis of microbial origin would be implied by the presence of the algal extracts in the icing system.

The study of lipophilic bacteria was also carried out in the present study. This bacteria group is reported to increase lipid hydrolysis development, a mechanism with a notable deteriorative effect on fish quality during chilled storage by facilitating protein denaturation and lipid oxidation events [44]. The results obtained regarding the capacity of lipolytic bacteria to originate extracellular lipases with activity against lipid classes like TAGs and PLs in the current rainbow trout muscle are displayed in Figure 10. In agreement with the above-described results for other bacterial groups, the lipolytic bacteria count increased in all kinds of samples with the storage time. In Control samples, the population of these bacteria increased from 2.24 to 5.17 CFU·g^{-1} by taking into account the initial and end times of the storage period, respectively. However, this lipolytic bacterial development was partially inhibited by the addition of the algal extracts to the icing medium. Thus, an inhibitory effect was detected for the 8–16-day period (water extract of *C. myrica*), 12–16-day period (aqueous–ethanolic extract of *C. trinodis*), and day 12 (aqueous–ethanolic extract of *C. myrica* and aqueous extract of *C. trinodis*).

As a possible explanation, lipolytic bacteria would interfere with the hydrolysis of fats into fatty acids and glycerol by producing lipase enzymes [56]. The present inhibitory effect can be related to the stronger impact of the algal extracts in reducing the activity of lipolytic bacteria regarding fat hydrolysis and further oxidation. Furthermore, algal extracts may quench bacterial enzymatic reactions associated with fat oxidation in fish, according to the above-mentioned results regarding lipid hydrolysis (Figure 3) and oxidation (Figure 5) development.

The decreased microbial activity observed in the current research is in agreement with the polyphenol compound presence in all tested algal extracts. It has been demonstrated that extracts of brown seaweeds contain notable levels of phenolic compounds and the distinctive phlorotannin profile found in this kind of algae, identified for its antimicrobial properties, mainly comprised phloroglucinol, eckol, and dieckol [74]. Polyphenols and their antimicrobial mechanisms are assigned to the alteration of microbial cell permeability, leading to the loss of internal macromolecules and, additionally, to the modification of membrane functions, interruption of cellular integrity, and, ultimately, causing cell death [44,45,75]. However, it is important to remark that the efficiency of extracted algal compounds may differ according to the specific bacteria, the concentration of the compounds used, and the extraction condition. Regarding *Cystoseira* genus, a combined ethanol–water extract of *C. stricta* was present in the icing system employed for the 11-day storage of horse mackerel (*T. trachurus*) [32]; as a result, decreased microbial development (aerobe, psychrotroph, proteolytic, lipolytic, and *Enterobacteriaceae* counts) was detected. Different kinds of extracts (petroleum ether, dichloromethane, or methanol) from *C. myrica* and *C. trinodis* showed a remarkable antimicrobial behavior against different kinds of microorganisms (*Staphylococcus aureus, Streptococcus pyogenes, Pseudomonas aeruginosa, Escherichia coli, Candida albicans,* and *Cryptococcus neoformans*) [30]; the intensity of this effect was shown to vary according to the extracting solvent employed.

As for the present results, previous studies have already proved the microbial activity inhibition resulting from the use of different kinds of brown algal extracts during the chilled storage of marine species. Thus, ethanolic extracts of *B. bifurcata* led to the inhibited development of aerobes, psychrotrophs, lipolytic bacteria, proteolytic bacteria, and *Enterobacteriaceae* in megrim (*L. whiffiagonis*) during a 14-day chilled storage [47]. In a comparative study of aqueous, ethanolic, and ethanolic–aqueous extracts of *F. spiralis* employed as the icing condition during a 13-day storage of hake (*M. merluccius*) [71], contrary to the present results, the highest inhibitory effects on microbial development (aerobe, psychrotroph, lipolytic bacteria, and proteolytic bacteria counts) were detected in batches including ethanol extracts. A comparative study of aqueous and ethanolic extracts of *B. bifurcata* employed as the icing medium during the storage of hake (*M. merluccius*) for 13 days was carried out [63]; as a result, both batches showed an inhibitory effect on the microbial development (psychrotrophic and lipolytic counts), but no differences were detected between both kinds of extracts. The addition of ethanolic extracts of *U. pinnatifida* to the icing system provoked a decrease in the aerobe, lipolytic, and proteolytic bacteria counts in megrim (*L. whiffiagonis*) stored for 9 days under chilling conditions [48].

Regarding red macroalgae, a remarkable decrease in mesophilic and psychrophilic bacteria counts was proved during a 15-day chilled storage of Indian mackerel (*R. kanagurta*) by including a methanolic extract of the red alga *G. verrucosa* in the icing system employed [55]; the valuable effect was explained after the isolation and identification of bioactive agents (butylated hydroxytoluene, sulfurous acid, 1,2-propanediol, cyclononasiloxane, and tetracosamethylcyclo-dodecasiloxane) present in the algal methanolic extracts.

4. Conclusions

A preservative effect of horse mackerel quality was observed with the presence of both algal extracts in the icing media employed during the chilled storage. The inhibitory behavior ($p < 0.05$) on lipid oxidation development and microbial activity was found to be more important at advanced storage times. In general, water and water–ethanol extracts led to higher ($p < 0.05$) preservative effects than their counterpart ethanol extracts. Additionally, higher ($p < 0.05$) total polyphenol values were detected in water and water–ethanol extracts of both algae than in their counterpart extracts obtained only with ethanol. It is assumed that the presence of preservative hydrophilic compounds has been more important than that of lipophilic ones.

According to the results obtained, the presence of aqueous and aqueous ethanol extracts of both algae in the icing system can be considered a valuable method for preserving the quality of farmed rainbow trout during chilled storage. A strategy for the quality enhancement and commercialization of this species under chilling conditions is proposed by employing different kinds of extracts obtained from both *Cystoseira* species. This new, simple, and practical strategy aligns with the nowadays global interest in the search for new tools related to the replacement of synthetic preservative compounds by others obtained from natural sources in order to produce high-quality seafood and food in general.

Further research is envisaged to fully understand the specific benefits and optimal usage of both macroalgal extracts and scale up its employment to other marine species. According to the consumers' acceptance and nutritional value requirements, complementary studies regarding the effects on the sensory descriptors and changes in the protein fraction of fish muscle resulting in alga-treated fish ought to be developed. Additionally, further research should also include the isolation (HPLC and chromatographic technologies in general) and identification (mass spectrometry) of active molecules (polyphenols and others), which are involved in the inhibitory mechanisms of lipid oxidation and microbial development and are responsible for the preservative behavior detected. To optimize the use of algal extracts, future studies developing an optimized design (namely a response surface methodology study) and taking into account the independent variables of the algal treatment (concentration of the algal extracts and polarity of the extracting system, i.e., water–ethanol concentration ratio) should be conducted. In such cases, the kind of marine species (wild and farmed; fatty and lean) ought to be taken into account.

Author Contributions: Conceptualization, P.P., B.S. and S.P.A.; methodology, S.A.; data curation, S.A. and P.P.; writing—original draft preparation, P.P.; writing—review and editing, B.S. and S.P.A. All authors have read and agreed to the published version of the manuscript.

Funding: Part of this work was supported by Gorgan University of Agricultural Sciences and Natural Resources through student grant number 9521043103.

Institutional Review Board Statement: Not applicable.

Informed Consent Statement: Not applicable.

Data Availability Statement: The original contributions presented in the study are included in the article; further inquiries can be directed to the corresponding author.

Conflicts of Interest: The authors declare no conflicts of interest.

References

1. Singh, A.K. Emerging scope, technological up-scaling, challenges and governance of rainbow trout *Oncorhynchus mykiss* (Walbaum, 1792) production in Himalayan region. *Aquaculture* **2020**, *518*, 734826. [CrossRef]
2. Longo, A.; Kurta, K.; Vanhala, T.; Jeuthe, H.; de Koning, D.J.; Palaiokostas, C. Genetic diversity patterns in farmed rainbow trout (*Oncorhynchus mykiss*) populations using genome-wide SNP and haplotype data. *Anim. Gen.* **2024**, *55*, 87–98. [CrossRef]
3. FAO. *The State of World Fisheries and Aquaculture 2022. Towards Blue Transformation*; FAO: Rome, Italy, 2022; p. 46.
4. Duan, Z.; Zhou, Y.; Liu, W.; Shi, C.C.; Li, L.; Dong, Y.; Gao, Q.; Dong, S. Variations in flavor according to fish size in rainbow trout (*Oncorhynchus mykiss*). *Aquaculture* **2020**, *526*, 735398. [CrossRef]
5. Abedi, E.; Naseri, M.; Ghanbarian, G.A.; Vazirzadeh, A. Coverage of polyethylene film with essential oils of thyme (*Thymus daenensis Celak)* and savory (*Satureja bachtiarica Bunge*) for lipid oxidation control in rainbow trout (*Oncorhynchus mykiss*) fillets during short-term storage in the refrigerator. *J. Food Proc. Preserv.* **2016**, *40*, 483–491. [CrossRef]
6. Tavakoli, S.; Naseri, M.; Abedi, E.; Imani, A. Shelf-life enhancement of whole rainbow trout (*Oncorhynchus mykiss*) treated with Reshgak ice coverage. *Food Sci. Nutr.* **2018**, *6*, 953–961. [CrossRef] [PubMed]
7. Baptista, R.C.; Horita, C.N.; Sant'Ana, A.S. Natural products with preservative properties for enhancing the microbiological safety and extending the shelf-life of seafood: A review. *Food Res. Int.* **2020**, *127*, 108762. [CrossRef]
8. Ghali, A.; Dave, D.; Budge, S.; Brooks, M. Fish spoilage mechanisms and preservation: Review. *Amer. J. Appl. Sci.* **2010**, *7*, 859–877. [CrossRef]
9. Özoğul, Y. Methods for freshness quality and deterioration. In *Handbook of Seafood and Seafood Products Analysis*; Nollet, L., Toldrá, F., Eds.; CRC Press: Boca Raton, FL, USA; Taylor & Francis Group: Boca Raton, FL, USA, 2010; pp. 189–214, Chapter 13.
10. Amaral, R.A.; Pinto, C.A.; Lima, V.; Tavares, J.; Martins, A.P.; Fidalgo, L.G.; Silva, A.M.; Gil, M.M.; Teixeora, P.; Barbosa, J.; et al. Chemical-based methodologies to extend the shelf life of fresh fish—A Review. *Foods* **2021**, *10*, 2300. [CrossRef]
11. Russo, G.L.; Langellotti, A.L.; Torrieri, E.; Masi, P. Emerging technologies in seafood processing: An overview of innovations reshaping the aquatic food industry. *Comp. Rev. Food Sci. Food Saf.* **2024**, *23*, e13281. [CrossRef]
12. Mei, J.; Ma, X.; Xie, J. Review on natural preservatives for extending fish shelf life. *Foods* **2019**, *8*, 490. [CrossRef]
13. Gökoğlu, N. Novel natural food preservatives and applications in seafood preservation: A review. *J. Sci. Food Agric.* **2019**, *99*, 2068–2077. [CrossRef] [PubMed]
14. Losada, V.; Barros-Velázquez, J.; Gallardo, J.M.; Aubourg, S.P. Effect of advanced chilling methods on lipid damage during sardine (*Sardina pilchardus*) storage. *Eur. J. Lipid Sci. Technol.* **2004**, *106*, 844–850. [CrossRef]
15. Katsaros, G.; Koseki, S.; Ding, T.; Valdramidis, V.P. Application of innovative technologies to produce activated safe ice. *Curr. Opinion Food Sci.* **2021**, *40*, 198–203. [CrossRef]
16. Nirmal, N.P.; Benjakul, S. Effect of green tea extract in combination with ascorbic acid on the retardation of melanosis and quality changes of Pacific white shrimp during iced storage. *Food Bioprocess Technol.* **2012**, *5*, 2941–2951. [CrossRef]
17. García-Soto, B.; Fernández-No, I.C.; Barros-Velázquez, J.; Aubourg, S.P. Use of citric and lactic acids in ice to enhance quality of two fish species during on-board chilled storage. *Int. J. Refr.* **2014**, *40*, 390–397. [CrossRef]
18. Arulkumar, A.; Swain, B.; Paramasivam, S. Shelf life extension of sardines (*Sardinella albella*) using betel leaf (*Piper betle*) incorporated ice. *Food Bioprocess Technol.* **2020**, *13*, 1255–1260. [CrossRef]

19. Özyurt, G.; Kuley, E.; Balikçi, E.; Kaçar, Ç.; Gökdogan, S.; Etyemez, M.; Özogul, F. Effect of the icing with rosemary extract on the oxidative stability and biogenic amine formation in sardine (*Sardinella aurita*) during chilled storage. *Food Bioprocess Technol.* **2012**, *5*, 2777–2786. [CrossRef]
20. Nazeri, F.S.; Soltanizadeh, N.; Goli, S.A.H.; Mazaheri, S. Chemical stability of rainbow trout in icing medium containing pistachio (*Pistachia vera*) green hull extract during chilled storage. *J. Food Sci. Technol.* **2018**, *55*, 449–456. [CrossRef]
21. Apang, T.; Xavier, K.A.M.; Lekshmi, M.; Kannuchamy, N.; Layana, P.; Balange, A.K. *Garcinia* spp. extract incorporated icing medium as a natural preservative for shelf life enhancement of chilled Indian mackerel (*Rastrelliger kanagurta*). *LWT-Food Sci. Technol.* **2020**, *133*, 110086. [CrossRef]
22. Roohinejad, S.; Koubaa, M.; Barba, F.J.; Saljoughian, S.; Amid, M.; Grenier, R. Applications of seaweeds to develop new food products with enhanced shelf-life, quality and health-related beneficial properties. *Food Res. Int.* **2017**, *99*, 1066–1083. [CrossRef]
23. Sadeghi, A.; Rajabiyan, A.; Meygoli Nezhad, N.; Nabizade, N.; Alvani, A.; Zarei-Ahmady, A. A review on Persian Gulf brown algae as potential source for anticancer drugs. *Algal Res.* **2024**, *79*, 103446. [CrossRef]
24. Gupta, S.; Abu-Ghannam, N. Bioactive potential and possible health effects of edible brown seaweeds. *Trends Food Sci. Technol.* **2011**, *22*, 315–326. [CrossRef]
25. Andrade, P.B.; Barbosa, M.; Matos, R.P.; Lopes, G.; Vinholes, J.; Mouga, T.; Valentão, P. Valuable compounds in macroalgae extracts. *Food Chem.* **2013**, *138*, 1819–1828. [CrossRef] [PubMed]
26. Tierney, M.; Smyth, T.; Rai, D.; Soler-Vila, A.; Croft, A.; Brunton, N. Enrichment of phenol contents and antioxidant activity of Irish brown macroalgae using food-friendly techniques based on polarity and molecular size. *Food Chem.* **2013**, *139*, 753–761. [CrossRef]
27. Bruno de Sousa, C.; Gangadhar, K.N.; Macridachis, J.; Pavão, M.; Morais, T.R.; Campino, L.; Varela, J.; Lago, J.H.G. *Cystoseira algae* (*Fucaceae*): Update on their chemical entities and biological activities. *Tetrah. Asymm.* **2017**, *28*, 1486–1505. [CrossRef]
28. Hentati, F.; Delattre, C.; Ursu, A.V.; Desbrières, J.; Le Cerf, D.; Gardarin, C.; Abdelkafi, S.; Michaud, P.; Pierre, G. Structural characterization and antioxidant activity of water-soluble polysaccharides from the Tunisian brown seaweed *Cystoseira compressa*. *Carboh. Polym.* **2018**, *198*, 589–600. [CrossRef]
29. Sathya, R.; Kanaga, N.; Sankar, P.; Jeeva, S. Antioxidant properties of phlorotannins from brown seaweed *Cystoseira trinodis* (Forsskål) C. Agardh. *Arab. J. Chem.* **2017**, *10*, S2608–S2614. [CrossRef]
30. Begum, S.; Nyandoro, S.; Buriyo, A.; Makangara, J.; Munissi, J.; Duffy, S.; Avery, V.; Erdelyi, M. Bioactivities of extracts, debromolaurinterol and fucosterol from macroalgae species. *Tanzanian J. Sci.* **2018**, *44*, 104–116.
31. Hamdy, A.H.A.; Aboutabl, E.A.; Sameer, S.; Hussein, A.A.; Díaz-Marrero, A.R.; Darias, J.; Cueto, M. 3-Keto-22-epi-28-nor-cathasterone, a brassinosteroid-related metabolite from *Cystoseira myrica*. *Steroids* **2009**, *74*, 927–930. [CrossRef]
32. Oucif, H.; Miranda, J.M.; Ali Mehidi, S.; Abi-Ayad, S.M.; Barros-Velázquez, J.; Aubourg, S.P. Effectiveness of a combined ethanol-aqueous extract of alga *Cystoseira compressa* for the quality enhancement of a chilled fatty fish species. *Eur. Food Res. Technol.* **2018**, *244*, 291–299. [CrossRef]
33. Oucif, H.; Ali Mehidi, S.; Aubourg, S.P.; Abi-Ayad, S.M. Antioxidant effect of an aqueous extract of alga *Cystosiera stricta* during the frozen storage of Atlantic Chub mackerel (*Scomber colias*). *Bulg. Chem. Com.* **2020**, *52*, 227–233.
34. Rodríguez-Bernaldo de Quirós, A.; Frecha-Ferreiro, S.; Vidal-Pérez, A.M.; López-Hernández, J. Antioxidant compounds in edible brown seaweeds. *Eur. Food Res. Technol.* **2010**, *231*, 495–498. [CrossRef]
35. Ntzimani, A.; Angelakopoulos, R.; Semenoglou, I.; Dermesonlouoglou, E.; Tsiro, T.; Moutou, K.; Taoukis, P. Slurry ice as an alternative cooling medium for fish harvesting and transportation: Study of the effect on seabass flesh quality and shelf life. *Aquacult. Fish.* **2023**, *8*, 385–392. [CrossRef]
36. Rukunudin, I.H.; White, P.J.; Bern, C.J.; Bailey, T.B. A modified method for determining free fatty acids from small soybean oil sample sizes. *J. Amer. Oil Chem. Soc.* **1998**, *75*, 563–568. [CrossRef]
37. Bligh, E.; Dyer, W. A rapid method of total extraction and purification. *Can. J. Biochem. Physiol.* **1959**, *37*, 911–917. [CrossRef]
38. Herbes, S.E.; Allen, C.P. Lipid quantification of freshwater invertebrates: Method modification for microquantitation. *Can. J. Fish. Aquat. Sci.* **1983**, *40*, 1315–1317. [CrossRef]
39. Rahman, M.H.; Hossain, M.M.; Rahman, S.M.E.; Amin, M.R.; Oh, D.H. Evaluation of physicochemical deterioration and lipid oxidation of beef muscle affected by freeze-thaw cycles. *Korean J. Food Sci. Anim. Resour.* **2015**, *35*, 772–782. [CrossRef]
40. Buege, J.A.; Aust, S.D. Microsomal lipid peroxidation. *Meth. Enzym.* **1978**, *52*, 302–310.
41. Ben-Gigirey, B.; Vieites Baptista de Sousa, J.; Villa, T.; Barros-Velázquez, J. Changes in biogenic amines and microbiological analysis in albacore (*Thunnus alalunga*) muscle during frozen storage. *J. Food Prot.* **1998**, *61*, 608–615. [CrossRef]
42. Ben-Gigirey, B.; Vieites Baptista de Sousa, J.; Villa, T.; Barros-Velázquez, J. Histamine and cadaverine production by bacteria isolated from fresh and frozen albacore (*Thunnus alalunga*). *J. Food Prot.* **1999**, *62*, 933–939. [CrossRef]
43. Ben-Gigirey, B.; Vieites Baptista de Sousa, J.M.; Villa, T.G.; Barros-Velázquez, J. Characterization of biogenic amine-producing *Stenotrophomonas maltophilia* strains isolated from white muscle of fresh and frozen albacore tuna. *Int. J. Food Microb.* **2000**, *57*, 19–31. [CrossRef]

44. Eom, S.H.; Kim, Y.M.; Kim, S.K. Antimicrobial effect of phlorotannins from marine brown algae. *Food Chem. Toxicol.* **2012**, *50*, 3251–3255. [CrossRef] [PubMed]
45. Pérez, M.J.; Falqué, E.; Domínguez, H. Antimicrobial action of compounds from marine seaweed. *Mar. Drugs* **2016**, *14*, 52. [CrossRef]
46. López, A.; Rico, M.; Rivero, A.; Suárez de Tangil, M. The effects of solvents on the phenolic contents and antioxidant activity of *Stypocaulon scoparium* algae extracts. *Food Chem.* **2011**, *125*, 1104–1109. [CrossRef]
47. Miranda, J.M.; Ortiz, J.; Barros-Velázquez, J.; Aubourg, S.P. Quality enhancement of chilled fish by including alga *Bifurcaria bifurcata* extract in the icing medium. *Food Bioprocess Technol.* **2016**, *9*, 387–395. [CrossRef]
48. Campos, C.A.; Miranda, J.M.; Trigo, M.; Barros-Velázquez, J.; Aubourg, S.P. Effect of alga *Undaria pinnatifida* ("wakame") extract on the quality evolution of chilled megrim (*Lepidorhombus whiffiagonis*). *Bulg. Chem. Comm.* **2019**, *51*, 137–143.
49. Kuda, T.; Ikemori, T. Minerals, polysaccharides and antioxidant properties of aqueous solutions obtained from macroalgal beach-casts in the Noto Peninsula, Ishikawa, Japan. *Food Chem.* **2009**, *112*, 575–581. [CrossRef]
50. Pereira, L.; Amado, A.; Critchley, A.; Van de Velde, F.; Ribeiro-Claro, P. Identification of selected seaweed polysaccharides (phycocolloides) by vibrational spectroscopy (FTIR-ATR and FT-Raman). *Food Hydroc.* **2009**, *23*, 1903–1909. [CrossRef]
51. Farvin, K.; Jacobsen, C. Phenolic compounds and antioxidant activities of selected species of seaweeds from Danish coast. *Food Chem.* **2013**, *138*, 1670–1681. [CrossRef]
52. Athukorala, Y.; Lee, K.W.; Song, C.; Ahn, C.B.; Shin, T.S.; Cha, Y.J.; Shahidi, F.; Jeon, Y.J. Potential antioxidant activity of marine red alga *Grateloupia filicina* extracts. *J. Food Lipids* **2003**, *10*, 251–265. [CrossRef]
53. Gram, L.; Huss, H.H. Microbiological spoilage of fish and fish products. *Int. J. Food Microb.* **1996**, *33*, 121–137. [CrossRef] [PubMed]
54. Gram, L.; Dalgaard, P. Fish spoilage bacteria—Problems and solutions. *Curr. Opinion Biotechnol.* **2002**, *13*, 262–266. [CrossRef] [PubMed]
55. Arulkumar, A.; Paramasivam, S.; Miranda, J.M. Combined effect of icing medium and red alga *Gracilaria verrucosa* on shelf life extension of Indian mackerel (*Rastrelliger kanagurta*). *Food Bioprocess Technol.* **2018**, *11*, 1911–1922. [CrossRef]
56. Sikorski, Z.E.; Kolakowski, E. Endogenous enzyme activity and seafood quality: Influence of chilling, freezing, and other environmental factors. In *Seafood Enzymes. Utilization and Influence on Postharvest Seafood Quality*; Haard, N.F., Simpson, B.K., Eds.; Marcel Dekker: New York, NY, USA, 2000; pp. 451–487.
57. Aubourg, S.P.; Quitral, V.; Larraín, M.A.; Rodríguez, A.; Gómez, J.; Maier, L.; Vinagre, J. Autolytic degradation and microbiological activity in farmed Coho salmon (*Oncorhynchus kisutch*) during chilled storage. *Food Chem.* **2007**, *104*, 369–375. [CrossRef]
58. Whittle, K.; Hardy, R.; Hobbs, G. Chilled fish and fishery products. In *Chilled Foods. The State of the Art*; Gormley, T., Ed.; Elsevier Applied Science: Ney York, NY, USA, 1990; pp. 87–116.
59. Olafsdóttir, G.; Martinsdóttir, E.; Oehlenschläger, J.; Dalgaard, P.; Jensen, B.; Undeland, I.; Mackie, I.; Henehan, G.; Nielsen, J.; Nilsen, H. Methods to evaluate fish freshness in research and industry. *Trends Food Sci. Technol.* **1997**, *8*, 258–265. [CrossRef]
60. Aubourg, S.P. Damage detection in horse mackerel (*Trachurus trachurus*) during chilled storage. *J. Am. Oil Chem. Soc.* **2001**, *78*, 857–862. [CrossRef]
61. Labuza, T. Kinetics of lipid oxidation in foods. *CRC Crit. Rev. Food Technol.* **1971**, *2*, 355–405. [CrossRef]
62. Miyashita, K.; Tagagi, T. Study on the oxidative rate and prooxidant activity of free fatty acids. *J. Am. Oil Chem. Soc.* **1986**, *63*, 1380–1384. [CrossRef]
63. Miranda, J.M.; Zhang, B.; Barros-Velázquez, J.; Aubourg, S.P. Preservative effect of aqueous and ethanolic extracts of the macroalga *Bifurcaria bifurcata* on the quality of chilled hake (*Merluccius merluccius*). *Molecules* **2021**, *26*, 3774. [CrossRef]
64. Rustad, T. Lipid oxidation. In *Handbook of Seafood and Seafood Products Analysis*; Nollet, L.M., Toldrá, F., Eds.; CRC Press: Boca Raton, FL, USA; Francis and Taylor Group: Boca Raton, FL, USA, 2010; pp. 87–95.
65. Pokorný, J. Browning from lipid-protein interactions. *Prog. Food Nutr. Sci.* **1981**, *5*, 421–428.
66. Howell, N.K. Interaction of proteins with small molecules. In *Ingredient Interactions—Effects on Food Quality*; Gaonkar, A., Ed.; Marcel Dekker: New York, NY, USA, 1995; pp. 269–289.
67. Aubourg, S.P.; Medina, I.; Pérez-Martín, R. A comparison between conventional and fluorescence detection methods of cooking-induced damage to tuna fish lipids. *Z. Lebensm. Unters. Forsch.* **1995**, *200*, 252–255. [CrossRef] [PubMed]
68. Anthony, K.P.; Deolu-Sobogun, S.A.; Saleh, M.A. Comprehensive assessment of antioxidant activity of essential oils. *J. Food Sci.* **2012**, *77*, C839–C843. [CrossRef] [PubMed]
69. Maqsood, S.; Benjakul, S.; Shahidi, F. Emerging role of phenolic compounds as natural food additives in fish and fish products. *Crit. Rev. Food Sci. Nutr.* **2013**, *53*, 162–179. [CrossRef]
70. Valls, R.; Piovetti, L.; Banaigst, B.; Praud, A. Secondary metabolites from Morocco brown algae of the genus *Cystoseira*. *Phytochem.* **1993**, *32*, 961–966. [CrossRef]

71. Barros-Velázquez, J.; Miranda, J.M.; Ezquerra-Brauer, J.M.; Aubourg, S.P. Impact of icing systems with aqueous, ethanolic and ethanolic-aqueous extracts of alga *Fucus spiralis* on microbial and biochemical quality of chilled hake (*Merluccius merluccius*). *Int. J. Food Sci. Technol.* **2016**, *51*, 2081–2089. [CrossRef]
72. Bensid, A.; Ucar, Y.; Bendeddouche, B.; Özogul, F. Effect of the icing with thyme, oregano and clove extracts on quality parameters of gutted and beheaded anchovy (*Engraulis encrasicholus*) during chilled storage. *Food Chem.* **2014**, *145*, 681–686. [CrossRef]
73. Liao, X.; Shen, W.; Wang, Y.; Bai, L.; Ding, T. Microbial contamination, community diversity and cross-contamination risk of food-contact ice. *Food Res. Int.* **2023**, *164*, 112335. [CrossRef]
74. Del Juncal-Guzmán, D.; Camacho-González, C.E.; López-Cárdenas, F.G.; Sáyago-Ayerdi, S.G.; Sánchez-Burgos, J.A. Immune system: Inflammatory response. In *Marine Phenolic Compounds*; Pérez-Correa, J.R., Mateos, R., Domínguez, H., Eds.; Elsevier: Cambridge, MA, USA, 2023; pp. 415–429.
75. Pradhan, B.; Nayak, R.; Bhuyan, P.P.; Patra, S.; Behera, C.; Sahoo, S.; Ki, J.S.; Quarta, A.; Ragusa, A.; Jena, M. Algal phlorotannins as novel antibacterial agents with reference to the antioxidant modulation: Current advances and future directions. *Mar. Drugs* **2022**, *20*, 403. [CrossRef]

foods

MDPI

Article

Assessment of Lipid Peroxidation Products in Adult Formulas: GC-MS Determination of Carbonyl and Volatile Compounds Under Different Storage Conditions

Jorge Antonio Custodio-Mendoza [1,*], Alexandra Rangel Silva [2], Marcin A. Kurek [1], Paulo Joaquim Almeida [2], João Rodrigo Santos [2], José António Rodrigues [2] and Antonia María Carro [3,4,5,*]

1 Department of Technique and Food Development, Institute of Human Nutrition Sciences, Warsaw University of Life Sciences (WULS-SGGW), 02-776 Warszawa, Poland; marcin_kurek@sggw.edu.pl
2 Laboratório Associado para a Química Verde (REQUIMTE/LAQV)—Departamento de Química e Bioquímica, Faculdade de Ciências, Universidade do Porto, 4169-007 Porto, Portugal; up201504466@edu.fe.up.pt (A.R.S.); pjalmeid@fc.up.pt (P.J.A.); jrodrigosantos@fc.up.pt (J.R.S.); jarodrig@fc.up.pt (J.A.R.)
3 Department of Analytical Chemistry, Nutrition and Food Science, University of Santiago de Compostela, 15782 Santiago de Compostela, Spain
4 Health Research Institute of Santiago de Compostela (IDIS), University of Santiago de Compostela, 15782 Santiago de Compostela, Spain
5 Instituto de Materiais (iMATUS), University of Santiago de Compostela, 15782 Santiago de Compostela, Spain
* Correspondence: jorge_custodio-mendoza@sggw.edu.pl (J.A.C.-M.); tuchi.carro@usc.es (A.M.C.)

Abstract: The occurrence of carbonyl compounds and volatile organic compounds (VOCs) in adult formulas is a critical issue in product safety and quality. This research manuscript reports the determination of targeted and untargeted carbonyl compounds and VOCs in adult formulas stored at different temperatures (room temperature, 4 °C, and 60 °C) over one month. Gas chromatography-mass spectrometry was utilized for the sample analysis. Ultrasound-assisted dispersive liquid–liquid microextraction at 60 °C for 20 min facilitated the extraction of six carbonyl compounds, while headspace solid-phase microextraction (HS-SPME) was employed for the determination of untargeted VOCs using a DVB/CAR/PDMS fiber, involving 15 min of equilibration and 45 min of extraction at 40 °C with magnetic stirring. Analytical features of the methods were assessed according to Food and Drug Administration guidelines, and good limits of detection and quantitation, linearity, accuracy, and precision were achieved. Notably, the highest levels of carbonyl compounds were found in high-protein formulas, with quantifiable levels of malondialdehyde, acrolein, and formaldehyde detected and quantified in 80% of samples. Additionally, significant levels of VOCs such as hexanal and 2-heptanone were found in samples stored at elevated temperatures. These findings suggest the importance of protein content and storage conditions in the levels of carbonyl compounds and VOCs found in adult formulas, with implications for consumer safety and quality control.

Keywords: dispersive liquid–liquid microextraction; enteral nutrition formula; food safety; solid-phase microextraction; risk of exposure; stability analysis; volatile organic compounds

1. Introduction

Adult formulas are specially designed nutritional products for enteral feeding, providing essential nutrients to individuals who may have difficulty consuming regular food [1,2]. These formulas play a vital role in the nutrition and health of various adult populations, particularly those with chronic illnesses, disabilities, or other health problems requiring specialized dietary support [2,3]. For instance, patients recovering from surgery, individuals with malabsorption issues, and the elderly often rely on these products to meet their nutritional needs and improve their overall health outcomes [1–3]. Stringent regulatory standards govern the manufacturing of adult formulas to ensure safety and efficacy [4]. Formulations can vary widely, including standard, high-protein, and specialized options tailored for specific health conditions [3,4]. Ultra-high temperature treatment, commonly used in adult formula processing, rapidly heats products to 135–150 °C for a few seconds to destroy harmful microorganisms and extend shelf life without refrigeration [5]. However, this process can also lead to the formation of heat-induced contaminants, such as lipid peroxidation products and advanced glycation end-products (AGEs), which may pose health risks to consumers [6].

Among the most concerning contaminants are malondialdehyde (MDA), formaldehyde (FCHO), acetaldehyde (ACE), acrolein (ACRL), methylglyoxal (MGO), diacetyl (DA) (the structural formula is provided in the Supplementary Materials, Figure S1), and other various volatile organic compounds (VOCs) [7]. These compounds are formed during lipid peroxidation as polyunsaturated fatty acids interact with reactive oxygen species [6,7]. This process follows a radical-driven pathway, forming lipid hydroperoxides as primary oxidation products [7]. These hydroperoxides decompose into more stable secondary products, including carbonyl compounds such as malondialdehyde, formaldehyde, acetaldehyde, acrolein, methylglyoxal, and diacetyl [7]. These compounds indicate oxidative degradation in lipid-containing foods like adult formula [6,7]. Lipid peroxidation can continue when the formulas are stored along with other degradation processes, potentially affecting their nutritional quality and safety [6,7]. Monitoring the presence and concentrations of these contaminants is crucial due to their toxicological profiles and classifications by health authorities, including the International Agency for Research on Cancer (IARC), which identifies several of these compounds as potential carcinogens [7]. For instance, FCHO is classified as a carcinogen with sufficient evidence of its carcinogenicity to humans (Group 1) with a tolerable daily intake (TDI) set at 150 μg/kg body weight (bw)/day. ACRL is classified as probably carcinogenic to humans (Group 2A), with a lower TDI of 7.5 μg/kg bw/day. ACE is categorized as possibly carcinogenic to humans (Group 2B), suggesting it may be carcinogenic, with an Acceptable Daily Intake (ADI) of 185 μg/day. MDA falls into the group of substances with insufficient evidence for their carcinogenicity (Group 3). However, it possesses a Threshold of Toxicological Concern set by the International Programme on Chemical Safety (IPCS) of 30 μg/kg bw/day. MGO is also in Group 3 and does not have a specific TDI. Finally, while unclassified by the IARC, DA has a notable ADI of 900 μg/kg bw/day [7–12].

Given the potential health implications of these contaminants, the need for reliable analytical methods to assess their levels in adult formulas is paramount [7]. Traditional methods for determining lipid peroxidation products in foods, such as the thiobarbituric acid (TBA) test and peroxide value measurement, are widely used due to their simplicity and accessibility [7,13,14]. However, these methods have several limitations: they are non-specific (which can lead to overestimation), time-consuming, and require harsh conditions and large amounts of solvents, which have a notable environmental impact [13,14]. To address these limitations, chromatographic methods such as high-performance liquid chromatography (HPLC) and gas chromatography (GC), paired with various detection systems (UV, Fluorescence Detection (FLD), Flame Ionization Detection (FID), and Mass Spectrometry (MS)), have been developed [7,13–20]. These methods allow for the specific determination of lipid peroxidation products, including aldehydes, ketones, malondialdehyde, and dicarbonyl compounds, with higher sensitivity and selectivity than traditional methods [7]. However, to detect specific compounds like AGEs and other lipid peroxidation products effectively, chemical derivatization (e.g., with hydrazine, TBA, or o-phenylene diamine) is often required to enhance instrument response, depending on the detector system [13,17–20]. Microextraction techniques have further advanced the sensitivity and specificity of carbonyl compounds and VOCs analysis by offering a more sustainable approach due to the reduced sample size and waste generation and the reduction (Dispersive Liquid–Liquid Microextraction, DLLME) or non-use (Solid Phase Microextraction, SPME) of organic solvent [7]. In this context, head-space (HS) SPME has been used to determine carbonyl compounds associated with lipid peroxidation in infant formula, milk, and dairy products [7,20,21]. VOC determination without derivatization has also been reported using

HS-SPME [7]. DLLME has been applied to measure FCHO, MDA, and ACRL [7,16,22] in beverages. To our knowledge, there is no analytical method for assessing lipid peroxidation products in adult formula products.

In this study, we aim to apply innovative microextraction techniques to determine the concentrations of carbonyl compounds and to monitor VOCs in adult formulas. Specifically, we propose using DLLME for the targeted analysis of carbonyl and dicarbonyl compounds as potential markers of lipid peroxidation and HS-SPME for untargeted VOC determination, as it requires no derivatization. Through these methods, we present the first known data on the occurrence of carbonyl compounds associated with oxidative and storage stability in adult formulas, as far as we know. Additionally, we assess the potential risk of exposure for consumers to these contaminants, highlighting the need for ongoing monitoring to warrant the safety and quality of adult formula.

2. Materials and Methods

2.1. Chemicals and Reagents

Unless otherwise specified, all chemicals used were high purity (>98%). Acetaldehyde (ACE, CAS 75-07-0), Acetonitrile (ACN, CAS 75-05-8), Chloroform ($CHCl_3$, 35%, CAS 67-66-3), Deuterated Acetaldehyde (ACEd4, CAS 1632-89-9), Deuterated Acetone (ACOd6, CAS 666-52-4), Diacetyl (DA, CAS 431-03-8), 2,4-Dinitrophenylhydrazine (DNPH, CAS 119-26-6), Formaldehyde (FCHO, CAS 50-00-0), Malondialdehyde (MDA, CAS 643-12-9), and Methylglyoxal (MGO, 40% CAS 78-98-8) were all supplied by Merck (Darmstadt, Germany). Acrolein (ACRL, CAS 107-02-8) was purchased from LGC Standards (London, UK).

SPME fibers with 50/30 μm divinylbenzene/Carboxen/Polydimethylsiloxane (DVB/CAR/PDMS) and 65 μm PDMS/DVB were also obtained from Merck. All fibers were conditioned to ensure optimal performance per the manufacturer's instructions before their initial use. Manual sampling was performed using a manual holder sourced from Merck. This study utilized various pieces of laboratory equipment, including a Centromix II-BL Centrifuge from J. P. Selecta (Barcelona, Spain), a Basic 20 pH meter from Crison Instruments (Barcelona, Spain), a 2510EMTH ultrasonic bath from Branson Ultrasonics (Danbury, CT, USA), and a Reax Top vortex mixer from Instruments GmbH & Co. (Schwalbach, Germany) to conduct all experiments.

2.2. Sample Selection and Stability Study Design

In this study, 12 commercially available adult nutritional formulas were selected from supermarkets and parapharmacies in Santiago de Compostela, Spain. The formulas were categorized into three groups: high-protein formulations (HP, n = 5), standard formulations (SAF, n = 4), and specialized formulations (SF, n = 3) encompassing an enteral formula for dysphagia and amylase resistance, a carbohydrate module formula, and an enteral powder for administration via tube or orally. All samples were kept in their original packaging before the occurrence study. From these, two unflavored standards and two unflavored high-protein adult formulas with similar compositions were selected for the storage stability study.

The stability study aimed to evaluate the effects of different storage temperatures on the formulations over 30 days. Unflavored samples, including high-protein (n = 2) and standard formulation (n = 2), were selected for this study. All samples were transferred into amber vials and the caps were covered with parafilm to prevent contamination. The formulas were stored in their original powdered form and were prepared according to the manufacturer's instructions by dissolving them in drinking water before analysis. Three storage conditions were tested: room temperature, 4 °C, and 60 °C. These conditions were selected to replicate typical domestic storage situations throughout the year. The samples were stored in the dark, using amber vials covered with aluminum foil and placed inside cardboard boxes to minimize light exposure. The four samples were analyzed on specified days (0, 1, 7, 14, 21, and 30). In the untargeted VOC determination, focusing on determining VOC changes during storage, two unflavored samples (also used in the target analysis) from the standard formulation category were selected for HS-SPME-GC-MS analysis after 30 days of storage, as they were the only formulations available in individual packaging of small portions. All analyses were conducted in triplicate to ensure reliable data.

2.3. Ultrasound-Assisted Dispersive Liquid–Liquid Microextraction

To extract FCHO, MDA, ACE, ACRL, MGO, and DA from the adult formula, a previously developed method to extract MDA, ACRL, and 4-hydroxy-2-nonenal in beverages was used (Custodio-Mendoza et al., 2022) with modifications [22]. The adult formula was prepared following the manufacturer's instructions: 15% *w*/*v* in drinking water at 40 °C, with gentle stirring until fully dissolved. A 0.5 mL aliquot of the solution was transferred to a 2.5 mL falcon tube, and 1.3 mL of ACN was added. The mixture was vortexed for 1 min to combine the phases, followed by centrifugation for 2 min at 1634 g to precipitate proteins. The supernatant was transferred to a clean tube, and 90 µL of CHCl3 was added. The mixture was homogenized by three cycles of charging and discharging using a Pasteur pipette, then rapidly transferred to a conical tube containing 4 mL of ultrapure water and 0.5 mL of DNPH solution (0.5 g/L in 2 M HCl). The system was incubated in an ultrasonic water bath at 60 °C for 20 min, followed by centrifugation for 2 min at 2146× *g* to separate the extractant phase. The drop containing the carbonyl-DNPH derivatives was collected with a microsyringe and directly injected into the GC-MS for target analysis.

2.4. Head-Space Solid Phase Microextraction

The VOCs were extracted from the adult formula following the method described by Clarke et al., (2019) with modifications [21]. Briefly, the adult formula was prepared as previously outlined. A 2 mL aliquot of the sample solution was placed in a 6 mL vial with a magnetic stirrer and sealed with a septum cap. The sample was equilibrated in a water bath with gentle stirring for 15 min at 40 °C. After the equilibration period, the SPME fiber was exposed, and extraction started for 45 min at the same temperature. Following extraction, the analytes were desorbed at 270 °C for 5 min in the GC-MS injection port.

2.5. Gas Chromatography-Mass Spectrometry

The targeted and untargeted analyses were performed using a GC–MS system (7890B-5977B, Agilent Technologies, Santa Clara, CA, USA) with a J&W HP-5MS column (30 m × 0.25 mm ID × 0.25 µm, Agilent) and a carrier gas flow rate of 1.5 mL/min. The transfer line was set to 280 °C, connecting the column to an electron impact (EI^+) source at 250 °C with 70 eV, and a single quadrupole mass analyzer was maintained at 120 °C.

For the targeted analysis, the injector temperature was set to 245 °C using an ultra-inert double taper liner in splitless mode. The temperature program began at 100 °C, increased at 100 °C/min to 230 °C (held for 3.3 min), then ramped at 35 °C/min to a final temperature of 280 °C, maintained for 2 min. The total analysis time for this targeted method was 8 min. Initially, full scans were performed over a 50–300 m/z range for each hydrazone, allowing the identification of the most intense ions. Single ion monitoring mode was then used with the quantifier and qualifier ions summarized in Table S1 (Supplementary Material).

For the untargeted analysis, the oven temperature started at 35 °C, held for 5 min, and increased at 5 °C/min to 280 °C (held for 10 min), resulting in a total run time of 64 min. The detector was turned off after 60 min. The injector was set to 270 °C using a straight ultra-inert liner (5190-4048) with a 0.755 mm inner diameter in splitless mode. Scan acquisition mode covered a 40–400 m/z range. Preliminary compound identification was performed using the NIST Mass Spectral Search Program (Version 2.2) by comparing experimental and reference mass spectra in the NIST/EPA/NIH Mass Spectral Library. Compound identification was based on matching fragmentation patterns and relative ion intensities, with only matches having a probabilistic score above 80% reported.

2.6. Analytical Validation

The analytical validation for matrix and analyte extension of the previously published DLLME-GC-MS method was conducted per FDA guidelines [23]. The acceptability criteria included several key components. First, the method demonstrated selectivity by providing specific retention times and identifying qualitative and quantitative ions for each analyte, ensuring accurate differentiation between compounds in the matrix. Linearity was assessed using a standard addition method with internal standard calibration within six concentration levels across a concentration range of 0.5 to 3 μg/mL for each analyte in triplicate, with strong linear relationships confirmed by high determination coefficients (r^2). The limit of detection (LOD) was calculated as the mean of the blank signal plus 3.3 times the standard deviation of this measurement. Similarly, the limit of quantitation (LOQ) was determined as the mean of the blank signal plus 10 times the standard deviation of the blank. The method's accuracy was evaluated at three concentration levels, with each level tested in quintuplicate, and recovery rates were calculated to assess the closeness of results to the true values. Finally, precision was evaluated by intraday and interday measurements at three concentration levels, each tested in quintuplicate, ensuring consistent results over time and under different conditions.

2.7. Statistical Analysis

Statistical analyses were conducted using Excel and Statistica (version 13.3) software. A box-and-whisker plot was used to present the occurrence data, showing the distribution of carbonyl compound concentrations for each sample. This type of plot displays the data's minimum, first-quartile, median, third-quartile, and maximum values. The "box" part represents the interquartile range (IQR), which covers the middle 50% of values. The line inside the box marks the median, indicating the central value of the data set. The "whiskers" extend to the minimum and maximum values, showing the full spread of the data, while any outliers are shown as individual points outside the whiskers. This plot helps to visualize the range, central tendency, and variability in carbonyl compound concentrations across samples. A heatmap was generated to present the storage stability data, visually representing the data trends over time and across different storage conditions. The heatmap illustrated the relative stability of the analytes, with color gradients indicating the intensity of the responses. Darker shades represented higher concentrations, while lighter shades indicated lower concentrations, allowing for quick identification of significant changes in analyte levels.

3. Results and Discussion

3.1. Method Performance—Matrix and Analyte Extension

The US-DLLME-GC-MS method was previously developed in our lab to determine MDA and ACRL in beverages [22]. Herein, we studied the method's applicability to four additional carbonyl compounds (FCHO, ACE, MGO, and DMGO) in adult formula. We used a standard adult formula as a blank sample to ensure the method's suitability for the simultaneous extraction of these six analytes from the adult formula. First, a kinetic study was performed in triplicate to assess the formation of carbonyl-DNPH derivatives (Figure 1). MDA and ACRL reached equilibrium after 5 min of ultrasound treatment, FCHO after 10 min, while ACE, MGO, and DMGO required 20 min.

To ensure all analytes reached equilibrium and were successfully extracted as their corresponding DNPH derivatives, we set the ultrasound incubation time to 20 min. The analytical performance of the method was evaluated under these conditions (Table 1).

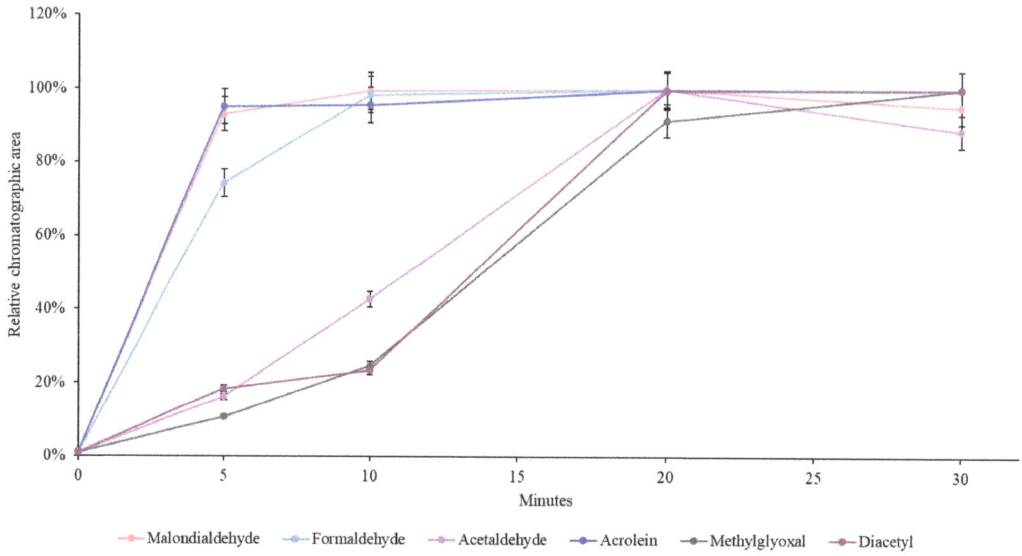

Figure 1. Kinetic analysis of DNPH derivatization and extraction of carbonyl compounds from adult formula using US-DLLME-GC-MS.

The method's specificity is based on specific ions for each analyte and internal standards eluting at specific retention times previously described in this manuscript (Section 2.5). Figure 2 confirms the absence of interfering signals in the retention time region for all analytes and the internal standards. The LOD ranged from 8 to 89 ng/mL, and the LOQ ranged from 0.061 to 0.671 μg/mL. Standard addition calibrations showed excellent linearity, with determination coefficients of $\geq$0.9990. The method's precision, assessed at concentrations of 0.7 μg/mL (QC1), 1.0 μg/mL (QC2), and 2.0 μg/mL (QC3), demonstrated good intraday precision with relative standard deviations (RSD) between 1.3% and 5.8%, and interday precision with RSD between 0.9% and 4.6%. Similarly, method accuracy at the same concentrations showed recoveries ranging from 98.0% to 102.6%.

Table S2 compares the analytical performance of this method with other methods for detecting carbonyl compounds in infant formula, milk, and dairy products. Single HPLC-UV determinations of MDA have been reported, using either the TBARS test or as a TBA-MDA derivative after liquid–liquid extraction or dilution. Despite being feasible and reproducible, with acceptable RSD values, these methods exhibit poorer linearity and lower recovery rates than those in this study. Similarly, single UV determinations of FCHO in milk have shown higher LOD and LOQ values than those reported here. Single MDA determination in infant formula using HPLC-MS without derivatization has been reported to have significantly lower detection limits than other HPLC methods. When paired with derivatization, HPLC-MS methods achieve even further reductions in detection limits, as reported for MGO and DA determination in baby food by Kocadağlı & Gökmen (2014), and in milk and dairy products reported by Zhang et al., 2022 [18,24]. In a previous study, we identified five carbonyl compounds, including MDA, ACRL, MGO, and DA, in infant formula using Gas-Diffusion Microextraction (GDME) with o-PDA derivatization via HPLC-UV; this method, though accurate and precise, had higher detection limits [19]. Additionally, we previously reported the development of an HS-SPME-GC-MS method for seven carbonyl and dicarbonyl compounds (including FCHO, ACE, MGO, DA, and MDA) as PFPH derivatives from infant formula, achieving detection limits similar to or slightly lower than those reported here. However, this method had higher RSD values [20].

Table 1. Analytical characteristics of matrix and analyte extension study for the US-DLLME-GC-MS method.

Analyte	LOD	LOQ	Slope	Intercept	r^2	Intraday Precision (n = 5) %RSD			Interday Precision (n = 5) %RSD			Accuracy (n = 5) % Recovery		
	μg/mL	μg/mL	$\times 10^{-3}$			QC1	QC2	QC3	QC1	QC2	QC3	QC1	QC2	QC3
Formaldehyde	0.022	0.671	24.90	2.3749	0.9997	3.9	2.6	2.9	2.4	1.1	1.5	102.6	98.5	98.5
Malondialdehyde	0.008	0.195	2.979	0.027	0.9999	4.7	1.9	2.2	4.3	1.7	1.1	100.9	98.5	99.4
Acetaldehyde	0.089	0.340	3.968	0.0981	0.9999	2.0	5.8	0.4	4.6	0.9	1.2	98.0	99.2	100.0
Acrolein	0.017	0.061	1.999	0.0067	0.9998	5.5	4.2	2.7	3.1	3.5	1.5	100.4	99.4	98.6
Methylglyoxal	0.083	0.229	0.574	0.0499	0.9990	3.4	2.8	6.9	2.4	0.9	3.4	98.0	95.8	93.1
Diacetyl	0.065	0.184	1.552	0.0889	0.9997	1.3	2.1	1.5	1.7	2.2	1.0	97.2	97.2	98.2

LOD, limit of detection; LOQ, limit of quantification; n, number of replications; RSD, relative standard deviation; QC1, low-level quality control sample spiked at 0.7 μg/mL; QC2, mid-level quality control sample spiked at 1.0 μg/mL; QC3, high-level quality control sample spiked at 2.0 μg/mL.

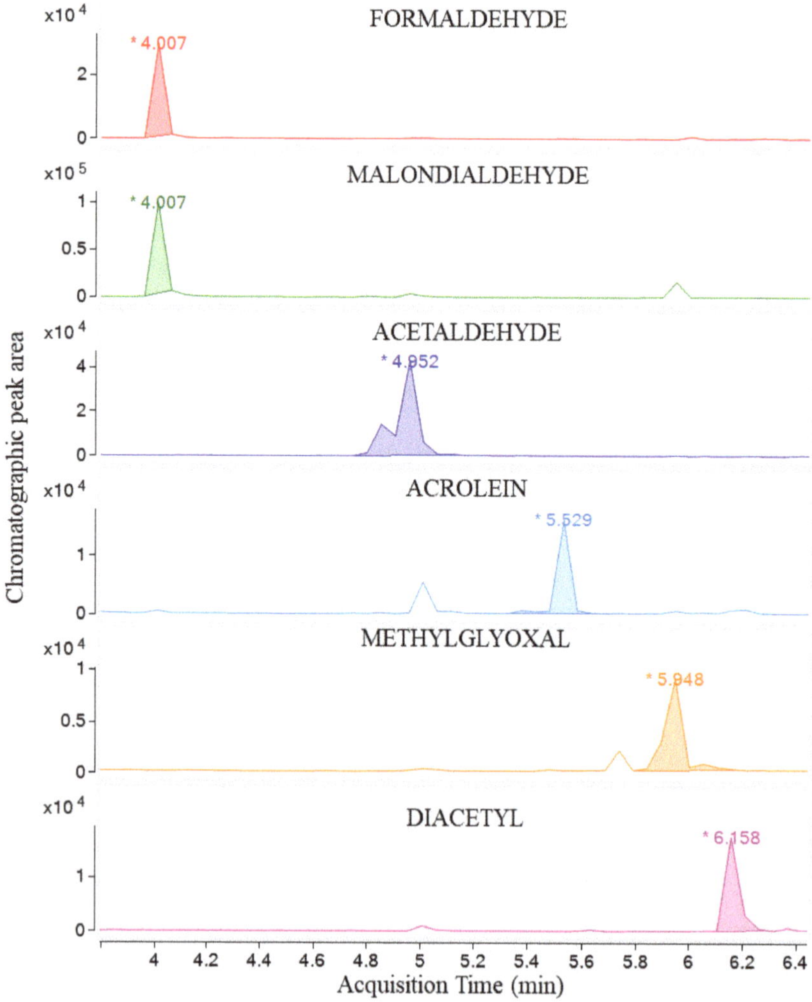

Figure 2. Chromatogram of standard adult formula spiked with 1 μg/mL of carbonyl compounds.

Overall, while combining derivatization with HPLC-MS produced the lowest detection limits, the extended method reported here achieved similar or lower detection limits compared to previous GC-MS and HPLC-UV methods for milk, dairy products, and infant formula, with more accurate recoveries (closer to 100%) and comparable precision, proving the suitability of this method for simultaneous monitoring of MDA, FCHO, ACE, ACRL, MGO, and DA in adult formulas.

3.2. Occurrence of Carbonyl Compounds in Adult Formulas

The validated US-DLLME-GC-MS method was applied to analyze carbonyl compounds in a set of adult formula samples, with results summarized in Table 2.

Table 2. Occurrence of carbonyl compounds in adult formula.

Adult Formula	Malondialdehyde μg/mL	±SD	Formaldehyde μg/mL	±SD	Acetaldehyde μg/mL	±SD	Acrolein μg/mL	±SD	Methylglyoxal μg/mL	±SD	Diacetyl μg/mL	±SD
HP1	2.38	0.39	2.86	0.90	2.16	0.31	2.29	0.73	1.50	0.28	ND	-
HP2	0.98	0.05	0.94	0.03	1.90	0.12	0.24	0.00	ND	-	ND	-
HP3	ND	-	ND	-	ND	-	ND	-	ND	-	ND	-
HP4	0.53	0.02	1.20	0.50	1.08	0.67	2.39	0.94	1.89	0.77	2.84	0.98
HP5	2.89	0.56	2.60	0.70	ND	-	2.80	0.17	0.61	0.08	ND	-
SAF1	2.65	0.82	0.90	0.11	1.63	0.45	1.27	0.70	1.65	0.77	ND	-
SAF2	1.73	0.54	1.56	0.31	2.91	0.23	1.43	0.35	2.69	0.82	0.97	0.27
SAF3	ND	-	ND	-	ND	-	ND	-	ND	-	ND	-
SAF4	ND	-	ND	-	ND	-	ND	-	ND	-	ND	-
SP1	2.09	0.75	1.65	0.75	1.16	0.24	2.49	0.76	2.02	0.52	2.11	0.49
SP2	2.00	0.80	1.48	0.51	0.82	0.53	1.89	0.42	2.46	0.93	1.67	0.96
SP3	ND	-	2.39	0.31	0.79	0.24	2.53	0.56	1.16	0.14	1.88	0.30

SD, standard deviation; HP, high-protein adult formula; SAF, standard adult formula; SP, special formulated adult formula; ND, non-determined as below limit of quantification.

All target analytes were detected, although not always at quantifiable levels; therefore, a box-and-whisker chart showing only quantifiable samples is presented for clearer comparison (Figure 3). In high-protein adult formulas, 80% of samples were quantifiable for MDA (0.53–2.89 μg/mL), FCHO (0.94–2.86 μg/mL), and ACRL (0.24–2.80 μg/mL). Additionally, 60% of samples were quantifiable for ACE (1.8–2.16 μg/mL) and MGO (0.61–1.89 μg/mL), with one sample quantifiable for DA at 2.94 μg/mL. In standard adult formulas, 50% of samples were quantifiable for MDA (1.73–2.65 μg/mL), FCHO (0.9–1.56 μg/mL), ACE (1.63–2.91 μg/mL), ACRL (1.27–1.43 μg/mL), and MGO (1.65–2.69 μg/mL), with only one sample quantifiable for DA at 0.97 μg/mL. For specialized formulations, only one enteral powder sample, intended for oral or tube administration, was quantifiable for MDA. All analytes in this sample type were quantifiable, with levels ranging from 2.0–2.09 μg/mL for MDA, 1.48–2.39 μg/mL for FCHO, 0.79–1.16 μg/mL for ACE, 1.89–2.49 μg/mL for ACRL, 1.16–2.46 μg/mL for MGO, and 1.67–2.11 μg/mL for DA.

To our knowledge, no studies have reported the occurrence levels of these compounds, specifically in adult formula. However, Kang et al. (2010) conducted short- and long-term trials to assess the effects of commercial enteral nutritional supports on the nutrition and health of stroke patients. They observed that MDA was significantly elevated in chronic stroke patients compared to healthy individuals after TBA derivatization and HPLC-UV [25].

Regarding similar foods, Bessaire et al. (2018) reported similar FCHO levels in milk powders after DNPH derivatization and HPLC-tandem mass spectrometry [26]. Custodio-Mendoza et al., (2024) observed lower levels of MDA, FCHO, and ACE in both powdered and liquid starter and follow-up infant formula after pentafluorophenyl hydrazine derivatization during HS-SPME extraction and GC-MS determination [19]. Similar ACRL, MGO, and DA levels were found in powdered starter formulas and ACRL levels in follow-up formulas after o-PDA derivatization during GDME extraction and HPLC-UV determination [20]. Similar MDA content in infant formula was reported by Pozzo et al. using TBA derivatization and HPLC-UV-FLD [27] and Cesa using the TBARS test [28]. Akıllıoğlu et al. with their microwave-assisted hydrolysis and HPLC-MS determination [15], as well as Kocadağlı et al. [24], reported comparable MGO results in the analysis of infant formulas, supporting these findings.

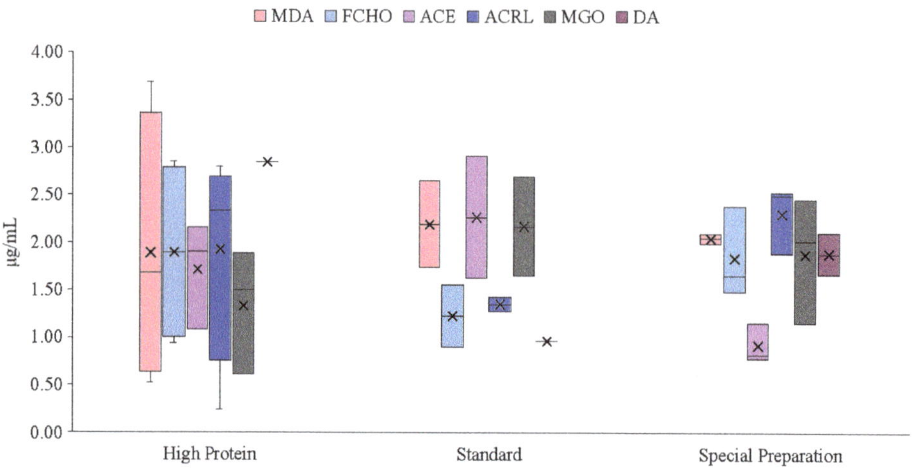

Figure 3. Box-and-whisker plot showing the occurrence of carbonyl compounds in adult formula samples. MDA, malondialdehyde; FCHO, formaldehyde; ACE, acetaldehyde; ACRL, acrolein; MGO, methylglyoxal; DA, diacetyl.

These findings suggest that MDA, FCHO, ACRL, ACE, MGO, and DA concentrations vary significantly across adult formula samples. The data suggest that high-protein and specialized formulations may be more susceptible to carbonyl compound formation than standard formulas. This variability highlights the importance of monitoring carbonyl compound levels in these products as potential indicators of oxidation which could affect product quality and safety.

3.3. *Temperature-Dependent Variations of Carbonyl Compound Levels in Adult Formulas During Storage*

Four samples were selected in the target stability study using the US-DLLME-GC-MS method. They included two standard adult formulas and two high-protein adult formulas from the same brand, differing only in protein content. Carbonyl compounds were detected in all samples, though quantification was only possible in HP samples at all time periods and in SAF at later stages. Results are summarized in Table 3.

In HP formulas, MDA levels increased over time under different storage conditions. At 4 °C, MDA remained relatively stable, detectable only in one HP sample, showing a slight increase at the end of 30 days. In the other HP sample, MDA became quantifiable by day 30. At room temperature, MDA accumulated more, reaching 2.01 μg/mL and 0.37 μg/mL after 30 days. Storage at 60 °C showed a similar trend, with MDA reaching comparable levels. MGO also showed accumulation over time, becoming detectable in HP samples after 7 days at 4 °C and increasing to 0.64 μg/mL and 0.98 μg/mL after 30 days. MGO levels rose further at higher temperatures, reaching 1.52–1.98 μg/mL after 30 days at room temperature and 2.54–2.78 μg/mL after 30 days at 60 °C. FCHO accumulated more gradually at 4 °C, becoming quantifiable only after 30 days at 0.5 μg/mL in one HP sample, while the other sample showed a higher range (0.94–1.93 μg/mL). FCHO accumulation intensified with higher storage temperatures, reaching up to 2.73 μg/mL at 60 °C after 30 days. In contrast, ACE and ACRL showed only slight and relatively stable increases across all temperatures.

Table 3. Variations of carbonyl content in adult formulas during 1 month of storage at different temperatures.

	Analyte	4 °C					Room Temperature					60 °C				
		1	7	14	21	30	1	7	14	21	30	1	7	14	21	30
HP2	MDA	1.22 ± 0.41	1.19 ± 0.99	1.54 ± 0.83	1.49 ± 0.95	1.41 ± 0.40	0.86 ± 0.09	0.85 ± 0.06	1.17 ± 0.10	1.99 ± 0.04	2.01 ± 0.04	0.97 ± 0.08	1.01 ± 0.04	1.91 ± 0.06	1.91 ± 0.16	2.23 ± 0.16
	FCHO	0.95 ± 0.45	1.14 ± 0.63	1.43 ± 0.33	1.64 ± 0.13	1.93 ± 0.22	0.94 ± 0.04	1.24 ± 0.05	1.51 ± 0.03	1.93 ± 0.09	2.03 ± 0.04	0.95 ± 0.04	1.24 ± 0.04	1.74 ± 0.09	2.13 ± 0.05	2.73 ± 0.05
	ACE	1.92 ± 0.27	2.06 ± 0.03	2.08 ± 0.25	2.53 ± 0.60	2.37 ± 0.65	2.03 ± 0.23	2.07 ± 0.20	2.19 ± 0.59	2.62 ± 0.40	2.83 ± 0.03	1.96 ± 0.07	2.00 ± 0.09	2.46 ± 0.04	2.47 ± 0.16	2.94 ± 0.23
	ACRL	0.24 ± 0.53	0.24 ± 0.05	0.38 ± 0.02	0.43 ± 0.01	0.44 ± 0.01	0.26 ± 0.06	0.22 ± 0.06	0.39 ± 0.06	0.40 ± 0.03	0.45 ± 0.04	0.24 ± 0.08	0.34 ± 0.04	0.33 ± 0.06	0.42 ± 0.06	0.56 ± 0.08
	MGO	ND	0.31 ± 0.08	0.48 ± 0.05	0.57 ± 0.02	0.64 ± 0.03	ND	0.32 ± 0.04	1.22 ± 0.03	1.75 ± 0.06	1.98 ± 0.01	0.23 ± 0.05	1.49 ± 0.01	1.60 ± 0.03	2.08 ± 0.09	2.78 ± 0.06
	DA	ND	ND	ND	ND	ND	ND	ND	ND	ND	ND	ND	ND	ND	ND	ND
HP3	MDA	ND	ND	ND	ND	0.28 ± 0.04	ND	ND	ND	ND	0.37 ± 0.03	ND	ND	0.30 ± 0.04	0.30 ± 0.06	0.40 ± 0.02
	FCHO	ND	ND	ND	ND	0.50 ± 0.08	0.77 ± 0.01	0.76 ± 0.02	0.80 ± 0.04	0.83 ± 0.08	0.92 ± 0.01	0.84 ± 0.07	0.88 ± 0.09	0.92 ± 0.09	0.96 ± 0.05	0.97 ± 0.07
	ACE	0.99 ± 0.01	1.44 ± 0.25	1.32 ± 0.27	1.36 ± 0.15	1.38 ± 0.17	1.13 ± 0.06	1.51 ± 0.12	1.81 ± 0.10	1.95 ± 0.33	2.39 ± 0.25	1.03 ± 0.31	1.65 ± 0.37	1.91 ± 0.28	2.15 ± 0.25	2.57 ± 0.17
	ACRL	ND	ND	ND	ND	ND	ND	ND	ND	ND	ND	ND	ND	ND	ND	ND
	MGO	ND	0.35 ± 0.02	0.67 ± 0.17	0.71 ± 0.15	0.98 ± 0.10	0.65 ± 0.03	1.05 ± 0.25	1.19 ± 0.13	1.23 ± 0.44	1.52 ± 0.27	0.82 ± 0.04	1.01 ± 0.04	1.43 ± 0.05	2.34 ± 0.09	2.54 ± 0.07
	DA	ND	ND	ND	ND	ND	ND	ND	ND	ND	ND	ND	ND	ND	ND	ND
SAF3	MDA	ND	ND	ND	ND	0.28 ± 0.06	ND	ND	0.28 ± 0.04	0.48 ± 0.05	0.58 ± 0.03	ND	ND	0.36 ± 0.01	0.57 ± 0.03	0.88 ± 0.09
	FCHO	ND	ND	ND	ND	0.71 ± 0.02	ND	ND	1.26 ± 0.01	0.13 ± 0.01	1.79 ± 0.09	ND	0.93 ± 0.09	1.41 ± 0.05	1.55 ± 0.05	1.79 ± 0.02
	ACE	ND	ND	ND	ND	0.37 ± 0.08	ND	ND	ND	ND	0.57 ± 0.08	ND	ND	ND	0.55 ± 0.05	0.62 ± 0.01
	ACRL	ND	ND	ND	ND	ND	ND	ND	ND	ND	ND	ND	ND	ND	ND	ND
	MGO	ND	ND	ND	ND	ND	ND	ND	ND	ND	ND	ND	ND	ND	ND	ND
	DA	ND	ND	ND	ND	ND	ND	ND	ND	ND	ND	ND	ND	ND	ND	ND
SAF4	MDA	ND	ND	ND	ND	ND	ND	ND	ND	ND	ND	ND	ND	ND	ND	0.32 ± 0.02
	FCHO	ND	ND	ND	ND	ND	ND	ND	ND	ND	ND	ND	ND	ND	ND	0.62 ± 0.02
	ACE	ND	ND	1.12 ± 0.01	1.26 ± 0.07	1.38 ± 0.06	ND	ND	1.81 ± 0.23	1.85 ± 0.27	1.94 ± 0.44	ND	ND	1.91 ± 0.09	2.15 ± 0.27	2.27 ± 0.57
	ACRL	ND	ND	ND	ND	ND	ND	ND	ND	ND	ND	ND	ND	ND	ND	ND
	MGO	ND	ND	ND	0.25 ± 0.01	0.71 ± 0.08	ND	ND	ND	0.88 ± 0.02	1.23 ± 0.05	ND	ND	1.43 ± 0.25	1.54 ± 0.43	1.94 ± 0.13
	DA	ND	ND	ND	ND	ND	ND	ND	ND	ND	ND	ND	ND	ND	ND	ND

HP, high-protein adult formula; SAF, standard adult formula; MDA, malondialdehyde; FCHO, formaldehyde; ACE, acetaldehyde; ACRL, acrolein; MGO, methylglyoxal; DA, diacetyl; ND, non-determined as it was below the limit of quantification.

In SAF samples, most compounds were undetectable until 7 days at both 4 °C and room temperature. MDA was below the limit of quantification in one SAF sample throughout the study, and ACRL and DA were undetectable in all SAF samples across temperatures. At 4 °C, only ACE was quantifiable in one sample after 14 days, rising to 0.37–1.38 μg/mL after 30 days. MGO was quantifiable in one SAF sample after 21 and 30 days at 0.25 and 0.71 μg/mL, respectively. In the other SAF sample, MDA, FCHO, and ACE were quantifiable only after 30 days, with concentrations of 0.28, 0.71, and 0.37 μg/mL, respectively. MDA and FCHO became quantifiable at room temperature in one SAF sample after 14 days, reaching 0.58 and 1.79 μg/mL by day 30. ACE was quantifiable in both SAF samples after 30 days, with concentrations of 0.57–1.94 μg/mL, while MGO reached 1.23 μg/mL in one sample after 21 days. At 60 °C, MDA became quantifiable after 14 days, reaching 0.32–0.88 μg/mL after 30 days. FCHO was quantifiable in one sample after 7 days, accumulating from 0.93 to 1.79 μg/mL by day 30. ACE became quantifiable in both samples after 14 and 21 days, reaching 0.62–2.27 μg/mL after 30 days, while MGO was only quantifiable in one sample after 14 days, reaching 1.43–1.94 μg/mL.

Although no data is available on the variations in carbonyl content in adult formula during one month of storage, our findings are consistent with those reported by other researchers in infant formula, milk, and dairy products. Bessaire et al. (2018) noted temperature-related differences in FCHO levels in milk powders, finding that FCHO levels were lower at room temperature than at 60 °C [26]. Cesa et al., (2015) reported a significant increase in MDA content in infant formula stored for two weeks at 55 °C [29]. Jia et al. (2018) found significant differences in content of advanced glycation end products (AGE), such as methylglyoxal and diacetyl, in milk-based infant formulas stored at 50 °C [30]. Liu and Li also observed similar variations in MGO after 60 days of storage [31]. Notably, Cheng et al. reported that even moderate increases in temperature can significantly contribute to the formation of AGEs, reaching detectable levels in infant formula milk powders, consistent with our results [32].

These findings suggest carbonyl compounds accumulate more over time in HP than in SAF, particularly at higher storage temperatures. The increase of MDA, MGO, and FCHO levels in HP formula indicates that higher protein content may promote oxidative reactions during storage, as reported in other foods [33]. Elevated temperatures further accelerate this process, with the most significant accumulation occurring at 60 °C. This implies that protein content and storage conditions (temperature and duration) are critical factors in adult formulas' stability and potential degradation.

3.4. Temperature-Dependent Variations of VOC Levels in Adult Formulas During Storage

In the untargeted stability study using HS-SPME-GC-MS, identification was based on comparing the mass spectrometry (MS) patterns with those in the NIST library. The results are summarized in Table S3 (Supplementary Materials), and heatmaps were created using the chromatographic peak areas obtained from a set of new SAF samples. These samples were stored for 30 days at 4 °C, room temperature, and 60 °C to facilitate comparison, as quantification was not possible (Figure 4).

A total of 40 compounds were observed, including 11 aldehydes, 10 ketones, 9 alcohols, and 2 esters, among others. Notably, hexanal, 2-heptanone, heptanal, 2-heptenal, 2-octenal, nonanal, and 2-hydroxybenzaldehyde were present in both new SAF samples. In contrast, pentyl acetate and 1-octen-3-ol were found in only one sample, while benzaldehyde, 2-pentylfuran, and octanal were found in the other sample.

Hexanal appears stable at 4 °C, with a slight increase noted in one sample stored at room temperature. However, both samples showed a significant increase in hexanal levels after 30 days at 60 °C. The other compounds in the fresh samples seem stable at both 4 °C and room temperature. In contrast, most compounds exhibited some increase when stored at 60 °C, except for nonanal and 2-hydroxybenzaldehyde, which remained stable at all temperatures. Moreover, compounds such as 1-hexanol, octanal, 1-octanol, 2-methyldecaline, and 17-octadecenoic acid appeared after 30 days of storage at all studied temperatures. Other compounds, including 1-pentanol, 1-heptanol, 2-nonanone, 1-decanol, 1-octanol, 2-decanone, decanal, γ-octalactone, 2-decenal, cyclodecanone, 6-undecanone, hexanoic acid pentyl ester, 2-undecanone, 2-undecenal, and 2-butyl-2,7-octadien-1-ol, were only present after 30 days of storage at 60 °C.

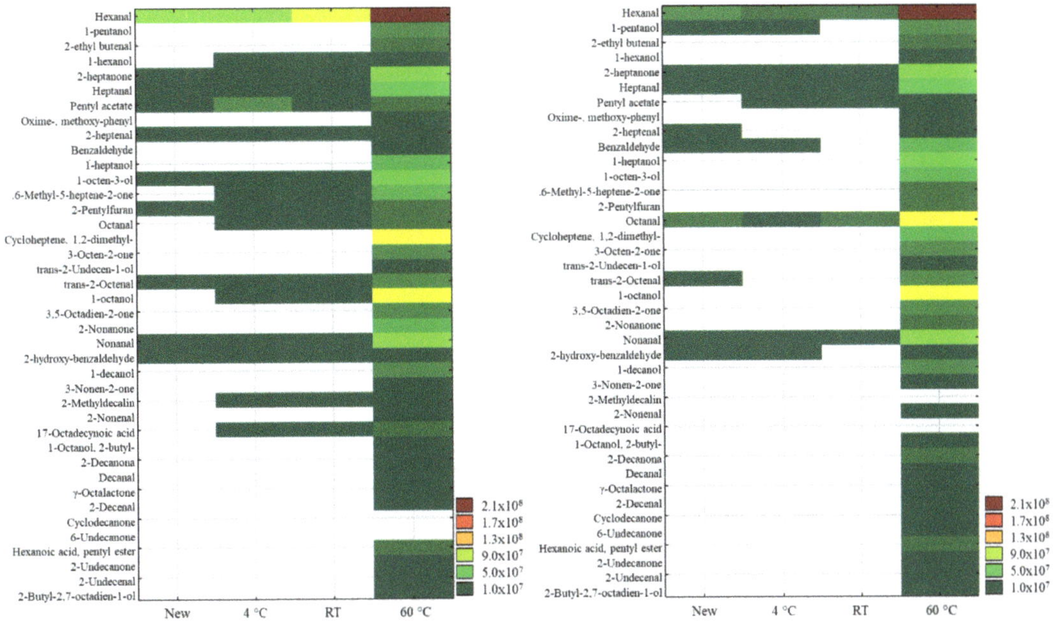

Figure 4. Heatmap of VOCs identified in adult formula after 1-month storage. Standard adult formula 3 (**left**), standard adult formula 4 (**right**).

Similar VOCs, including aldehydes, ketones, alcohols, and volatile acids, were reported by Jia et al. [30] in their analysis of milk-based infant formulas, as well as by Hausner et al., (2009) in their characterization of infant formulas and breast milk [34]. Li, Zhang, and Wang also noted the formation of aldehydes and ketones in milk powders stored at various temperatures, which aligns with our findings [35]. Li et al., noticed the presence of aldehydes and ketones that influence the sensory properties of infant formulas [36]. They also found that long-term storage (up to one year) at room temperature leads to significant variations in volatile compounds, similar to our observations, with aldehydes and ketones being the most prevalent, significantly impacting the sensory attributes of infant formulas [36].

3.5. *Risk of Exposure Assessment Based on Adult Formula Consumption*

In this study, we calculated the estimated consumption of adult formula to assess potential exposure to contaminants. The analysis accounted for differences in caloric needs between males and females to avoid gender bias and individual activity levels, considering the role of adult formula as a meal replacement and the highest carbonyl content identified in the occurrence study [37]. Based on average daily caloric requirements for males and females by activity level (Table S4, Supplementary Materials) and noting that 100 mL of adult formula provides between 100–479 kcal when prepared according to the manufacturer's instructions, we summarized the average daily intake for males weighing 70–90 kg and females weighing 55–75 kg in Table 4, expressed as μg/kg of body weight per day. We compared the average daily intakes with the tolerable or acceptable daily intakes for carbonyl compounds (see Introduction section) [38].

Table 4. Average daily intake of carbonyl compounds for males and females by activity level.

Level of Physical Activity	Malondialdehyde		Formaldehyde		Acetaldehyde		Acrolein		Methylglyoxal [a]		Diacetyl	
	µg/Kg bw/day		µg/Kg bw/day		µg/Kg bw/day		µg/Kg bw/day		µg/Kg bw/day		µg/Kg bw/day	
	M	F	M	F	M	F	M	F	M	F	M	F
Low	53.5 *	55.3 *	52.9	69.8	53.8	55.7	51.8 *	53.6 *	49.8	51.5	52.5	67.7
Moderate	58.9 *	61.4 *	58.3	60.8	59.3	61.8	57.1 *	59.5 *	54.9	57.1	57.9	60.3
High	67.9 *	67.5 *	67.2	66.8	68.3	67.9	65.7 *	65.4 *	63.2	62.8	66.7	66.3

bw, body weight; M, male; F, female; * above tolerable daily intake (or analog exposure level); [a] compared to glyoxal equivalent due to lack of exposure level.

The risk of exposure assessment indicates that, when consuming high-protein adult formula as a meal replacement, adult women with low physical activity may experience slightly higher exposure to FCHO and DA. In contrast, for individuals with moderate to high physical activity levels, exposure to these contaminants is similar for both males and females. While exposure to most of the targeted compounds across the studied weight range and activity levels remains below the tolerable daily intake (or equivalent exposure level), both males and females would be exposed to levels of malondialdehyde twice the toxicological threshold of concern and approximately eight times the tolerable daily intake for acrolein.

4. Conclusions

This study highlights the need for a reliable analytical method to assess lipid peroxidation products in adult formula, a gap not addressed in prior research. By adopting a previously established method for detecting MDA and ACRL in beverages, we successfully extended it to determine six key analytes—MDA, ACRL, FCHO, ACE, MGO, and DA—in adult formulas. The method showed excellent specificity, precision, and accuracy, achieving low limits of detection and quantification.

To our knowledge, this is the first occurrence study of carbonyl compounds in adult formulas. Remarkably, high-protein formulas showed the highest detectability, with 80% of samples quantifiable for MDA, FCHO, and ACRL and 60% quantifiable for ACE and MGO. Standard formulas had a 50% quantification rate for most compounds, while only one special formulation sample, intended for oral or tube feeding, had quantifiable levels of all analytes.

The analysis of various adult formula samples revealed significant variations in the levels of carbonyl compounds when stored at different temperatures using HS-SPME-GC-MS. High-protein adult formulas consistently exhibited higher concentrations of these compounds than standard formulas. This suggests that higher protein content may enhance the formation of carbonyl compounds during storage, particularly under elevated temperatures. These findings suggest that the protein content and storage conditions—such as temperature and duration—are critical factors influencing the stability and degradation of these products.

The untargeted determination of VOCs by HS-SPME-GC-MS identified a diverse range of compounds, including various aldehydes, ketones, alcohols, and volatile acids. The results indicated that many VOCs accumulate over time, particularly in high-protein adult formulas stored at elevated temperatures. Notably, hexanal, 2-heptanone, and other compounds showed significant increases, suggesting that storage conditions critically affect the stability of these products. The presence of these VOCs, many of which are associated with sensory attributes, emphasizes the importance of understanding their impact on product quality. The findings also suggest that monitoring VOC levels can serve as an indicator of oxidation processes in adult formulas.

Our findings suggest that, once opened, adult formula should be stored in its original packaging, protected from light and moisture, and kept away from heat sources such as ovens or central heating. Formulas with higher protein content should ideally be stored in a refrigerator, particularly during the summer or in locations where the room temperature exceeds 25 °C.

Furthermore, we assessed the potential risk of exposure to these contaminants based on the estimated consumption of adult formula. Our results indicate that males and females, across all activity levels, may exceed safety thresholds for MDA and acrolein when consuming high-protein formulas. This underscores the importance of continuously monitoring carbonyl compounds in adult formulas, as high levels may compromise product quality and safety.

Supplementary Materials: The following supporting information can be downloaded at: https://www.mdpi.com/article/10.3390/foods13233752/s1, Figure S1. Structural formula, molecular formula, and molecular weight of target carbonyl compounds. Table S1: Retention time, quantifier and qualifier ions for the US-DLLME-GC-MS method. Table S2: Analytical methods for malondialdehyde, acrolein, and α-dicarbonyl compounds in infant and adult formulae, milk, and dairy products. Table S3: Volatile compounds identified in adult formulas. Table S4. Interval of daily caloric request.

Author Contributions: J.A.C.-M.: Conceptualization, Investigation, Validation, Formal analysis, Visualization, Writing—original draft, Writing—review and editing, Supervision; A.R.S.: Investigation and Validation, Formal analysis, Data curation, Writing—original draft; M.A.K.: Formal analysis, Writing—review and editing, Project administration and Funding acquisition; P.J.A.: Writing—review and editing, Project administration and Funding acquisition. J.R.S.: Writing—review and editing, Project administration and Funding acquisition. J.A.R.: Writing—review and editing, Project administration and Funding acquisition. A.M.C.: Conceptualization, Formal Analysis, Writing—review and editing, Resources, Project administration and Funding acquisition, Supervision. All authors have read and agreed to the published version of the manuscript.

Funding: This work was supported by the Polish National Science Center within the project 2023/07/X/NZ9/01113 as part of the MINIATURA-7 program. It also received support from the Portuguese government through The Foundation for Science and Technology (FCT/MCTES) under the project UIDP/50006/2020 (DOI 10.54499/UIDP/50006/2020) and national funds.

Institutional Review Board Statement: Not applicable.

Informed Consent Statement: Not applicable.

Data Availability Statement: The data presented in this study are available on request from the corresponding author due to privacy.

Acknowledgments: The article is based upon work from the Sample Preparation Study Group and Network, supported by the Division of Analytical Chemistry of the European Chemical Society. We further acknowledge support from FCT/MCTES (LA/P/0008/2020 DOI 10.54499/LA/P/0008/2020, UIDP/50006/2020 DOI 10.54499/UIDP/50006/2020, and UIDB/50006/2020 DOI 10.54499/UIDB/50006/2020). ARS acknowledges her PhD grant (ref. 2021.05227.BD) supported by FCT, and JRS acknowledges the program DL 57/2016—Norma transitória (ref. SFRH/BPD/76544/2011), also supported by FCT.

Conflicts of Interest: The authors declare no conflicts of interest. The funders had no role in the study's design; in the collection, analyses, or interpretation of data; in the writing of the manuscript; or in the decision to publish the results.

References

1. Kozeniecki, M.; Fritzshall, R. Enteral nutrition for adults in the hospital setting. *Nutr. Clin. Pract.* **2015**, *30*, 634–651. [CrossRef] [PubMed]
2. Parker, E.K.; Flood, V.; Halaki, M.; Wearne, C.; Anderson, G.; Gomes, L.; Kohn, M. A standard enteral formula versus an iso-caloric lower carbohydrate/high fat enteral formula in the hospital management of adolescent and young adults admitted with anorexia nervosa: A randomised controlled trial. *J. Eat. Disord.* **2021**, *9*, 160. [CrossRef] [PubMed]
3. Ojo, O.; Adegboye, A.R.A.; Ojo, O.O.; Wang, X.; Brooke, J. An evaluation of the nutritional value and physical properties of blenderised enteral nutrition formula: A systematic review and meta-analysis. *Nutrients* **2020**, *12*, 1840. [CrossRef] [PubMed]
4. Sinha, S.; Lath, G.; Rao, S. Safety of enteral nutrition practices: Overcoming the contamination challenges. *Indian J. Crit. Care Med.* **2020**, *24*, 709. [CrossRef]

5. Yang, H.; Xu, L.L.; Hou, L.; Xu, T.C.; Ye, S.H. Application of the global stability index method to shelf-life prediction and physiochemical characteristics analysis of enteral feeding formula during storage. *J. Food Eng.* **2023**, *348*, 111433. [CrossRef]
6. Vistoli, G.; De Maddis, D.; Cipak, A.; Zarkovic, N.; Carini, M.; Aldini, G. Advanced glycoxidation and lipoxidation end products (AGEs and ALEs): An overview of their mechanisms of formation. *Free Radic. Res.* **2013**, *47*, 3–27. [CrossRef]
7. Custodio-Mendoza, J.A.; Ares-Fuentes, A.M.; Carro, A.M. Innovative solutions for food analysis: Microextraction techniques in lipid peroxidation product detection. *Separations* **2023**, *10*, 531. [CrossRef]
8. Center for Drug Evaluation and Research. *M7(R2) Addendum: Application of the Principles of the ICH M7 Guideline*; U.S. Food and Drug Administration: Silver Spring, MD, USA, 2021. Available online: https://www.fda.gov/media/157451/download (accessed on 29 October 2024).
9. Gomes, R.; Meek, M.K. *Acrolein. Concise International Chemical Assessment Document 43*; World Health Organization: Geneva, Switzerland, 2022. Available online: https://www.who.int/ipcs/publications/cicad/en/cicad43.pdf (accessed on 29 October 2024).
10. Clark, S.; Winter, C.K. Diacetyl in foods: A review of safety and sensory characteristics. *Compr. Rev. Food Sci. Food Saf.* **2015**, *14*, 634–643. [CrossRef]
11. Papastergiadis, A.; Fatouh, A.; Jacxsens, L.; Lachat, C.; Shrestha, K.; Daelman, J.; De Meulenaer, B. Exposure assessment of malondialdehyde, 4-hydroxy-2-(E)-nonenal and 4-hydroxy-2-(E)-hexenal through specific foods available in Belgium. *Food Chem. Toxicol.* **2014**, *73*, 51–58. [CrossRef]
12. European Food Safety Authority (EFSA). Opinion of the Scientific Panel on food additives, flavourings, processing aids and materials in contact with food (AFC) related to use of formaldehyde as a preservative during the manufacture and preparation of food additives. *EFSA J.* **2007**, *5*, 415. [CrossRef]
13. Zhou, X.; Zhang, Z.; Liu, X.; Wu, D.; Ding, Y.; Li, G.; Wu, Y. Typical reactive carbonyl compounds in food products: Formation, influence on food quality, and detection methods. *Compr. Rev. Food Sci. Food Saf.* **2020**, *19*, 503–529. [CrossRef] [PubMed]
14. Martysiak-Żurowska, D. Content malondialdehyde (MDA) in infant formulae and follow-on formulae. *Pol. J. Food Nutr. Sci.* **2006**, *15*, 323–328.
15. Akıllıoğlu, H.G.; Lund, M.N. Quantification of advanced glycation end products and amino acid cross-links in foods by high-resolution mass spectrometry: Applicability of acid hydrolysis. *Food Chem.* **2022**, *366*, 130601. [CrossRef] [PubMed]
16. Nascimento, C.F.; Brasil, M.A.; Costa, S.P.; Pinto, P.C.; Saraiva, M.L.M.; Rocha, F.R. Exploitation of pulsed flows for on-line dispersive liquid–liquid microextraction: Spectrophotometric determination of formaldehyde in milk. *Talanta* **2015**, *144*, 1189–1194. [CrossRef] [PubMed]
17. Altomare, A.; Baron, G.; Gianazza, E.; Banfi, C.; Carini, M.; Aldini, G. Lipid peroxidation derived reactive carbonyl species in free and conjugated forms as an index of lipid peroxidation: Limits and perspectives. *Redox Biol.* **2021**, *42*, 101899. [CrossRef] [PubMed]
18. Zhang, J.; Wei, F.; Zhang, T.; Cui, M.; Peng, B.; Zhang, Y.; Wang, S. Simultaneous determination of seven α-dicarbonyl compounds in milk and milk products based on an LC–MS/MS method with matrix-matched calibration. *Food Anal. Methods* **2022**, *15*, 1652–1662. [CrossRef]
19. Custodio-Mendoza, J.A.; Muñoz-Menendez, L.; España-Fariñas, M.P.; Valente, I.M.; Rodrigues, J.A.; Almeida, P.J.; Carro, A.M. Simultaneous determination of carbonyl compounds related to thermal treatment and oxidative stability of infant formulas by gas-diffusion microextraction and high-performance liquid chromatography with ultraviolet detection. *Anal. Chim. Acta* **2024**, *1288*, 342164. [CrossRef]
20. Custodio-Mendoza, J.A.; Blanco, A.L.; Ares-Fuentes, A.M.; Díaz, A.M.C. Green infant formula analysis: Optimizing headspace solid-phase microextraction of carbonyl compounds associated with lipid peroxidation using GC-MS and pentafluorophenylhydrazine derivatization. *Talanta* **2024**, *273*, 125816. [CrossRef]
21. Clarke, H.J.; Mannion, D.T.; O'Sullivan, M.G.; Kerry, J.P.; Kilcawley, K.N. Development of a headspace solid-phase microextraction gas chromatography mass spectrometry method for the quantification of volatiles associated with lipid oxidation in whole milk powder using response surface methodology. *Food Chem.* **2019**, *292*, 75–80. [CrossRef]
22. Custodio-Mendoza, J.A.; Caamaño-Fernandez, C.; Lage, M.A.; Almeida, P.J.; Lorenzo, R.A.; Carro, A.M. GC–MS determination of malondialdehyde, acrolein, and 4-hydroxy-2-nonenal by ultrasound-assisted dispersive liquid-liquid microextraction in beverages. *Food Chem.* **2022**, *384*, 132530. [CrossRef]
23. U.S. Food and Drug Administration. *Guidelines for the Validation of Chemical Methods in Food, Feed, Cosmetics, and Veterinary Products*; Food Drug Admin: Silver Spring, MD, USA, 2019; pp. 1–39. Available online: https://www.fda.gov/media/81810/download (accessed on 29 October 2024).
24. Kocadağlı, T.; Gökmen, V. Investigation of α-dicarbonyl compounds in baby foods by high-performance liquid chromatography coupled with electrospray ionization mass spectrometry. *J. Agric. Food Chem.* **2016**, *87*, 26–32. [CrossRef] [PubMed]
25. Kang, Y.; Lee, H.S.; Paik, N.J.; Kim, W.S.; Yang, M. Evaluation of enteral formulas for nutrition, health, and quality of life among stroke patients. *Nutr. Res. Pract.* **2010**, *4*, 393–399. [CrossRef] [PubMed]
26. Bessaire, T.; Savoy, M.C.; Tarres, A.; Mujahid, C.; Goldmann, T.; Perrin, I.; Mottier, P. Artefact formation of formaldehyde in milk powders: Impact of analytical conditions. *Food Control* **2018**, *93*, 23–31. [CrossRef]
27. Pozzo, L.; Cirrincione, S.; Russo, R.; Karamać, M.; Amarowicz, R.; Coscia, A.; Giribaldi, M. Comparison of oxidative status of human milk, human milk fortifiers, and preterm infant formulas. *Foods* **2019**, *8*, 458. [CrossRef] [PubMed]

28. Cesa, S. Malondialdehyde contents in infant milk formulas. *J. Agric. Food Chem.* **2004**, *52*, 2119–2122. [CrossRef]
29. Cesa, S.; Casadei, M.A.; Cerreto, F.; Paolicelli, P. Infant milk formulas: Effect of storage conditions on the stability of powdered products towards autoxidation. *Foods* **2015**, *4*, 487–500. [CrossRef]
30. Jia, H.X.; Chen, W.L.; Qi, X.Y.; Su, M.Y. The stability of milk-based infant formulas during accelerated storage. *CyTA-J. Food* **2019**, *17*, 96–104. [CrossRef]
31. Liu, H.; Li, J. Changes in glyoxal and methylglyoxal content in the fried dough twist during frying and storage. *Eur. Food Res. Technol.* **2014**, *238*, 323–331. [CrossRef]
32. Cheng, H.; Zhu, R.G.; Erichsen, H.; Sorensen, J.; Petersen, M.A.; Skibsted, L.H. High temperature storage of infant formula milk powder for prediction of storage stability at ambient conditions. *Int. Dairy J.* **2017**, *73*, 166–174. [CrossRef]
33. Guo, Y.; Cai, W.; Tu, K.; Tu, S.; Wang, S.; Zhu, X.; Zhang, W. Infrared and Raman spectroscopic characterization of structural changes in albumin, globulin, glutelin, and prolamin during rice aging. *J. Agric. Food Chem.* **2013**, *61*, 185–192. [CrossRef]
34. Hausner, H.; Philipsen, M.; Skov, T.H.; Petersen, M.A.; Bredie, W.L.P. Characterization of the volatile composition and variations between infant formulas and mother's milk. *Chemosens. Percept.* **2009**, *2*, 79–93. [CrossRef]
35. Li, Y.; Zhang, L.; Wang, W. Formation of aldehyde and ketone compounds during production and storage of milk powder. *Molecules* **2012**, *17*, 9900–9911. [CrossRef] [PubMed]
36. Li, Y.; Wang, H.; Li, R.; Liu, G.; Zhong, K.; Gao, L.; Wang, S. Monitoring volatile changes in infant formula during long-term storage at room temperature. *Curr. Res. Food Sci.* **2023**, *7*, 100645. [CrossRef] [PubMed]
37. Arnold, M.J.; Harding, M.C.; Conley, A.T. Dietary guidelines for Americans 2020–2025: Recommendations from the US Departments of Agriculture and Health and Human Services. *Am. Fam. Physician* **2021**, *104*, 533–536. Available online: https://www.dietaryguidelines.gov/sites/default/files/2020-12/Dietary_Guidelines_for_Americans_2020-2025.pdf (accessed on 29 October 2024).
38. Boon, P.E.; Pustjens, A.M.; Te Biesebeek, J.D.; Brust, G.M.H.; Castenmiller, J.J.M. Dietary intake and risk assessment of elements for 1- and 2-year-old children in the Netherlands. *Food Chem. Toxicol.* **2022**, *161*, 112810. [CrossRef]

foods

MDPI

Article

Chemical Composition, Biological Activity, and Application of *Rosa damascena* Essential Oil as an Antimicrobial Agent in Minimally Processed Eggplant Inoculated with *Salmonella enterica*

Andrea Verešová [1], Milena D. Vukic [2], Nenad L. Vukovic [2], Margarita Terentjeva [3], Zhaojun Ban [4], Li Li [5], Alessandro Bianchi [6], Ján Kollár [7], Rania Ben Saad [8], Anis Ben Hsouna [8,9], Joel Horacio Elizondo-Luévano [10], Maciej Ireneusz Kluz [11], Natália Čmiková [1], Stefania Garzoli [12] and Miroslava Kačániová [1,11,*]

[1] Institute of Horticulture, Faculty of Horticulture and Landscape Engineering, Slovak University of Agriculture, Tr. A. Hlinku 2, 94976 Nitra, Slovakia; andrea.veresova1979@gmail.com (A.V.); n.cmikova@gmail.com (N.Č.)

[2] Department of Chemistry, Faculty of Science, University of Kragujevac, 34000 Kragujevac, Serbia; milena.vukic@pmf.kg.ac.rs (M.D.V.); nvchem@yahoo.com (N.L.V.)

[3] Faculty of Veterinary Medicine, Latvia University of Life Sciences and Technologies, LV-3001 Jelgava, Latvia; margarita.terentjeva@llu.lv

[4] Zhejiang Provincial Key Laboratory of Chemical and Biological Processing Technology of Farm Products, Zhejiang Provincial Collaborative Innovation Center of Agricultural Biological Resources Biochemical Manufacturing, School of Biological and Chemical Engineering, Zhejiang University of Science and Technology, Hangzhou 310023, China; banzhaojun@zust.edu.cn

[5] Key Laboratory for Agro-Products Postharvest Handling of Ministry of Agriculture and Rural Affairs, College of Biosystems Engineering and Food Science, Zhejiang University, Hangzhou 310058, China; lili1984@zju.edu.cn

[6] Department of Agriculture, Food and Environment, University of Pisa, Via del Borghetto 80, 56124 Pisa, Italy; alessandro.bianchi@phd.unipi.it

[7] Institute of Landscape Architecture, Faculty of Horticulture and Landscape Engineering, Slovak University of Agriculture, Tulipánová 7, 94976 Nitra, Slovakia; jan.kollar@uniag.sk

[8] Laboratory of Biotechnology and Plant Improvement, Centre of Biotechnology of Sfax, B.P "1177", Sfax 3018, Tunisia; raniabensaad@gmail.com (R.B.S.); benhsounanis@gmail.com (A.B.H.)

[9] Department of Environmental Sciences and Nutrition, Higher Institute of Applied Sciences and Technology of Mahdia, University of Monastir, Monastir 5000, Tunisia

[10] Department of Chemistry, Faculty of Biological Sciences, Universidad Autónoma de Nuevo León, San Nicolás de los Garza C.P. 64455, Nuevo León, Mexico; joel.elizondolv@uanl.edu.mx

[11] School of Medical & Health Sciences, University of Economics and Human Sciences in Warsaw, Okopowa 59, 01-043 Warszawa, Poland; m.kluz@vizja.pl

[12] Department of Chemistry and Technologies of Drug, Sapienza University, P. le Aldo Moro, 5, 00185 Rome, Italy; stefania.garzoli@uniroma1.it

* Correspondence: miroslava.kacaniova@gmail.com or m.kacaniova@vizja.pl

Abstract: *Rosa damascena* is mostly grown for its usage in the food, medical, and perfume industries, while it is also used as an attractive plant in parks, gardens, and homes. The use of *R. damascena* essential oil may yield new results in relation to the antimicrobial activity of essential oils and their use mainly in extending the shelf life of foods. This study investigates the chemical composition and antimicrobial properties of *Rosa damascena* essential oil (RDEO) using gas chromatography–mass spectrometry (GC-MS) and various bioassays to explore its potential applications in food preservation and microorganism growth control. The GC-MS analysis revealed that RDEO is predominantly composed of phenylethyl alcohol (70%), which is known for its antimicrobial and aromatic properties. Additionally, other significant constituents were identified, including nerol, citronellol, and geraniol, which may contribute to the EOs overall bioactivity. The antimicrobial activity was assessed through the minimal inhibition concentration against five *Candida* yeast strains, four Gram-positive, and four Gram-negative bacteria, including biofilm-forming *Salmonella enterica*. Determination of minimum inhibitory concentrations (MIC) revealed the strongest effects of RDEO's on Gram-negative species, with MIC_{50} values as low as 0.250 mg/mL for *S. enterica*. Moreover, an in situ assessment utilizing fruit and vegetable models demonstrated that the vapor phase of RDEO significantly suppressed microbial growth, with the most substantial reductions observed on kiwi and banana models. As a result of our study, the antimicrobial effect of RDEO on the microbiota of sous vide processed eggplant was detected, as well as an inhibitory effect on *S. enterica* during storage. The insecticidal activity against *Megabruchidius dorsalis* Fahreus, 1839, was also studied in this work and the best insecticidal activity was found at the highest concentrations. These results suggest that RDEO has the potential to serve as a natural antimicrobial agent in food preservation and safety applications, providing an alternative to synthetic preservatives.

Keywords: phytochemicals; food spoilage microorganisms; rose absolute; antimicrobial activity; insecticidal activity; food model

1. Introduction

A growing amount of produce purchased by consumers is minimally processed, especially organically produced fruits and vegetables [1]. Consumers are increasingly choosing convenient, ready-to-use produce that maintains fresh quality and contains only natural ingredients [2,3]. To extend the shelf life of fresh fruits and vegetables, the growth of microbial populations must be controlled [4]. Several postharvest procedures, such as washing and removing damaged tissues, are employed to reduce initial high microbial counts. However, this processing can negatively impact the plant tissues' metabolism and may reduce the shelf life of the fruits or vegetables [5].

Fruits and vegetables have a short shelf life due to weight loss and decay, primarily caused by fungal activity. However, they are becoming increasingly recognized for their health benefits due to their nutrient and fiber content. To extend the shelf life of fruits and vegetables, essential oils (EOs) can be incorporated into protective biofilms or applied in modified atmosphere packaging (MAP) [6]. To offer consumers fresh produce more efficiently, the food industry introduced the concept of "fresh-cut food", which refers to produce that has been physically altered but remains fresh. Fresh-cut food is ready to eat and appealing due to its fresh-like flavor, appearance, and taste [7]. It also offers health benefits, requiring less preparation time and effort. Since these products are pre-cleaned and cut, consumers save time in preparation. Fresh-cut fruits and vegetables (FCFV) are more convenient and easier to transport and consume than whole produce. However, there are significant challenges in maintaining their quality, freshness, and microbiological safety [5]. Although the studies show that the consumers are opting for fresh-cut products due to their health benefits, there has also been an increase in foodborne illnesses caused by bacteria such as *Salmonella enterica* (ex. Kauffmann and Edwards 1952), *Escherichia coli* (Migula 1895), *Shigella* spp., *Campylobacter* spp., *Listeria monocytogenes* (Murray et al. 1926), *Staphylococcus aureus* Rosenbach 1884, *Yersinia* spp., and *Bacillus cereus* Frankland and Frankland 1887 [8].

The use of the EOs as potent antimicrobial agents in minimally processed foods has been expected to exhibit a significant effect on safety of fresh produce and, hence, has attracted the attention of research [9]. Low storage temperatures and decreased pH enhance the antibacterial activity of EOs in fruit and vegetable-based foods [9]. Researchers have employed EOs in various studies to enhance the antimicrobial properties of fruits and vegetables, showing promising results [10–15]. The strong antibacterial activity of *Rosa damascena* EO and absolute was demonstrated by Ulusoy et al. [16] against strains of *Pectobacterium carotovorum* (Jones 1901), *Pseudomonas aeruginosa* (Schroeter 1872), *Bacillus subtilis* (Ehrenberg 1835), *Staphylococcus aureus* Rosenbach 1884, and *Chromobacterium violaceum* Bergonzini 1880. *Chromobacterium violaceum* was found to be the most susceptible to rose absolute and EO, while *E. coli* also showed sensitivity to rose EO. However, none of the microorganisms were affected by the antibacterial properties of the hydrosol [16].

Both Gram-positive and Gram-negative bacteria were also susceptible to the antibacterial properties of rose absolute [16].

Rosa damascena Mill., commonly known as the Damask rose, is a fragrant shrub belonging to the *Rosa* genus. It has over 18,000 varieties and approximately 200 species. Most roses are shrubs, found across the Northern Hemisphere's temperate and subtropical zones [17,18]. On an industrial scale, *R. centifolia* L., *R. gallica* L., and *R. damascena* Mill. are primarily used for their fragrance and flavor. Among these, *R. damascena* is regarded as the highest quality EO producer [19]. While primarily cultivated in Bulgaria, Turkey, and Iran, this plant is also grown in China, India, Libya, Morocco, South Italy, South France, South Russia, and Ukraine [20,21]. The main products derived from *R. damascena* include rose water (hydrosol), essential oil, absolute, and concrete. The use of rose EO dates back to ancient Persia. The practice of distilling roses for their EO likely began in Persia during the late seventh century A.D. and spread to Ottoman territories [22,23]. As one or more synergists can provide the desired fragrance and flavor without adversely affecting the food, the use of EOs in the food industry and consumer goods is expected to increase in the future [24]. *Rosa damascena* is being used more and more in laboratories and industries to obtain essential oil because of its high product yield, low organic solvents, quick processing time, and cheap maintenance expenses [25].

The prevalence of *Salmonella* spp. in plants poses a significant contamination risk in the food industry. Additionally, *Salmonella* strains can form biofilms on various food contact surfaces, such as processing equipment, knives and cutting boards, leading to cross-contamination of vegetables [26,27]. Cleaning and sanitation practices, often involving chemical disinfectants like chlorinated water [28,29], are fundamental in reducing *Salmonella* infections. However, the use of chemical disinfectants raises concerns about sustainability and safety. Therefore, it is essential to develop "green washing solutions" as effective alternatives to chemical disinfectants [30].

It is well known that the bacterial growth phase can influence their sensitivity to antimicrobial treatments [31]. Therefore, the effectiveness of antibacterial agents could be affected by bacterial cell physiology. Understanding whether bacterial cells in different growth phases are more susceptible or resistant to treatments is crucial [32].

EO, their distillation co-products (hydrosols), and plant extracts have been studied for their potential use in the food industry. These natural alternatives have demonstrated antimicrobial, antioxidant, and food preservation properties [33,34]. Some EOs have even shown a potential in reducing *Salmonella* spp. on fruits and vegetables [35–38]. Biofilm formation occurs when bacteria adhere to surfaces, such as food processing equipment, and secrete extracellular polymeric substances, providing structural protection to bacterial communities [39]. Bacteria in biofilms are more resistant to the human immune system, fungicides, antibiotics, and environmental stress than planktonic bacteria [40,41]. Organic matter left after food processing creates favorable conditions for biofilm formation, which poses cross-contamination risks [42,43]. In fact, bacterial adhesion and biofilm formation on processing equipment surfaces were identified as primary contributors to cross-contamination in trace analyses of *Salmonella* contamination [44,45].

Insect control in post-harvest crops and stored grains has also received significant attention. Various methods are used to protect crops from pests after harvest [46]. Chemical pesticides are often employed in agricultural fields and storage facilities to prevent and control pests. However, excessive use of chemical pesticides can result in resistance, environmental harm, and toxicity to non-target species [47]. Plant-derived bioinsecticides, such as EOs, may offer a safe and effective alternative to chemical pesticides [48]. Numerous studies have demonstrated the potential of plant-derived compounds, including EOs, in the management of pets in stored grain [49,50].

Inspired by the positive biological features of *Rosa damascena* Mill. and its metabolites, this work provides an extensive analysis of the chemical composition and biological profile of *R. damascena* EO derived from fresh flowers (RDEO). The therapeutic benefits of *R. damascena* motivated a comprehensive study of RDEO's bioactive effects, focusing on its antibacterial (in vitro and in situ), antibiofilm, and insecticidal properties. This study also explores the potential use of RDEO as a green storage protectant, specifically for handling preserved foods like sous vide eggplant contaminated with *Salmonella enterica*.

2. Materials and Methods

2.1. Essential Oil

The absolute 100% *Rosa damascena* Mill. essential oil (RDEO) used in this study was purchased from Hanus s.r.o. (Nitra, Slovakia). *R. damascena* was cultivated in Turkey. The first step in the process was the production of rose concretion, obtained by extracting the flowers of *R. damascena*. This fragrant waxy substance was then dissolved in pure ethanol and cooled. The frozen wax was filtered off, and the ethanol was fully evaporated from the solution, resulting in rose absolute.

2.2. GC/MS Examination of the Volatile Components in RDEO

The analysis on the volatile content of the *Rosa damascena* essential oil (RDEO) sample was conducted using an Agilent Technologies (Santa Clara, CA, USA) 6890N gas chromatograph. Prior to analysis, the EO was dissolved in a 10% hexane solution. The oven was set to operate at 50 °C and increased by 4 °C/min to 70 °C (holding for 2 min), then increased by 5 °C/min to 120 °C (holding for 1 min), and finally increased by 5 °C/min to 290 °C. The entire run took 52 min, with a 1 mL injection volume. Data collection was set to start after a 3.2 min solvent delay. Additionally, the MS quadrupole and MS ion source temperatures were 150 °C and 230 °C, respectively. The investigated sample was injected in split mode with a split ratio of 40.8:1. The carrier gas, helium 5.0, was used and flowed at a rate of 1 µL/min. Data were gathered in scan mode using electron impact mass spectrometry (EI-MS; 70 eV) in the 35–550 m/z range. To identify volatile compounds, retention indices were established experimentally using n-alkanes (C_7–C_{35}). These indices were then compared with those available in the NIST and Wiley databases. Compound percentages (amounts >0.1%) were calculated from their GC peak regions [51,52].

2.3. Microorganisms Tested for Antimicrobial Activity

The assessed EO antibacterial activity was evaluated using the following strains of bacteria: *Bacillus cereus* CCM 7934, *Listeria monocytogenes* CCM 4699, *Staphylococcus aureus* subsp. *aureus* CCM 4423, and *Streptococcus pneumoniae* CCM 4501, which are examples of Gram-positive bacteria; *Salmonella enterica* subsp. *enterica* CCM 3807, *Serratia marcescens* CCM 8588, *Shigella sonnei* CCM 4421, *Yersinia enterocolitica* CCM 7204T, which are examples of Gram-negative bacteria; and yeasts including *Candida albicans* CCM 8186, *Candida glabrata* CCM 8270, *Candida krusei* CCM 8271, *Candida parapsilosis* CCM 8260, and *Candida tropicalis* CCM 8223. For this experiment, all Gram-positive and Gram-negative bacterial species and yeasts were obtained from the Czech Collection of Microorganisms (CCM), which is kept in Brno, Czech Republic. From milk production, the biofilm-forming Gram-negative *Salmonella enterica* was isolated and sequenced to evaluate the antibacterial and antibiofilm activity. Before analysis, the bacterial and yeast inoculums were cultivated for 24 h at 37 °C and 25 °C, respectively, in Mueller–Hinton Broth (MHB, Oxoid, Basingstoke, UK) and Sabouraud Dextrose Broth (SDB, Oxoid, Basingstoke, UK). On the day of the experiment, the optical density of the bacterial and yeast inoculum was set at the 0.5 McFarland standard [53].

2.4. Minimal Inhibition Concentration (MIC)

The minimal inhibitory concentration values (MIC_{50} and MIC_{90}) were calculated. The microbial inoculum was initially added to 50 µL of a 96-well microtiter plate. Different concentrations of RDEO [54] (10 mg/mL to 0.00488 mg/mL in MHB) were then added. The negative and positive controls were prepared using MHB and SDB with EO (at the appropriate concentration) and MHB and SDB with inoculum, respectively. After the incubation period, absorbance at 570 nm was measured using a spectrophotometer (Glomax,

Promega Inc., Madison, WI, USA). The MIC_{50} and MIC_{90} values correspond to the lowest concentrations of RDEO that inhibit 50% and 90% of bacterial growth, respectively. The test was performed in triplicate.

2.5. Examination of the Fruit and Vegetables In Situ

To evaluate the antimicrobial properties of RDEO in situ, a variety of commercial kiwi, banana, eggplant, and pumpkin substrates were used, along with specific strains of yeast and Gram-positive and Gram-negative bacteria [55]. The substrates were chopped into 0.5 mm pieces, cleaned, and then placed into 60 mm Petri plates containing the bacteria. RDEO samples at concentrations of 500, 250, 125, and 62.5 μg/L were dispersed using ethyl acetate. Filter sheets made of ethyl acetate were used as controls. The Petri plates were sealed and incubated for seven days at 37 °C. The volume density of the bacteria and yeasts was calculated using the ImageJ program, and conventional techniques were employed to measure the in situ growth of microbial colonies [54].

2.6. Assay for Antibiofilm

2.6.1. Crystal Violet Assay

In the crystal violet study, Kačániová et al. [56] incubated bacterial suspensions in Mueller–Hinton Broth at 37 °C to determine the minimum biofilm inhibitory concentration (MBIC). A microtiter plate containing an inoculum was treated with two-fold dilutions of EO (from 100 mg/mL to 0.049 mg/mL). Wells were cleaned, stained, and their absorbance at 570 nm was measured after a 24 h period. The doses that inhibit 100%, 50%, and 90% of biofilm formation were determined to be $MBIC_{50}$ and $MBIC_{90}$, respectively.

2.6.2. Applying the MALDI-TOF MS Biotyper to Detect the Development of Biofilms

A Bruker Daltonics MALDI-TOF MicroFlex (Bruker Daltonics GmbH & Co. KG, Bremen, Germany) instrument was used to measure the protein degradation that occurs during the biofilm-forming process. Stainless steel and glass slides were placed into 50 mL polypropylene tubes containing 20 mL of Mueller–Hinton Broth (MHB) and 100 μL of *Salmonella enterica* biofilm-forming bacterial inoculum. Experimental tubes were filled with EO to a concentration of 0.1%, while control tubes were left undisturbed. Biofilms were removed from glass and plastic surfaces by shaking at 170× *g* for three days and then incubated at 37 °C for five, seven, nine, twelve, and fourteen days. Additionally, planktonic cells from control samples used for the RDEO experiment were investigated. Dendrograms based on the estimation of Euclidean distance using 19 typical global spectra were produced and protein spectra were obtained using MALDI-TOF in the linear positive mode [56].

2.7. Kinetics Growth Measurement

To develop growth curves for *Salmonella enterica*, optical density (OD) at 850 nm was measured using a personal bioreactor (RTS-1, Biosan, Riga, Latvia). The strain was first cultured for 24 h on Mueller–Hinton Agar (MHA; Oxoid, Basingstoke, UK) at 37 °C. One colony was then transferred to a 30 mL sealed tube containing Mueller–Hinton Broth (MHB) and incubated at 37 °C until the OD850 reached 1, representing the log phase of bacterial growth. The bioreactor operated at 2000 rpm with directional changes every second. The temperature was gradually raised to 50 °C, with OD readings recorded every 5 min over a 20 min period. This procedure was then repeated at 55 °C, 60 °C, and 65 °C. In a separate experiment, 1% EO was introduced to the culture at the point when OD850 reached 1, marking the log phase. OD readings were recorded for both control (no EO) and experimental (with EO) conditions. Notably, the bioreactor is calibrated for microorganisms of 0.4–0.8 × 1–3 μm in size.

2.8. Antimicrobial Action of Sous Vide in the Eggplant Model

The two and a half kg of eggplant samples used in this investigation came from an authorized dealer in the Slovak Republic. Utilizing a sterile knife, the eggplant (*Solanum melongena*) was cut into five-gram halves after cooling and was then transported to the microbiological laboratory. The study included a total of 480 five-gram samples: 3 raw samples, 240 treated and control samples on day 1, and 240 treated and control samples on day 7. The chopped eggplant samples were treated with a 1% v/w RDEO solution dissolved in rapeseed oil, and each was vacuum packed separately with a Concept vacuum packer. Both vacuum-packed and unpackaged samples were employed as control samples. Samples containing 100 µL of *S. enterica* and 1% v/w RDEO were intended to imitate the presence of *S. enterica* without harming the eggplant. Before the samples were vacuum packed, they were stimulated for approximately a min [57].

We had the following data accessible to us while we were assessing it:

(i) Control: New eggplant samples were kept at 4 °C in polyethylene bags. Following this, they underwent treatments at 50 to 65 °C for 5 to 20 min.
(ii) Control + vacuum: Fresh eggplant samples were treated for 5 to 25 min at 50–65 °C after being vacuum packed in polyethylene bags and kept at 4 °C.
(iii) EO: Collected eggplant samples received vacuum packaging, 1% RDEO treatment, and preservation at 4 °C. After that, they were cooked for 5 to 25 min at 50 to 65 °C.
(iv) *Salmonella*: Fresh eggplant samples that were vacuum packed and treated with *S. enterica* were kept at 4 °C before being exposed to the bacterium for 5 to 25 min at 50 to 65 °C.
(v) *Salmonella* + EO: Vacuum-packed fresh eggplant samples treated with *S. enterica* and containing 1% RDEO were held at 4 °C before being treated for 5 to 25 min at 50 to 65 °C.

On the first day, a raw, uncooked eggplant sample was used as the control. All of the samples were macerated for a full day after the EO from the first set of samples and the *S. enterica* from the second group of tests were applied, gently mixed, and integrated. The samples were created using a CASO SV1000 sous vide device, produced by a company in Arnsberg, Germany. To prepare the samples for sous vide cooking, they were divided into groups and cooked at a specific temperature for a predetermined amount of time while being closely monitored.

The high-barrier polyethylene vacuum packaging bags are made of an impermeable substance that is resistant to moisture and extremely high or low temperatures (−30 °C to 100 °C), with a thickness between 40 and 200 microns. The data sheet also states that they taste and smell good, have a very long shelf life, and are free of bisphenol A and all plasticizers, including microplastics. They can last for several years when kept in freezers and refrigeration cases.

2.9. Microbiological Analyses of Eggplant Samples

Microbiological studies were carried out from day 0 to day 7. After heating, a portion of the samples were assessed one and seven days later. After weighing five grams, the eggplant samples were put in an aseptic stomacher bag. The materials were diluted to 10^{-1} in 45 mL of peptone water and then homogenized using a stomacher for two min. Subsequently, a standard pre-dried plate count agar medium was covered with 0.1 mL of an aliquot pipetted from an appropriate dilution. The samples were homogenized in the GFL 3031 shaking incubator (GFL, Burgwedel, Germany) for 30 min. The following microorganism populations were tested: Coliform bacteria were cultivated on Violet Red Bile Lactose Agar (VRBL, Oxoid, Basingstoke, UK) at 37 °C for 24 to 48 h and Total Viable Counts (TVCs) on Plate Count Agar (PCA, Oxoid, Basingstoke, UK) at 30 °C for 48 to 72 h [57].

2.10. Identification of Microbial Strains by MALDI-TOF MS Biotyper

The MALDI-TOF (Matrix-Assisted Laser Desorption/Ionization Time of Flight) MS Biotyper (Bruker, Daltonics, Bremen, Germany) and reference libraries were used for determining the eggplant samples. Once a stock solution had been prepared, it transformed into a substance that was organic. 50% acetonitrile, 47.5% water, and 2.5% trifluoroacetic acid made up the standard solution. 1 mL of stock solution was made by mixing 500 µL of pure 100% acetonitrile, 475 µL of filtered water, and 25 µL of pure 10% trifluoroacetic acid. In a 250 µL Eppendorf flask, the "HCCA matrix portioned" was created and homogenized with the organic solvent. The matrix materials were from Vrable, Slovakia-based Aloqence Science. The samples were created in accordance with previous suggestions [57].

2.11. Insect-Related Activity

The model organism utilized for assessing the insecticidal effectiveness of RDEO was *Megabruchidius dorsalis* Fahreus, 1839. Petri plates were used to hold groups of fifty *M. dorsalis* insects, and each plate was covered with sterile filter paper. To achieve concentrations of 100%, 50%, 25%, 12.5%, 6.25%, and 3.125%, RDEO was diluted with 0.1% polysorbate. After saturating sterile filter paper disks with 100 µL of each RDEO concentration, the plates were covered with parafilm and left at room temperature for an entire day. A 100 µL 0.1% polysorbate solution was given to the control group. After one full day, counts of live and dead insects were recorded. Three distinct studies successfully replicated this experimental setup [56].

2.12. Statistical Analysis

The data are presented as mean values ± standard deviation (SD), and each evaluation was carried out in triplicate. A one-way ANOVA was conducted, and then Tukey's HSD test was run at $p \leq 0.05$ of significance (CoStat version 6.451, CoHort Software, Pacific Grove, CA, USA). Finally, the JMP Pro 17.0 software program (SAS Institute, Cary, NC, USA) was used for the graphic elaborations.

3. Results

3.1. Chemical Composition of Rosa Damascena Essential Oil

The GC/MS technique was used to investigate the chemical composition of the RDEO volatile oil. The results, which show the percentage distribution of detected compounds, are presented in Table 1. A total of six volatiles were found, accounting for 99.9% of the EO composition. The findings suggest that the examined EO was primarily characterized by a significantly elevated percentage (70%) of phenylethyl alcohol.

Table 1. Volatile constituents of RDEO.

No	RI (lit.) [a]	RI (calc.) [b]	Compound [c]	% [d]
1	1108	1115	phenylethyl alcohol	70.0
2	1229	1223	nerol	3.7
3	1225	1227	citronellol	11.3
4	1252	1249	geraniol	7.1
5	1875	1873	1-nonadecene	3.1
6	1990	1899	nonadecane	4.7
			Total	99.9

[a] Literature values of retention indices on HP-5MS column; [b] Calculated values of retention indices on HP-5MS column; [c] identified compounds; [d] percentage amounts of identified compounds.

3.2. Minimal Inhibition Concentration

The broth microdilution method was used to determine the MIC_{50} and MIC_{90}, i.e., the minimum inhibitory concentrations. This was performed in order to gain a better understanding of the antibacterial activity of RDEO. Overall, RDEO had the strongest effect on inhibition of Gram-negative species. In particular, the lowest MIC_{50} values

(0.250 mg/mL) and MIC_{90} values (0.13 and 0.14 mg/mL) were observed for *S. enterica* and *S. sonnei*. For Gram-positive *S. pneumoniae* and *S. aureus*, MIC_{50} (0.249 and 0.274 μL/mL, respectively) and MIC_{90} (0.283 and 0.291 mg/mL, respectively) were determined. The results of inhibiting *S. enterica* biofilm formation using RDEO showed $MIBC_{50}$ values of 0.270 mg/mL and $MIBC_{90}$ values of 0.291 mg/mL. The yeast strains *C. albicans* and *C. tropicalis* showed the best MIC results, with MIC_{50} and MIC_{90} values of 0.363 and 0.369 mg/mL and 0.393 and 0.386 mg/mL, respectively. The detailed results of the minimum inhibitory concentration (MIC) assay are presented in Table 2.

Table 2. Minimal inhibition concentration and minimal biofilm inhibition concentration of RDEO in mg/mL.

Microorganism	MIC_{50}	MIC_{90}
Gram-negative bacteria		
Salmonella enterica CCM 3807	0.250 ± 0.015 [b]	0.277 ± 0.027 [b]
Serratia marcescens CCM 8588	0.264 ± 0.014 [b]	0.295 ± 0.006 [b]
Shigella sonnei CCM 4421	0.250 ± 0.015 [b]	0.282 ± 0.013 [b]
Yersinia enterocolitica CCM 7204T	0.254 ± 0.015 [b]	0.284 ± 0.010 [b]
Gram-positive bacteria		
Bacillus cereus CCM 7934	0.352 ± 0.013 [a]	0.382 ± 0.006 [a]
Listeria monocytogenes CCM 4699	0.353 ± 0.029 [a]	0.380 ± 0.003 [a]
Staphylococcus aureus CCM 4423	0.274 ± 0.022 [b]	0.291 ± 0.005 [b]
Streptococcus pneumoniae CCM 4501	0.249 ± 0.029 [b]	0.283 ± 0.007 [b]
Yeast		
Candida albicans CCM 8186	0.363 ± 0.016 [a]	0.393 ± 0.006 [a]
Candida tropicalis CCM 8264	0.369 ± 0.012 [a]	0.386 ± 0.011 [a]
Candida glabrata CCM 8270	0.372 ± 0.014 [a]	0.389 ± 0.011 [a]
Candida krusei CCM 8271	0.369 ± 0.016 [a]	0.389 ± 0.010 [a]
Candida parapsilosis CCM 8260	0.372 ± 0.015 [a]	0.387 ± 0.017 [a]
Biofilm forming bacteria (BFB)	$MIBC_{50}$	$MIBC_{90}$
Salmonella enterica	0.270 ± 0.016 [b]	0.291 ± 0.005 [b]

Data are the mean (±SD) of three samples. Different letters in each column refer to significant differences (Tukey, $p \leq 0.05$).

3.3. Antimicrobial Activity in Vapor Phase

In the following experiment, we carried out an in situ antimicrobial investigation using fruit and vegetables as food models. The pre-selected bacteria have also been used in the in vitro evaluation. Table 3 and Figure 1a, b showed the vapor phase data for the fruit model. Overall, compared to the in vitro tests, the vapor phase of RDEO showed higher efficacy in suppressing Gram-positive strains, following yeasts and Gram-negative in the fruit model. When RDEO was used for in situ assessment of *C. tropicalis* development on the kiwi model, the results indicated that the highest concentration used (500 μg/L) had the strongest inhibitory effect (88.48%). The maximum dose of RDEO tested (88.32 and 87.55%, respectively) significantly inhibited the growth of *C. albicans* and *C. krusei* in the kiwi model (Table 3). RDEO was the most effective at inhibiting biofilm-forming *S. enterica* (76.36%) at 500 μg/L, but even at the highest concentration (500 μg/L), a significant level of inhibition was maintained compared to the other tested microorganisms. Furthermore, EO showed significant antibacterial activity against *S. pneumoniae* (86.89%). Maximum antibacterial activity against *S. sonnei* was observed at a concentration of 500 μg/L (76.46%). RDEO showed modest inhibitory effects in the in situ evaluation of Gram-positive bacterial development on the banana model. Among the Gram-positive strains, RDEO showed the highest inhibitory activity against *B. cereus* (67.13%) at 62.5 μg/L. In contrast, at the highest concentrations tested, RDEO showed the greatest inhibitory effect on biofilm-forming *S. enterica* (57.46%) among the Gram-negative strains growing on the banana model (Table 3). The results showed that among the yeasts grown in the banana model, RDEO, when applied at the maximum concentration, was the most effective at inhibiting the growth of *C. albicans* (46.83%) and *C. krusei* (45.63%). Furthermore, RDEO at the highest concentration tested strongly inhibited *Y. enterocolitica*. In the case of Gram-negative bacteria and yeasts, pro-microbial growth was detected in some of the species at the lowest concentrations, indicating that the concentrations of EO promoted the growth of the individual microorganisms.

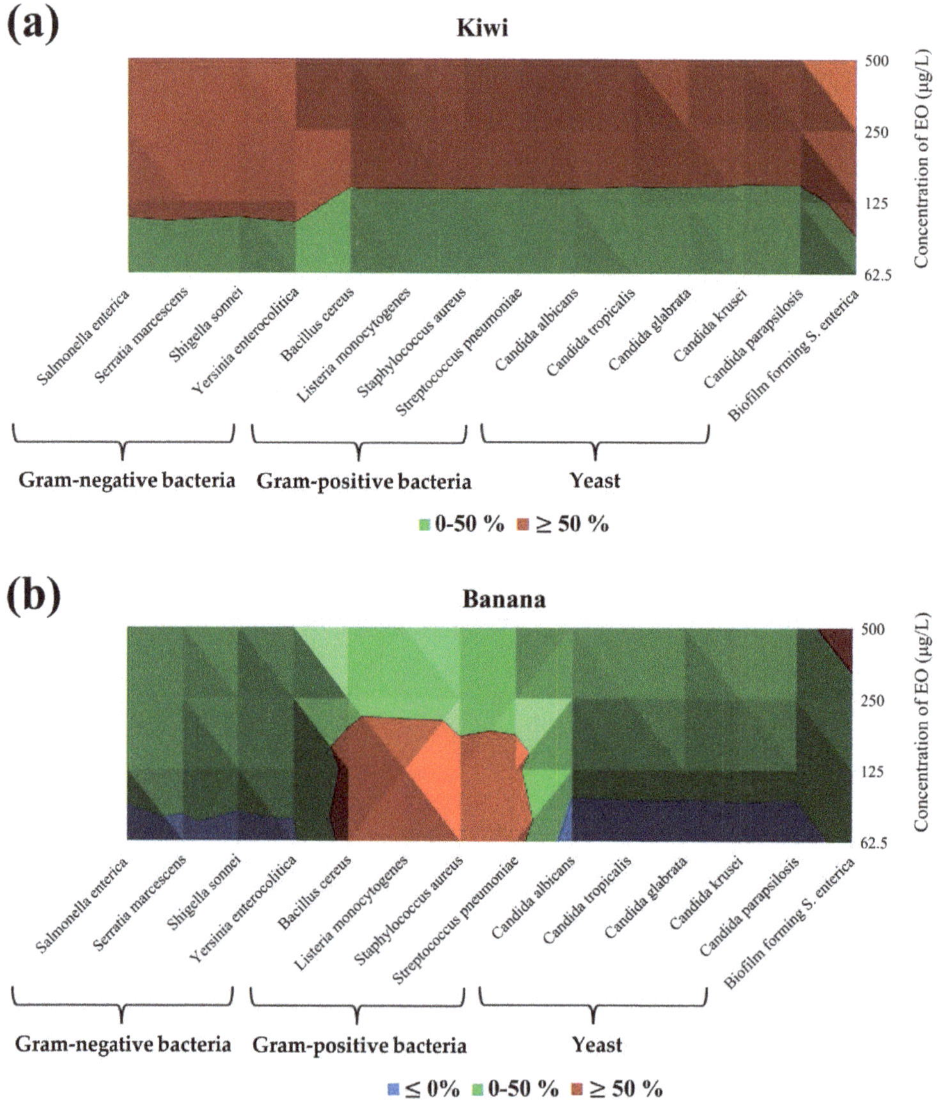

Figure 1. Isometric elaboration of the data of Table 4 (In situ analysis of the antimicrobial activity of the vapor phase of RDEO in fruits model): (**a**) Kiwi; (**b**) Banana. Blue $\leq$ 0%; Green: 0–50%; Red: $\geq$ 50%.

Table 3. In situ analysis of the antimicrobial activity (%) of the vapor phase of RDEO in fruits model.

Food Model	Microorganisms	Concentration of RDEO in μg/L			
		62.5	125	250	500
Kiwi					
Gram-negative	*Salmonella enterica*	37.10 ± 2.28 [b]	53.77 ± 1.06 [a]	66.59 ± 1.72 [a]	75.69 ± 1.05 [b]
	Serratia marcescens	35.63 ± 1.76 [b]	54.90 ± 1.68 [a]	66.07 ± 2.24 [a]	76.33 ± 2.19 [b]
	Shigella sonnei	34.97 ± 2.53 [b]	54.16 ± 0.82 [a]	65.66 ± 2.21 [a]	76.46 ± 2.10 [b]
	Yersinia enterocolitica	37.13 ± 1.30 [b]	55.25 ± 2.36 [a]	66.09 ± 2.33 [a]	76.15 ± 2.61 [b]
Gram-positive	*Bacillus cereus*	25.22 ± 2.18 [c]	46.29 ± 2.63 [b]	64.36 ± 3.14 [a]	85.50 ± 2.16 [a]
	Listeria monocytogenes	25.62 ± 2.92 [c]	46.13 ± 1.54 [b]	66.68 ± 2.84 [a]	86.56 ± 1.90 [a]
	Staphylococcus aureus	25.40 ± 0.55 [c]	46.33 ± 2.57 [b]	65.97 ± 2.60 [a]	85.05 ± 1.56 [a]
	Streptococcus pneumoniae	25.69 ± 2.20 [c]	45.88 ± 2.78 [b]	66.06 ± 2.31 [a]	86.89 ± 3.13 [a]
Yeast	*Candida albicans*	24.41 ± 2.51 [c]	46.14 ± 2.64 [b]	66.76 ± 2.92 [a]	88.32 ± 1.15 [a]
	Candida tropicalis	25.43 ± 2.14 [c]	45.48 ± 1.61 [b]	65.69 ± 2.94 [a]	88.48 ± 1.14 [a]
	Candida glabrata	24.80 ± 1.16 [c]	45.29 ± 1.91 [b]	67.06 ± 2.41 [a]	84.93 ± 1.70 [a]
	Candida krusei	25.70 ± 1.88 [c]	45.00 ± 2.19 [b]	64.85 ± 3.18 [a]	87.55 ± 2.80 [a]
	Candida parapsilosis	26.09 ± 3.05 [c]	45.21 ± 2.32 [b]	65.15 ± 0.51 [a]	86.37 ± 2.94 [a]
BFB	*Salmonella enterica*	44.26 ± 1.65 [a]	55.30 ± 1.84 [a]	66.43 ± 1.70 [a]	76.38 ± 3.51 [b]
Banana					
Gram-negative	*Salmonella enterica*	−14.77 ± 2.11 [de]	13.63 ± 1.16 [d]	26.06 ± 1.68 [e]	36.44 ± 1.70 [cd]
	Serratia marcescens	−6.74 ± 0.97 [c]	10.73 ± 0.77 [d]	24.40 ± 2.09 [e]	36.84 ± 2.21 [cd]
	Shigella sonnei	−12.77 ± 3.90 [cd]	17.50 ± 1.67 [d]	26.76 ± 2.91 [de]	36.40 ± 2.30 [cd]
	Yersinia enterocolitica	−7.85 ± 1.08 [cd]	16.33 ± 2.57 [d]	35.73 ± 2.86 [c]	42.10 ± 1.40 [bc]
Gram-positive	*Bacillus cereus*	67.13 ± 2.29 [a]	56.68 ± 1.21 [ab]	45.47 ± 3.60 [a]	35.76 ± 1.83 [cd]
	Listeria monocytogenes	65.76 ± 4.21 [a]	63.52 ± 2.11 [a]	44.67 ± 1.06 [a]	36.03 ± 1.68 [cd]
	Staphylococcus aureus	66.08 ± 2.33 [a]	55.43 ± 2.30 [b]	44.08 ± 1.38 [ab]	33.63 ± 0.96 [d]
	Streptococcus pneumoniae	65.37 ± 2.28 [a]	54.33 ± 1.39 [b]	45.61 ± 2.31 [a]	35.66 ± 2.21 [cd]
Yeast	*Candida albicans*	−23.37 ± 2.96 [f]	14.60 ± 2.17 [d]	36.77 ± 2.20 [bc]	46.83 ± 1.86 [b]
	Candida tropicalis	−22.40 ± 1.10 [ef]	15.70 ± 2.95 [d]	36.81 ± 1.85 [bc]	45.43 ± 4.35 [b]
	Candida glabrata	−23.33 ± 1.40 [f]	16.39 ± 3.55 [d]	33.62 ± 2.72 [cd]	44.74 ± 2.63 [b]
	Candida krusei	−22.48 ± 4.12 [ef]	17.56 ± 3.96 [d]	37.11 ± 2.27 [bc]	45.63 ± 2.63 [b]
	Candida parapsilosis	−22.93 ± 1.00 [f]	16.39 ± 3.55 [d]	34.19 ± 3.38 [cd]	45.61 ± 2.32 [b]
BFB	*Salmonella enterica*	15.66 ± 3.32 [b]	35.39 ± 3.22 [c]	45.66 ± 3.32 [a]	57.46 ± 3.01 [a]

Data are the mean (± SD) of three samples. Different letters in each column (for each fruit type: kiwi and banana) refer to significant differences (Tukey, $p \leq 0.05$).

The results of the antibacterial activity on vegetables against all microorganisms observed in the vapor phase are reported in a subsequent section of our experiment (Table 4, Figure 2a,b). On the eggplant model, the best antimicrobial results were found against *S. pneumoniae* (85.62%) at the lowest concentration, followed by the bacteria *B. cereus* (85.13%), *S. aureus* (84.49%), and *L. monocytogenes* (82.41%), also at the lowest concentration of 62.5 μg/L (Table 4). Similarly, the lowest concentration of EO applied, 62.5 μg/L, showed the highest antimicrobial activity in the pumpkin model. The best results were obtained against *S. pneumoniae* (77.44%), followed by *S. enterica* with 76.62% inhibition and *L. monocytogenes* with 75.92% inhibition in the pumpkin model (Table 4). Satisfactory results were obtained against other Gram-positive and Gram-negative bacteria, while the lowest antimicrobial activity on the cucumber model was achieved by yeast against *C. parapsilosis* (58.62%) at the lowest RDEO concentration. In the case of the biofilm-forming bacterium *S. enterica*, pro-bacterial growth was detected at the lowest concentration.

Table 4. In situ analysis of the antimicrobial activity (%) of the vapor phase of RDEO in vegetable models.

Food Model	Microorganisms	Concentration of RDEO in μg/L			
		62.5	125	250	500
Eggplants					
Gram-negative	*Salmonella enterica*	76.34 ± 2.11 bc	65.20 ± 3.28 c	56.29 ± 2.46 bc	42.59 ± 1.94 b
	Serratia marcescens	76.03 ± 2.78 bc	65.87 ± 4.44 bc	56.09 ± 4.14 bc	45.03 ± 1.68 b
	Shigella sonnei	74.43 ± 3.49 c	66.08 ± 1.21 bc	56.86 ± 1.76 bc	47.19 ± 0.61 b
	Yersinia enterocolitica	86.29 ± 4.29 a	74.19 ± 0.64 a	63.15 ± 1.58 ab	57.79 ± 1.75 a
Gram-positive	*Bacillus cereus*	85.15 ± 0.51 a	75.93 ± 4.32 a	63.30 ± 2.24 ab	57.51 ± 1.69 a
	Listeria monocytogenes	82.41 ± 1.11 ab	63.37 ± 2.25 c	54.41 ± 3.81 c	43.52 ± 1.89 b
	Staphylococcus aureus	84.49 ± 1.86 a	73.75 ± 2.17 ab	64.76 ± 3.50 a	55.74 ± 0.93 a
	Streptococcus pneumoniae	85.62 ± 2.26 a	74.58 ± 3.93 a	67.87 ± 0.02 a	57.17 ± 1.22 a
Yeast	*Candida albicans*	56.14 ± 1.18 d	43.89 ± 0.58 d	35.70 ± 1.88 d	24.33 ± 2.26 c
	Candida tropicalis	54.74 ± 2.71 d	43.67 ± 2.06 d	35.37 ± 1.59 d	25.69 ± 3.53 c
	Candida glabrata	54.56 ± 2.01 d	46.03 ± 2.56 d	33.63 ± 1.11 d	23.96 ± 3.40 c
	Candida krusei	57.49 ± 0.62 d	43.47 ± 2.36 d	34.97 ± 3.41 d	25.33 ± 2.28 c
	Candida parapsilosis	54.86 ± 1.72 d	46.06 ± 2.23 d	35.66 ± 2.72 d	25.74 ± 2.10 c
BFB	*Salmonella enterica*	56.03 ± 2.31 d	44.20 ± 1.29 d	24.90 ± 2.73 e	−6.40 ± 1.65 d
Pumpkin					
Gram-negative	*Salmonella enterica*	76.62 ± 4.15 a	66.54 ± 1.07 a	55.66 ± 2.93 a	45.59 ± 2.33 a
	Serratia marcescens	65.28 ± 3.75 bc	55.88 ± 2.59 bc	44.34 ± 1.25 b	35.10 ± 1.68 b
	Shigella sonnei	72.96 ± 1.79 ab	64.37 ± 1.72 ab	56.81 ± 1.05 a	45.00 ± 1.25 a
	Yersinia enterocolitica	73.59 ± 4.31 ab	55.84 ± 4.51 bc	43.37 ± 2.07 b	34.00 ± 3.69 bc
Gram-positive	*Bacillus cereus*	75.15 ± 1.57 a	65.88 ± 4.05 a	55.34 ± 1.71 a	46.86 ± 1.14 a
	Listeria monocytogenes	75.92 ± 3.98 a	64.59 ± 1.97 ab	56.14 ± 3.71 a	44.12 ± 2.94 a
	Staphylococcus aureus	74.38 ± 3.17 a	62.95 ± 8.21 ab	56.06 ± 2.78 a	44.06 ± 0.55 a
	Streptococcus pneumoniae	77.44 ± 1.67 a	65.92 ± 0.58 a	56.50 ± 1.19 a	33.96 ± 0.71 bc
Yeast	*Candida albicans*	56.14 ± 1.19 d	46.32 ± 2.10 cd	32.66 ± 2.00 cd	24.80 ± 3.61 d
	Candida tropicalis	56.44 ± 1.63 cd	46.41 ± 0.64 cd	34.63 ± 1.00 c	27.42 ± 2.18 cd
	Candida glabrata	55.63 ± 1.23 d	44.70 ± 0.96 d	33.40 ± 1.45 cd	23.56 ± 2.11 d
	Candida krusei	55.94 ± 2.79 d	44.73 ± 3.81 d	35.06 ± 3.27 c	27.88 ± 1.10 cd
	Candida parapsilosis	58.60 ± 0.66 cd	46.38 ± 2.76 cd	35.29 ± 3.21 c	25.30 ± 3.36 d
BFB	*Salmonella enterica*	−29.52 ± 5.26 e	16.03 ± 1.68 e	26.77 ± 2.93 d	34.15 ± 3.43 bc

Data are the mean (±SD) of three samples. Different letters in each column (for each vegetable type: eggplants and pumpkin) refer to significant differences (Tukey, $p \leq 0.05$).

In conclusion, as shown in Figure 1a,b, the two used fruit models (kiwi and banana), had different trends. In kiwi, by increasing the concentration of RDEO used, the % inhibition tends to increase. For all tested microorganisms, concentrations of 250 μg/L and 500 μg/L lead to a % inhibition greater than 50%; indeed, this inhibition value was maintained with 125 μg/L of RDEO for Gram-negative bacteria and BFB *S. enterica* (Figure 1a).

In banana, on the other hand, the 50% inhibition was archived only for Gram-positive bacteria for concentrations less than 125 μg/L and for BFB *S. enterica* at 500 μg/L. For the other microorganisms (such as yeast and Gram-negative bacteria), the RDEO had not shown a significant effect on inhibition. Indeed, the growth of microorganisms was favored at concentrations of 62.5 μg/L (Figure 1b).

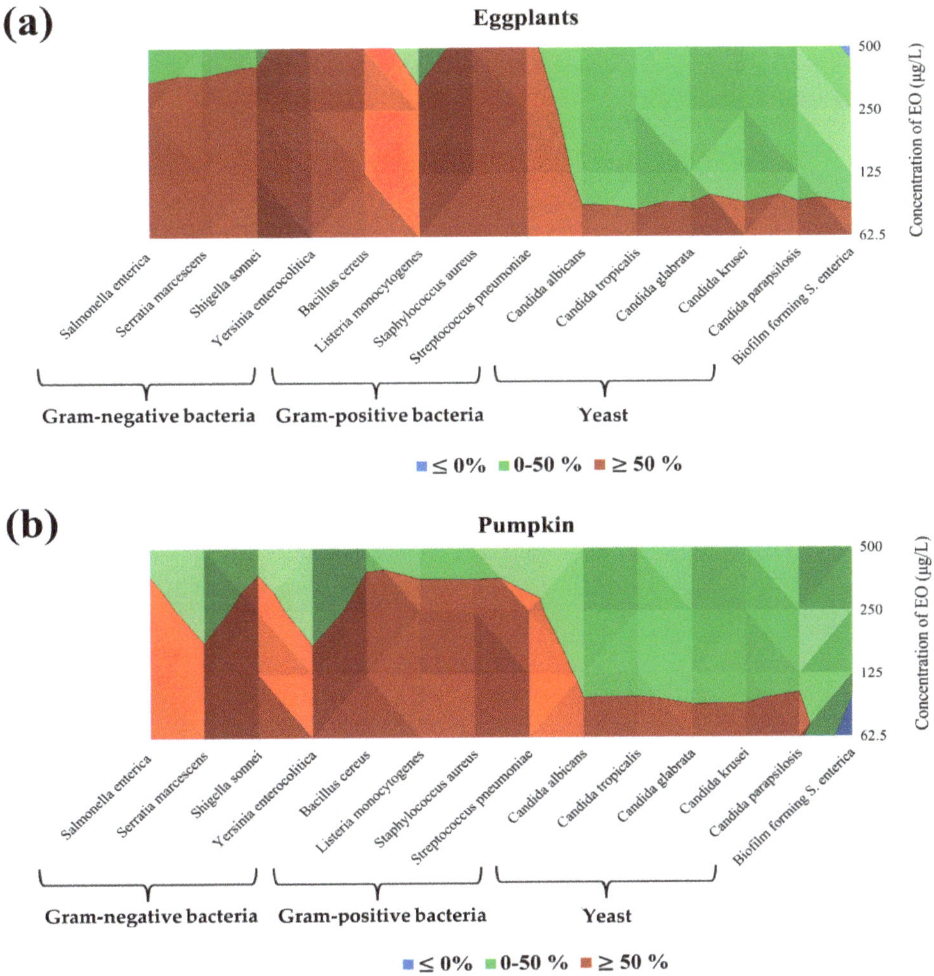

Figure 2. Isometric elaboration of the data of Table 5 (In situ analysis of the antimicrobial activity of the vapor phase of RDEO in vegetable models): (**a**) Eggplants; (**b**) Pumpkin. Blue $\leq 0\%$; Green: 0–50%; Red: $\geq 50\%$.

Table 5. Insecticidal activity of RDEO against *Megabruchidius dorsalis* (n = 50).

Concentration (%)	Number of Living Individuals	Number of Dead Individuals	Insecticidal Activity (%)
100	10	90	90.00 ± 0.00
50	20	80	80.00 ± 0.00
25	30	70	70.00 ± 0.00
12.5	50	50	50.00 ± 0.00
6.25	100	0	0.00 ± 0.00
3.125	100	0	0.00 ± 0.00
Control group	100	0	0.00 ± 0.00

On the contrary, in the vegetal models (eggplants and pumpkin) in Figure 2a,b, the similar trend has been seen. In fact, in this case, the inhibition percentage increased with decreasing the RDEO concentrations, indicating that a greater effect occurs at low concentrations (62.5 µg/L).

In eggplants, the application of RDEO showed the major effect. Gram-positive bacteria and Gram-negative bacteria were inhibited at concentrations lower than 250 µg/L, while for yeast and BFB *S. enterica* only at a concentration of 62.5 µg/L (Figure 2a).

In pumpkin, all Gram-positive bacteria and Gram-negative bacteria were inhibited at concentrations lower than 125 µg/L but at a concentration of 62.5 µg/L. Meanwhile, the RDEO promoted growth of BFB S. enterica (Figure 2b).

3.4. Antibiofilm Activity of Rosa Damascena

Figure 3A–F showed the spectra of the developmental stages of *S. enterica* biofilm throughout the experiment. RDEO was added to the experimental groups. The spectra were sorted into pairs according to their growth stage on different surfaces, except for the spectra of planktonic cells, which were obtained from the culture medium. Mass spectra of *S. enterica* on day 3 of the experiment are shown (Figure 3A). Spectra of control planktonic cells and experimental spectra with both stainless steel and glass are similar. On day 5 of the experiment (Figure 3B), a difference can be observed between the control planktonic cells and the experimental group with stainless steel, while the experimental group with glass is comparable. On day 7 of the experiment (Figure 3C), there was a change in the spectra in both experimental groups compared to the control planktonic cells. The same trend can be observed on day 9–14 of the experiment (Figure 3D,F).

The results suggest that distinct stages of biofilm development could be distinguished based on MSP distance using MALDI profiling. They were divided into different clusters (Figure 4). It can be observed that the planktonic stage of *S. enterica* showed the most significant similarity at day 3 with the experimental group in MSP distance. The control groups showed shorter MSP distances from the planktonic cells than the experimental groups on the following days.

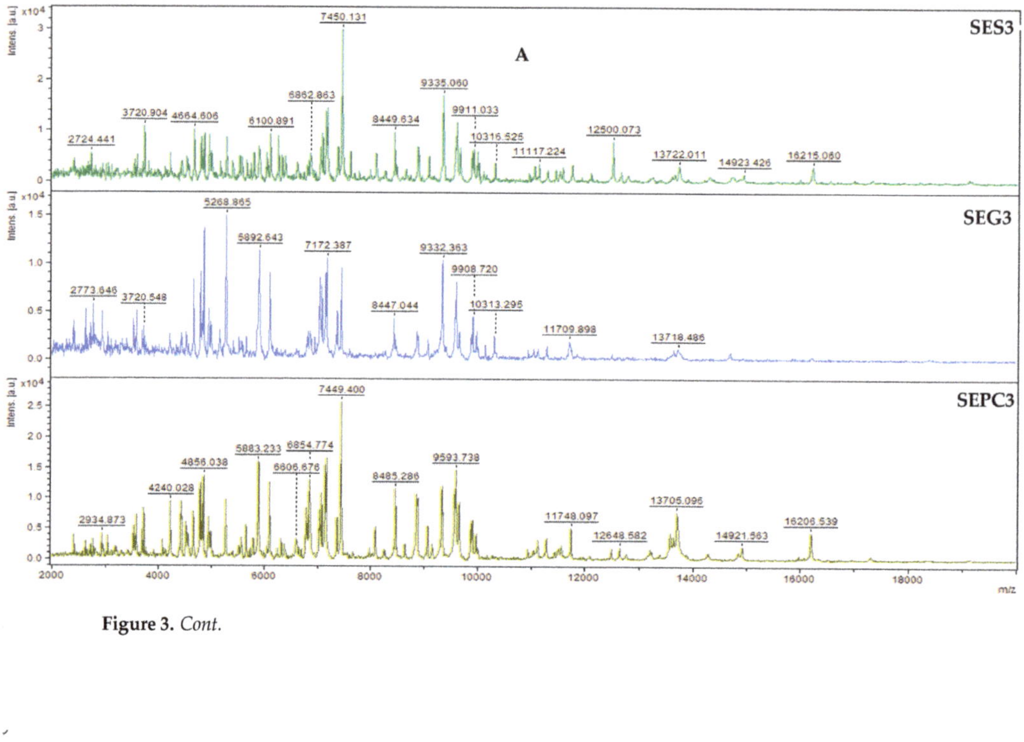

Figure 3. *Cont.*

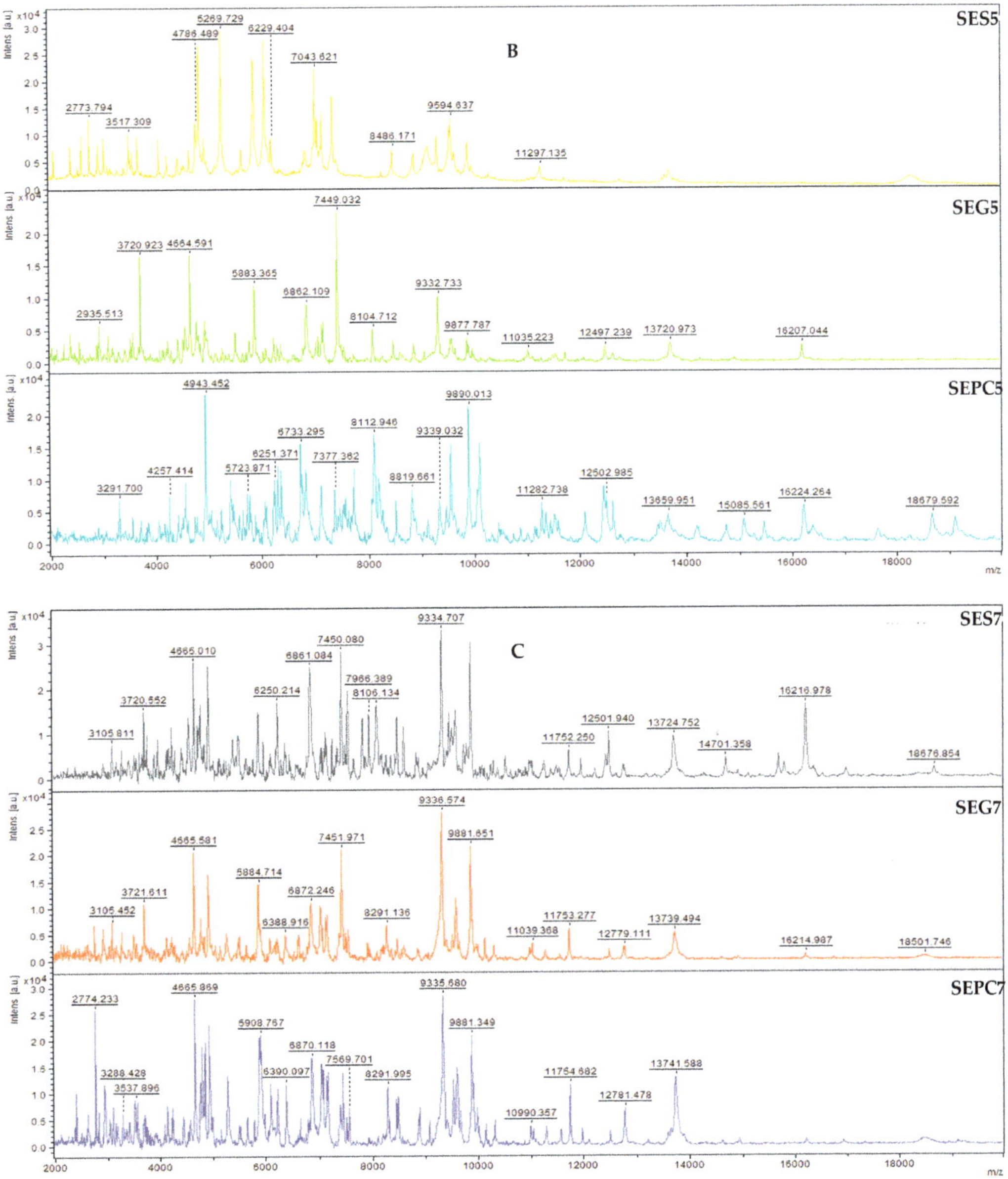

Figure 3. *Cont.*

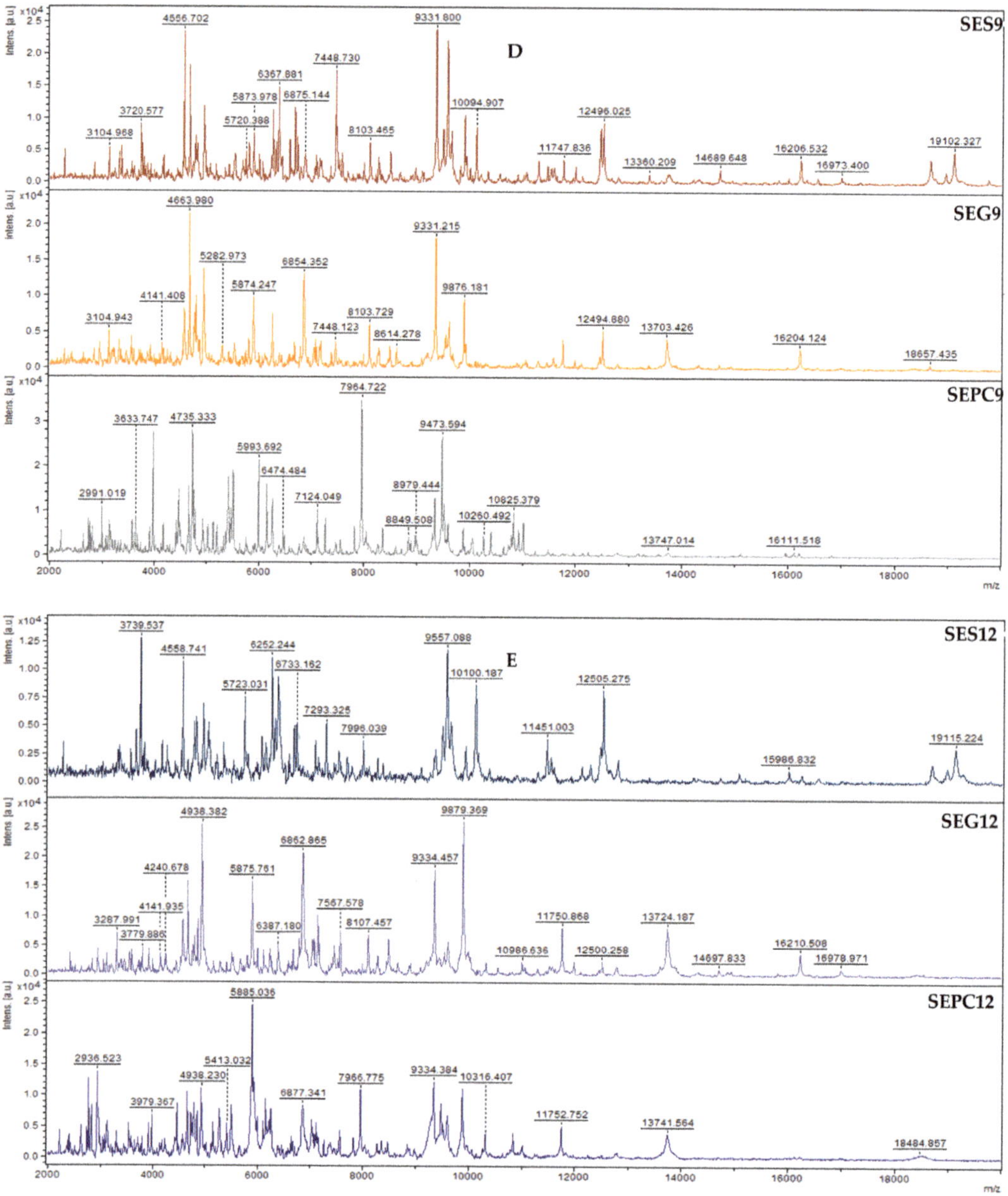

Figure 3. *Cont.*

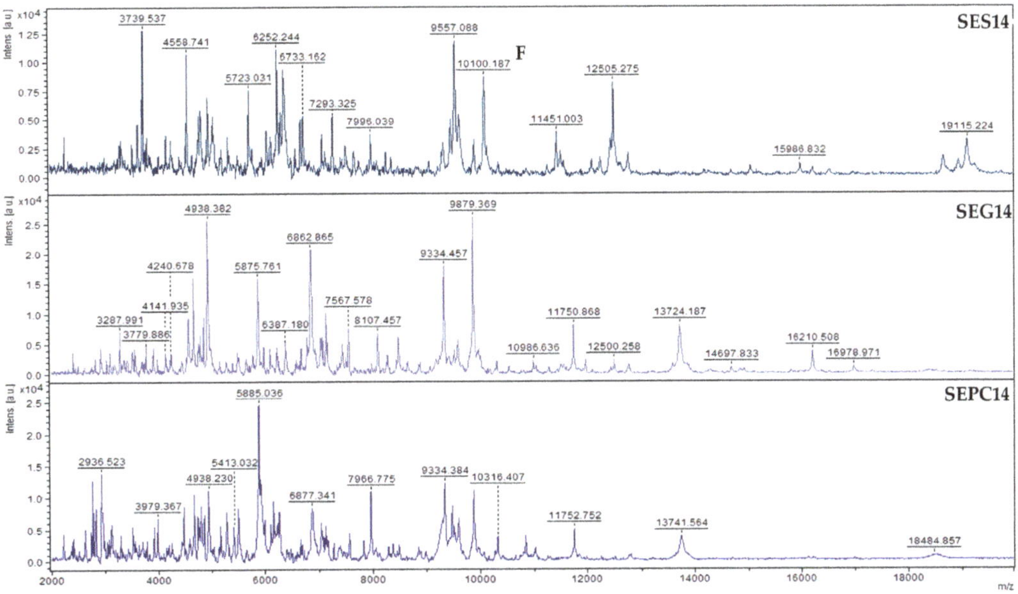

Figure 3. Representative MALDI-TOF mass spectra of *S. enterica*: (**A**) 3rd day; (**B**) 5th day; (**C**) 7th day; (**D**) 9th day; (**E**) 12th day; (**F**) 14th day. SE = *S. enterica*; G = glass; S = stainless steel; and PC = planktonic cells.

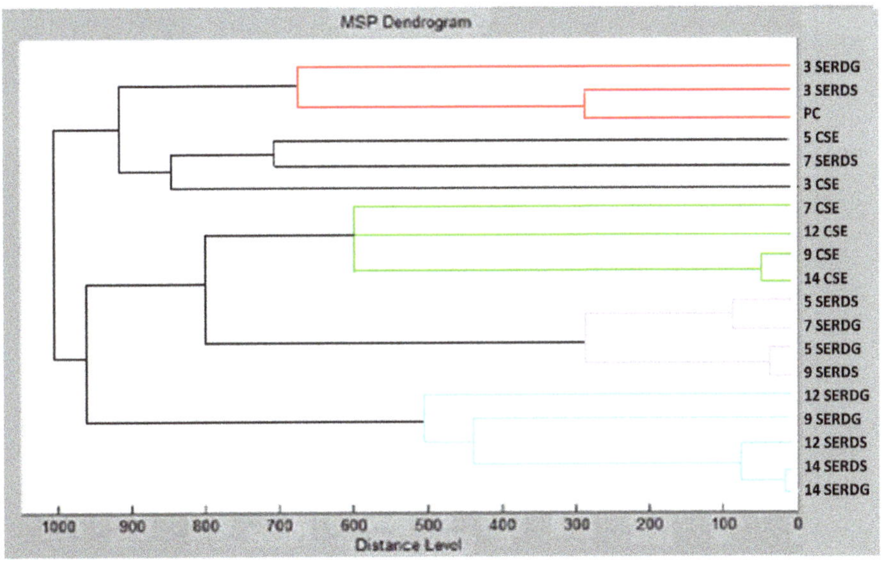

Figure 4. Dendrogram of *S. enterica* generated using MSPs of the planktonic cells and the control. SE = *S. enterica*; C = glass; S = stainless steel; and PC = planktonic cells.

3.5. Kinetic Growth of *S. enterica*

Figure 5 illustrates the growth of *S. enterica* over time, comparing the optical density (OD850) under two conditions: with 1% EO added at OD 1 (log phase) and without EO (control), alongside the temperature profile of the experiment. In the control group (without

EO), *S. enterica* demonstrated continuous growth during the initial 3 h, reaching a peak OD of approximately 1.0. As the temperature increased beyond 40 °C after 3 h, bacterial growth stabilized, followed by a gradual decline in OD, indicating inhibited growth at higher temperatures. In contrast, the addition of 1% EO at the log phase resulted in a similar growth trend up to 3 h, with the OD peaking near 1.0. However, beyond the 3 h mark, when the temperature exceeded 40 °C, the OD began to decrease significantly in the EO-treated group. The decline in OD was more pronounced compared to the control, suggesting that the presence of EO enhanced the inhibitory effects of elevated temperature on bacterial growth. By the end of the experiment (4.5 h), the OD in the EO-treated group had dropped substantially lower than that of the control, indicating a marked reduction in bacterial viability. These findings highlight the combined effect of temperature and EO on bacterial inhibition.

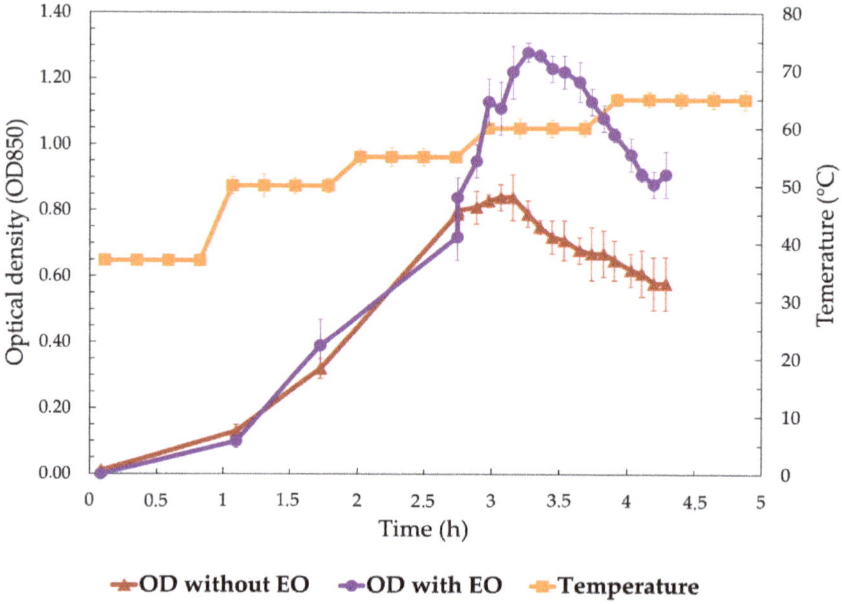

Figure 5. Optical density (OD850) of *S. enterica* over time with and without EO treatment at varying temperatures.

Figure 6 depicts the growth rate (μ) of *S. enterica* over time, under two experimental conditions: with 1% EO added at the log phase (OD 1) and without EO (control), alongside the corresponding temperature profile. In the control condition (no EO), the bacterial growth rate increased steadily until approximately 3 h, when it began to plateau at approximately 0.2 h. After this point, the growth rate began to decrease, coinciding with an increase in temperature above 40 °C. At the end of the 4.5 h period, the growth rate dropped to negative values, indicating inhibition of the bacteria due to the elevated temperature. In the EO-treated conditions, the growth rate during the first 3 h had a similar pattern and peaked at approximately 0.4 h, slightly higher than the control group. However, after 3 h, the growth rate dropped dramatically and fell below the control as the temperature continued to increase. This sharp decrease in growth rate suggests a significant inhibitory effect of EO, especially when the temperature exceeded 40 °C. At the end of the experiment, the growth rate in the EO-treated group reached a more negative value compared to the control, indicating a stronger inhibition of bacterial growth.

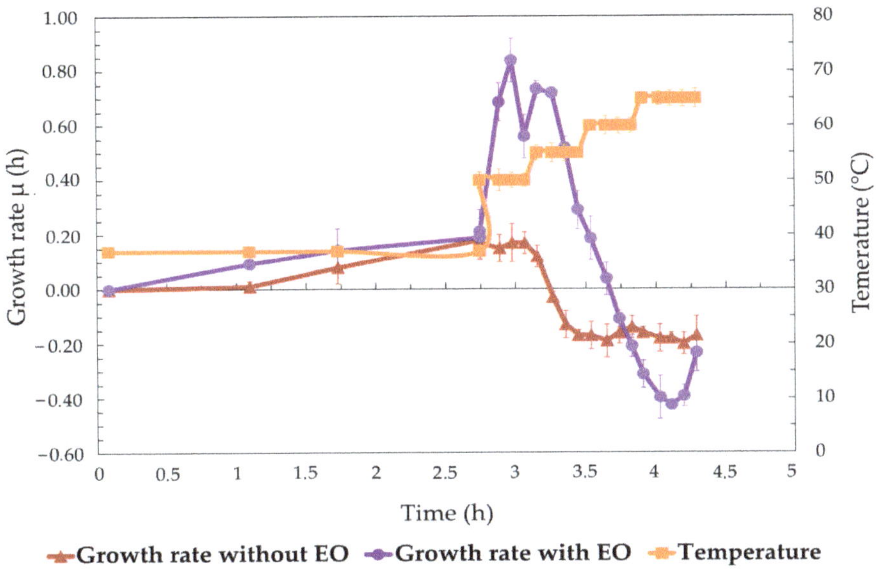

Figure 6. Growth rate (µ) of *S. enterica* over time with and without EO treatment at increasing temperatures.

These results demonstrate that EO in combination with increasing temperature significantly inhibits the growth of *S. enterica*.

3.6. Microbiological Quality of Sous Vide Eggplant

A microbiological examination was conducted on raw eggplant to verify the presence of *Salmonella enterica* on XLD agar. On day 0, no coliforms were detected and the total bacterial count (TBC) was 1.98 log CFU/g. The TBC of the first day and the seventh day of storage were used to evaluate the microbiological quality of the vacuum-packed eggplant (Figures 7 and 8). Within the control group, the range of the bacterial counts was from 1.18 to 2.45 log CFU/g and from 1.11 to 2.67 log CFU/g on the seventh day. TBC ranged from 1.19 to 2.26 log CFU/g on day 1 and 1.27 to 2.22 log CFU/g on day 7 in the vacuum-packed group. In the group that was vacuum packed and subjected to the RDEO treatment, the log CFU/g varied between 1.04 and 1.67 on the first day and between 1.06 and 1.43 on the seventh day. For TBC, the log CFU/g ranged from 1.22 to 2.32 on the first day and from 1.58 to 2.32 on the seventh day for the group treated with *S. enterica*. In the presence of *S. enterica*, TBC varied from 1.34 to 2.32 log CFU/g on day 1 and from 1.16 to 1.78 log CFU/g on day 7 after vacuum packaging and RDEO treatment. In general, there were fewer bacteria in the groups receiving RDEO and *S. enterica* inoculation.

Coliform bacteria count (CBC) was not identified in the RDEO group, in the vacuum-packed control group, or in the control group on the first day (Figure 8). Only the last two groups, where *S. enterica* was inoculated into the samples, revealed coliform bacteria on day 1. CBC ranged from 1.81 to 2.24 log CFU/g in the *S. enterica* inoculated group and from 1.37 to 1.95 log CFU/g in the RDEO with inoculation of *S. enterica* group. On the seventh day, the CBC in the control group ranged from 1.06 to 1.26 log CFU/g. Neither the RDEO group nor the vacuum-packed group contained coliforms after seven days of the experiment. In the group to which *S. enterica* was inoculated, the CBC ranged from 1.19 to 1.63 log CFU/g. In the treatment group receiving RDEO addition and inoculation with *S. enterica*, CBC ranged from 1.00 to 1.32 log CFU/g on day 7.

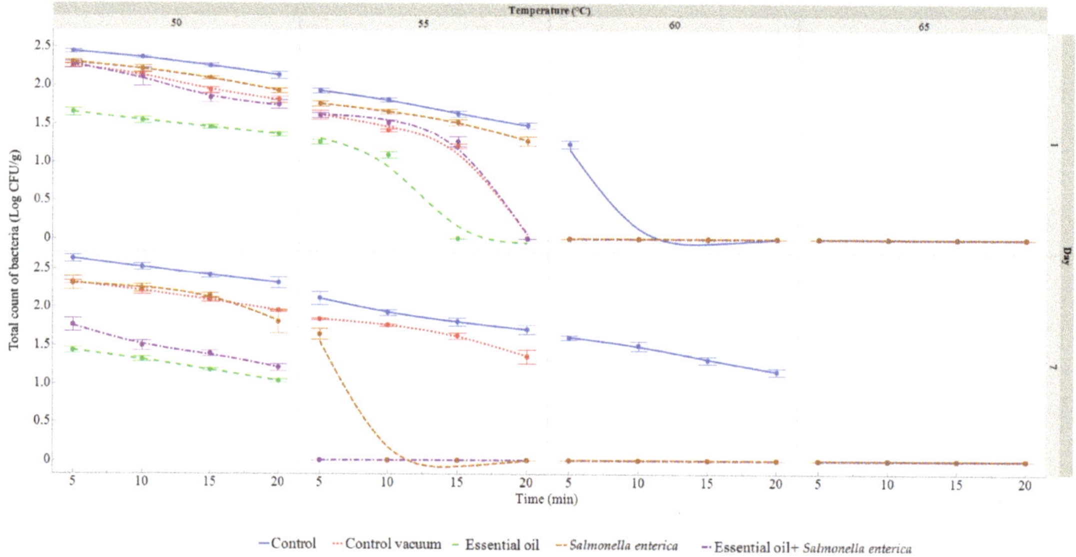

Figure 7. Total bacterial count (TBC) treated at temperatures ranging between 50 and 65 °C for durations of 5 to 20 min. (expressed in log CFU/g) on the first day and seventh day. Data are the mean (±SD) of three samples. Control: fresh samples were treated at 50–65 °C for 5 to 20 min after being packaged in polyethylene bags and kept at 4 °C. Control vacuum: fresh samples were treated at 50–65 °C for 5 to 20 min after being vacuum packed in polyethylene bags and kept at 4 °C. Ess. oil: vacuum-packed fresh samples were treated with 1% RDEO kept at 4 °C and treated for 5–25 min at 50–65 °C. *Salmonella*: vacuum-packed fresh samples were treated with *S. enterica* kept at 4 °C and treated for 5–20 min at 50–65 °C. *Salmonella* + EO: vacuum-packed fresh samples were treated with *S. enterica* and 1% RDEO was kept at 4 °C and treated for 5–20 min at 50–65 °C.

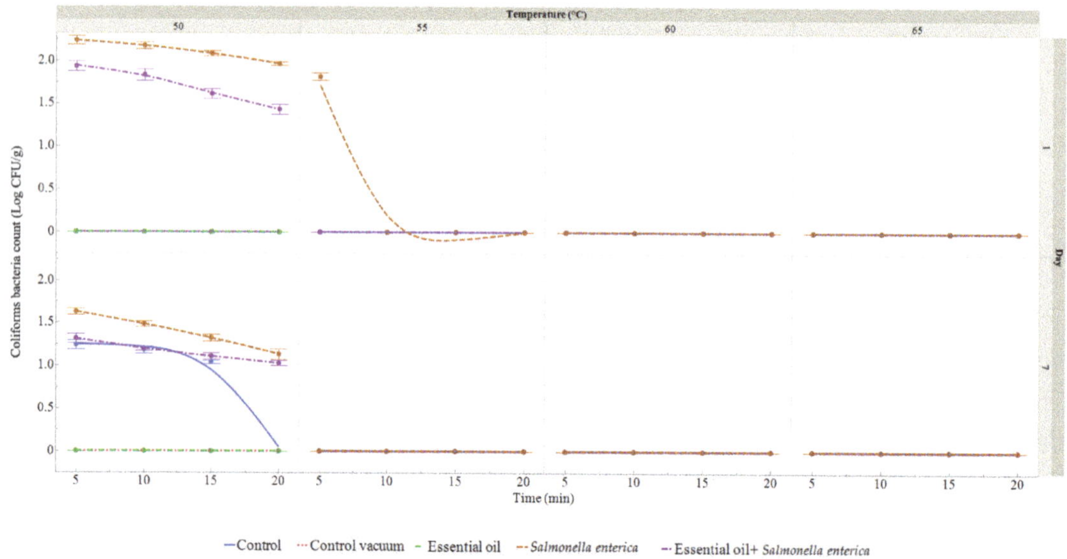

Figure 8. Coliform bacteria count (CBC) treated at temperatures ranging between 50 and 65 °C for durations of 5 to 20 min (expressed in log CFU/g) on the first and seventh day. Data are the mean (±SD) of three samples. Control: fresh samples were treated at 50–65 °C for 5 to 20 min after

being packaged in polyethylene bags and kept at 4 °C. Control vacuum: fresh sample were treated at 50–65 °C for 5 to 20 min after being vacuum packed in polyethylene bags and kept at 4 °C. Ess. oil: vacuum-packed fresh samples were treated with 1% RDEO kept at 4 °C and treated for 5–25 min at 50–65 °C. *Salmonella*: vacuum-packed fresh sample treated with *S. enterica* was kept at 4 °C and treated for 5–20 min at 50–65 °C. *Salmonella* + EO: vacuum-packed fresh sample treated with *S. enterica* and 1% RDEO was kept at 4 °C and treated for 5–20 min at 50–65 °C.

The microbial species, genera, and families that have been isolated since the first day of storage are displayed in Figure 9. A total of 246 isolates were identified using mass spectrometry; the scores in all categories might go up to 2. There were twenty-five species, eight genera, and six families among these isolates. Most of the discovered species belonged to the families *Burkholderiaceae* and *Bacillaceae*. On the first day of the experiment, the most frequently recovered species from the sous vide eggplant samples in the experimental groups were *Salmonella enterica* (12%), *Bacillus amyloliquefaciens* (10%), *Bacillus cereus*, *Bacillus licheniformis*, *Burkholderia cenocepacia*, and *Ralstonia pickettii* (8%).

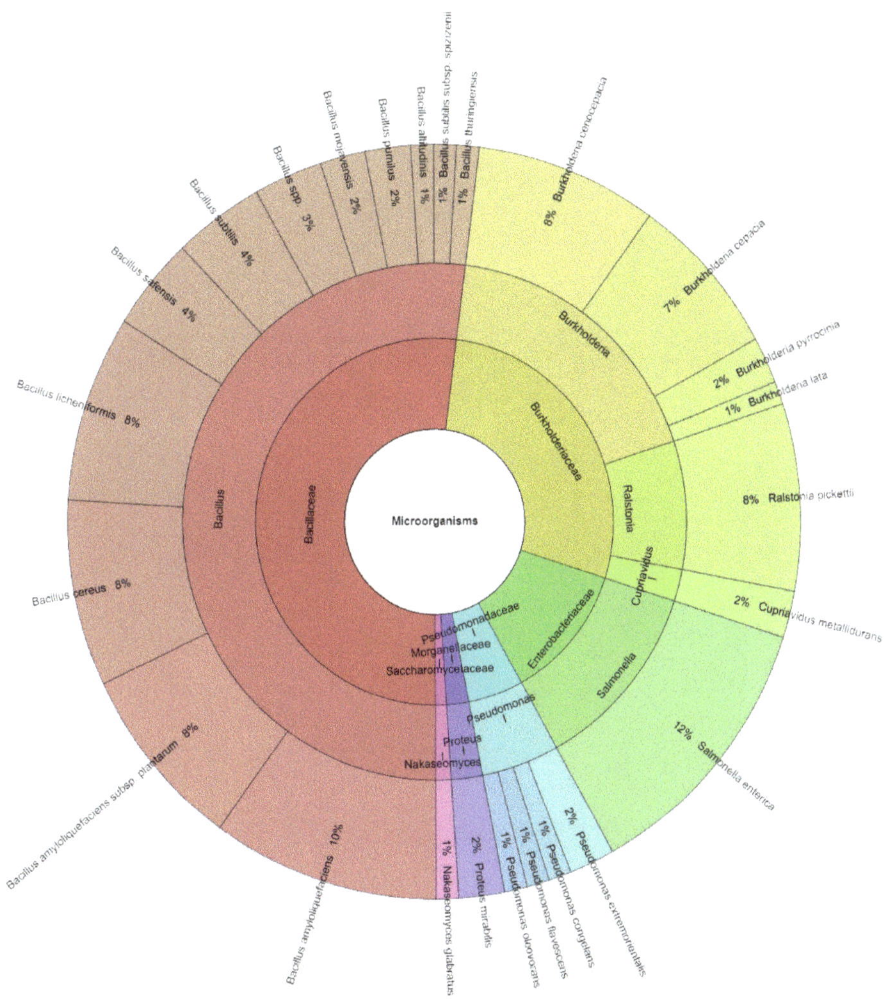

Figure 9. Krona chart: Species, genera, and families isolated from eggplant at the first day of storage.

The microbial species, genera, and families that were identified from sous vide-cooked samples of eggplant (*Solanum melongena*) after a seven-day period of preservation are displayed in Figure 10. A total of 363 isolates were found using mass spectrometry, and every group received a score of two or higher. In total, 24 species from 10 families and 14 genera were among these isolates. The Enterobacteriaceae and Pseudomonadaceae families contained the majority of the species. *S. enterica*, which was inoculated on eggplant, was the most commonly found isolate (18% of all isolates). After *S. enterica*, the most frequent bacteria were *Rhizobium radiobacter* (11%), *Stenotrophomonas maltophilia* (8%), *Acinetobacter calcoaceticus* (6%), *Acinetobacter baumannii*, *Klebsiella oxytoca*, and *Pseudomonas kilonensis*, each at 4%.

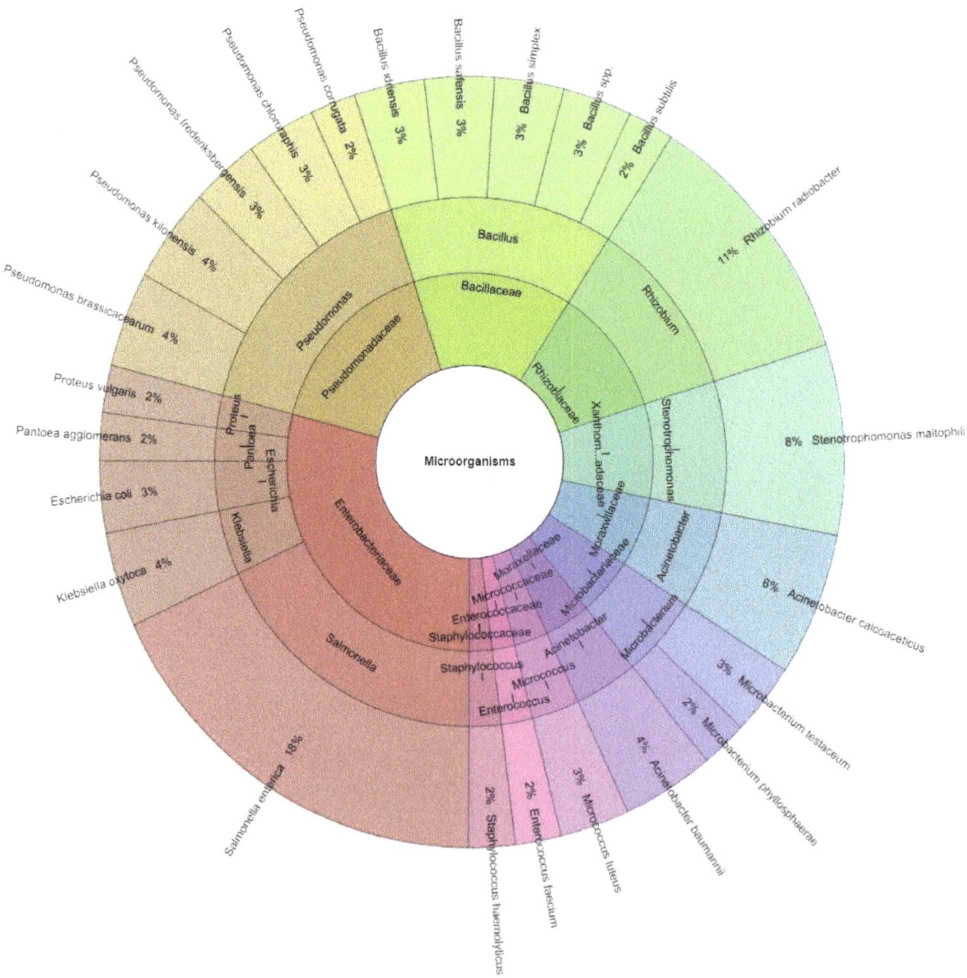

Figure 10. Krona chart: Species, genera, and families isolated from eggplant at the seventh day of storage.

3.7. Insecticidal Activity of RDEO

The evaluation of the insecticidal activity of RDEO against *M. dorsalis* is presented in Table 5. According to the results, the application of 50% and 100% of the tested EO produced the highest levels of insecticidal activity. RDEO, when applied at a concentration of 6.25% and 3.125%, did not show any significant repellent effect against *M. dorsalis*. It is noteworthy that a concentration of 12.5% had an effect on the *M. dorsalis* population (50%). However, a concentration of 25% was effective against 70% of the insects, respectively.

4. Discussion

Terpenes, glycosides, flavonoids, and anthocyanins were among the components of *Rosa damascena* that were extracted from its flowers, petals, and hips (seed pods) [58–60]. Citronellol, geraniol, and nerol—the three main components of rose EO—have been shown to possess antibacterial properties [61,62]. Thus, these compounds may act as mediators for the antibacterial properties of the EO. The high concentration of phenylethyl alcohol in rose absolute may contribute to its antibacterial qualities, as alcohols have long been recognized for their antibacterial properties [63]. In our study, the main component was phenylethyl alcohol (70.0%), followed by citronellol (11.3%), geraniol (7.1%), nonadecane (4.7%), nerol (3.7%), and 1-nonadecene (3.1%). The obtained results agree with the studies by Ulusoy et al. [16] and Aydinli et al. [64] who showed that the main compound of rose absolute EO is characterized by the high relative amount of phenylethyl alcohol that varies in the range from 72.73% to 78.38%, followed by a significant abundance of citronellol, nonadecane, geraniol, and nerol. On the other hand, Lei et al. [65] showed that rose water from Chine had a higher amount of phenylethyl alcohol (90.2%). The principal constituents of rose EO extracted from the central region of Iran have been identified as citronellol, nonadecane, and geraniol [66]. According to other studies, the main constituents of the essential oil of *Rosa damascena* EO collected from northern Iran were 1-nonadecene, hexatriacontane, n-tricosane, and geraniol [67]. The most representative bioactive compounds of the essential oil of *Rosa damascena* EO collected from southern Iran were reported to be nonadecane, heneicosane, docosane, citronellol, and 9-nonadecene [67]. The principal constituents of the *Rosa damascena* EO extracted from the Kashan region were β-citronellol (14.88–47.43%), nonadecane (10.5–40.5%), geraniol (5.5–18%), and heneicosane (7–14%) [68].

The results showed that RDEO exhibited the strongest antibacterial action against Gram-positive *Streptococcus pneumoniae*, Gram-negative *S. sonnei*, and yeasts such as *Candida parapsilosis*. It has been demonstrated that RDEO exhibits antibacterial action against the biofilm-forming Gram-negative bacteria *Salmonella enterica*. The wide range of antibacterial activity of *Rosa damascena* has been established. Hydrosol, absolute, and EOs are significant products that demonstrate these benefits [62]. Strong antibacterial efficacy against strains of *Escherichia coli, Pseudomonas aeruginosa, Bacillus subtilis, Staphylococcus aureus, Chromobacterium violaceum*, and *Erwinia carotovora* was reported by Ulusoy et al. [16] using EO and absolute. *C. violaceum* was the microorganism most susceptible to rose absolute and EO. Additionally, *E. coli* was sensitive to rose essential oil. Nevertheless, none of the microbes were susceptible to the antibacterial effects of hydrosol [16]. Both Gram-positive and Gram-negative bacteria were also susceptible to the antibacterial properties of rose absolute. In a different investigation, the antibacterial properties of the EO extracted from the petals of *R. damascena* were assessed against three strains of *Xanthomonas axonopodis* spp. *vesicatoria*. The growth of the investigated strains of *X. axonopodis vesicatoria* was significantly reduced by the essential oil of *R. damascena* EO from flowers [69]. The antibacterial properties of *R. damascena* flowers were investigated against 15 different species of bacteria, including *Aeromonas hydrophila, Bacillus cereus, Enterobacter aerogenes, Enterococcus faecalis, E. coli, Klebsiella pneumoniae, Mycobacterium smegmatis, Proteus vulgaris, P. aeruginosa, P. fluorescens, Salmonella* Enteritidis, *Salmonella* Typhimurium, *S. aureus*, and *Yersinia enterocolitica* [70]. A different study demonstrated the in vitro antibacterial activity of the EO from *R. damascena* against *Pseudomonas, S. aureus*, and *E. coli*. In this work, *R. damascena* showed antibacterial efficacy against *S. aureus* [61]. The antibacterial activity of EOs from several plants, including *R. damascena*, was also evaluated against the yeast *Candida albicans*, Gram-positive *S. aureus*, Gram-negative *E. coli*, and Gram-negative *P. aeruginosa*. At low doses, the studied essential oils demonstrated both bactericidal and inhibitory effects against all tested bacteria [71]. In a different study, rose EO was found to be less effective against

Gram-positive *S. aureus* compared to rifampin and gentamicin [72]. The same results were observed in our study. The antibacterial activity of the EO against this bacteria in Bulgaria and Saudi Arabia was greater than the lethal and inhibitory power of the EO against these bacteria [73]. Furthermore, in our work, we investigated the antimicrobial activity in the vapor phase on model fruits and vegetables. The results revealed that the inhibitory effect of RDEO at different concentrations acted on various microorganisms, depending on the model food. The best inhibitory effect was observed on kiwifruit against *C. tropicalis* yeast at the highest concentration, on banana against *B. cereus* at the lowest concentration, on eggplant against *S. pneumoniae*, and on pumpkin against *S. enterica* at the lowest concentrations. In some cases, pro-bacterial growth was demonstrated. A different study found antimicrobial effects in situ on fruit and vegetable models with similar results [54–57].

As far as we know, this is the first study to investigate the antibiofilm activity of RDEO. However, the current findings align with earlier research [74–76] that demonstrated the antibiofilm capabilities of various EOs against a range of foodborne pathogens on different surfaces. After employing RDEO to prevent the formation of *Salmonella enterica* biofilms, we obtained minimum inhibitory biofilm concentration (MIBC) values of 0.270 mg/mL and 0.291 mg/mL for $MIBC_{90}$ using the crystal violet assay. The MALDI-TOF MS investigations indicated changes in the protein composition of the biofilm-forming bacteria *S. enterica*. The increasing incidence of bacterial diseases can be attributed to the lack of effective treatment options, antibiotic resistance, and the tendency of bacteria to form biofilms [77]. Consequently, new approaches that emphasize innovative technologies such as natural antimicrobials have emerged to control bacterial biofilms and mitigate the resistance of *Salmonella*, a highly prevalent pathogen. The EOs are antibacterial substances extracted from various herbal plants using a range of methods. EOs have been shown to possess significant antimicrobial activity against planktonic microbes, in addition to their antibacterial properties against microorganisms embedded in biofilm [78–81]. All examined free EOs and their active ingredients successfully inhibited planktonic bacterial cells [82,83]. However, the biofilms required significantly higher concentrations of Eos, or their active components, than their planktonic counterparts, to achieve comparable or zero microbial reductions [75,84–89]. In several studies, even when EOs or their active components were applied at twice or even higher concentrations than their minimum inhibitory concentrations (MIC) against planktonic cells, the biofilms were not completely eradicated [85,88,90–96]. Moreover, longer exposure times and greater EO concentrations were necessary to eliminate the biofilms that had formed over extended periods [80,97].

Salmonella enterica bacteria are responsible for a variety of severe, potentially fatal diseases in both humans and animals across the globe [98]. *Salmonella* grows best at 37 °C and pH 7, with a range of 5–47 °C and pH 4–9 [99]. In our study, without EO, the bacteria followed this growth pattern, thriving near 37 °C. However, with 1% EO, the growth was significantly inhibited, as shown by lower optical density (OD) and growth rates. At higher temperatures, both treated and untreated bacteria experienced a decline, but the EO-treated samples showed a steeper drop, highlighting EO's stronger inhibitory effect even at suboptimal growth conditions.

The microbiological quality of eggplant was monitored during storage over 7 days. Eggplant was treated using the sous vide cooking method, while microbiological quality was assessed at different temperatures and times, in combination with *Salmonella enterica* inoculation and RDEO treatment. During the observation period, we found that increasing temperature and time positively affected the total number of microorganisms and the number of coliforms. RDEO treatment demonstrated an antimicrobial effect against *S. enterica* inoculation on eggplant. According to Zhou et al. [100], the shelf life of a green vegetable is determined by its ability to maintain an appealing appearance—a crisp, green texture with minimal browning or wetness—for consumers. After day 7 of storage, these products become more perishable than untreated materials due to off-flavors, tissue weakening, and microbial growth [101]. The relationship between the development of spoilage bacteria, their production of metabolites—particularly volatiles—and consumers' is very important. Additionally, the inherent composition of food may affect bacterial susceptibility and reduce the antibacterial effectiveness of EOs. Higher concentrations of free EOs or their active ingredients are often required to achieve equivalent antimicrobial effects in food compared to in vitro assays [102–105]. This could have unfavorable organoleptic effects and reduce food products' general appeal [106,107]. It was discovered that EOs or their active ingredients were more beneficial in foods with no or low fat content than in foods with high fat content [102,104,105]. Additionally, certain extrinsic factors like pH, temperature, and oxygen present may have an impact on the antibacterial action of EOs. Due to their higher hydrophobicity and easier dissolving in bacterial cell membranes, numerous EOs generally exhibited greater antibacterial action at low pH [102,103,108,109]. Some investigations found conflicting results regarding the effect of temperature on the antibacterial activity of EOs. Due to an increase in unsaturated phospholipids in the composition of cytoplasmic membranes and an increase in membrane fluidity, it has been discovered that lower temperatures (7 °C compared to 35 °C) boost the antibacterial action of the EOs or their active components [102]. According to Cava et al. [102], the enhanced fluidity makes the membrane connection weaker and facilitates the simpler dissolution of the EOs into it. In contrast, it was discovered that the antibacterial activity of carvacrol and cymene was more effective at 25 °C than it was at 15 and 4 °C [104]. The reason why bacterial cells were less sensitive to antimicrobial agents at lower temperatures remained unclear, but possible explanations included modifications to the fluidity and/or properties of the membrane, or the possibility that low temperatures would impact the synthesis of target sites in both EOs and bacterial cytoplasmic membranes, which in turn influences the susceptibility of microorganisms to EOs [104]. Additionally, EOs were reported to have greater antibacterial action in low-oxygen environments (vacuum and modified atmosphere packing) than in aerobic environments [110,111]. The microbiological quality of sous vide eggplant was associated with all the factors described above. That is, in our study, we investigated the effect of RDEO at 1% concentration as a function of temperature and time during storage for 7 days at 4 °C. In another study compared to our results, much higher numbers of microorganisms were obtained on eggplant treated with gamma irradiation and ascorbic acid [112]. On the first day of storage, the most frequently isolated species from sous vide eggplant were *Salmonella enterica, Bacillus amyloliquefaciens, Bacillus cereus, Bacillus licheniformis, Burkholderia cenocepacia,* and *Ralstonia pickettii,* and on the seventh day of storage they were *S. enterica, Rhizobium radiobacter, Stenotrophomonas maltophilia, Acinetobacter calcoaceticus, Acinetobacter baumannii, Klebsiella oxytoca,* and *Pseudomonas kilonensis.* In different study bacteria species namely, *Staphylococcus aureus, Bacillus subtilis, Micrococcus luteus,* and *Enterococcus faecalis,* and four fungal species, namely, *Saccharomyces, Fusarium, Aspergillus,* and *Rhizopus* were isolated from eggplant after preservative treatments [113].

The results indicated that applying 50% and 100% concentrations of the tested RDEO provided the highest insecticidal efficacy against *M. dorsalis*. This study represents the first investigation into the insecticidal activity of RDEO against *M. dorsalis*. Through a variety of mechanisms of action, such as contact activity, inhibition of molting and respiration, reduction in growth and fertility, cuticle destruction, and an impact on the octopamine pathway invertebrates, the EO exhibit a wide range of activity against insects and mites [114,115]. Geraniol and citronellol, the two main constituents of rose EO, shown contact actions against both adult and nymph *Tetranychus urticae*. On the adult and nymphal stages of *T. urticae*, the amounts of rose oil, geraniol, and citronellol that were applied, particularly after 96 h, had a strong fatal effect. At all concentrations, geraniol exhibited the greatest contact effect on *T. urticae* adults and nymphs, followed by citronellol and rose EO. According to Attia et al. [116], the primary constituents of the EO contribute to their acaricidal properties. EOs and monoterpenoids have been found to exhibit acaricidal efficacy against two-spotted mites [117–121]. The literature contains no reports on the insecticidal or acaricidal properties of rose EO. Nonetheless, research has been conducted on geraniol, an ingredient in several EOs.

5. Conclusions

The findings of this study confirm that *Rosa damascena* essential oil (RDEO) possesses potent antimicrobial and insecticidal properties, primarily attributed to its high concentration of phenylethyl alcohol. RDEO exhibited significant inhibitory effects against a variety of microorganisms, including clinically relevant Gram-positive and Gram-negative bacteria, as well as yeast strains. The antimicrobial activity was robustly demonstrated through minimum inhibitory concentration (MIC) assessments, indicating its effectiveness against biofilm-forming bacteria like *Salmonella enterica*. Additionally, the in situ analyses using fruit and vegetable models showcased the essential oil's ability to inhibit microbial growth in vapor phase applications, highlighting its practical utility in food preservation. Furthermore, the study revealed effective antibiofilm activity, emphasizing the potential of RDEO to mitigate biofilm formation, a critical challenge in food safety and public health. The results also suggest that sous vide conditions may enhance the antimicrobial efficacy of RDEO, offering a novel approach to integrating EOs in cooking processes to improve food safety. Moreover, the insecticidal activity tests support the role of the EOs, as a natural pest deterrent, underscoring its applicability in sustainable agricultural practices. Given the rising concerns over antibiotic resistance, RDEO represents a promising candidate for further exploration as a safe and effective antimicrobial agent across various industries. Future studies should focus on elucidating the specific mechanisms of action of RDEO's constituents and evaluating its effectiveness in real-world food systems and agricultural practices. Additionally, the research could investigate the formulation of RDEO with other natural preservatives to enhance its stability and efficacy and explore its potential synergistic effects in combination with conventional antimicrobial agents. By addressing these areas, RDEO could pave the way for innovative solutions to contemporary challenges in food safety and pest management.

Author Contributions: Conceptualization, A.V., M.D.V., N.L.V., A.B. and M.K.; methodology, A.V., M.D.V., N.L.V., M.T., Z.B., L.L., A.B., J.K., R.B.S., A.B.H., J.H.E.-L., M.I.K., N.Č., S.G. and M.K.; investigation, M.K.; writing—original draft, A.V., M.D.V., N.L.V., M.T., Z.B., L.L., A.B., J.K., R.B.S., A.B.H., J.H.E.-L., M.I.K., S.G. and M.K.; formal analyses, A.V., M.D.V., N.L.V., M.T., Z.B., L.L., A.B., J.K., R.B.S., A.B.H., J.H.E.-L., M.I.K., N.Č., S.G. and M.K; writing—review and editing, A.V., M.D.V., N.L.V., M.T., Z.B., A.B., J.K., R.B.S., A.B.H., J.H.E.-L., M.I.K., N.Č., S.G. and M.K; resources, M.K; supervision, M.K. and S.G.; project administration, M.K.; funding acquisition, M.K.; visualization, M.K.; software, A.B.; data curation, A.B. All authors have read and agreed to the published version of the manuscript.

Funding: This research was funded by the grant APVV-20-0058 "The potential of the essential oils from aromatic plants for medical use and food preservation". This research was supported by the grant VEGA 1/0059/24 "Chemical properties and biological activity (in vitro, in vivo, and in situ) of plant volatile mixtures, their main components and inclusion systems". This work was supported by the Ministry of Science, Technological Development, and Innovations of the Republic of Serbia, Agreement No. 451-03-66/2024-03/200122 and Agreement No. 451-03-65/2024-03/200122. This study was supported by CONAHCYT (Consejo Nacional de Humanidades, Ciencias y Tecnologías) of Mexico (Postdoctoral Scholarship I1200/331/2023).

Institutional Review Board Statement: Not applicable.

Informed Consent Statement: Not applicable.

Data Availability Statement: The original contributions presented in the study are included in the article, further inquiries can be directed to the corresponding author.

Conflicts of Interest: The authors declare no conflicts of interest.

References

1. Gundala, R.R.; Singh, A. What Motivates Consumers to Buy Organic Foods? Results of an Empirical Study in the United States. *PLoS ONE* **2021**, *16*, e0257288. [CrossRef] [PubMed]
2. Mesías, F.J.; Martín, A.; Hernández, A. Consumers' Growing Appetite for Natural Foods: Perceptions towards the Use of Natural Preservatives in Fresh Fruit. *Food Res. Int.* **2021**, *150*, 110749. [CrossRef] [PubMed]

3. Testa, R.; Schifani, G.; Migliore, G. Understanding Consumers' Convenience Orientation. An Exploratory Study of Fresh-Cut Fruit in Italy. *Sustainability* **2021**, *13*, 1027. [CrossRef]
4. Ramos, B.; Miller, F.A.; Brandão, T.R.S.; Teixeira, P.; Silva, C.L.M. Fresh Fruits and Vegetables—An Overview on Applied Methodologies to Improve Its Quality and Safety. *Innov. Food Sci. Emerg. Technol.* **2013**, *20*, 1–15. [CrossRef]
5. Giannakourou, M.C.; Tsironi, T.N. Application of Processing and Packaging Hurdles for Fresh-Cut Fruits and Vegetables Preservation. *Foods* **2021**, *10*, 830. [CrossRef] [PubMed]
6. Mastromatteo, M.; Lucera, A.; Sinigaglia, M.; Corbo, M.R. Combined Effects of Thymol, Carvacrol and Temperature on the Quality of Non Conventional Poultry Patties. *Meat Sci.* **2009**, *83*, 246–254. [CrossRef]
7. Hamdan, N.; Lee, C.H.; Wong, S.L.; Fauzi, C.E.N.C.A.; Zamri, N.M.A.; Lee, T.H. Prevention of Enzymatic Browning by Natural Extracts and Genome-Editing: A Review on Recent Progress. *Molecules* **2022**, *27*, 1101. [CrossRef]
8. Iwu, C.D.; Okoh, A.I. Preharvest Transmission Routes of Fresh Produce Associated Bacterial Pathogens with Outbreak Potentials: A Review. *Int. J. Environ. Res. Public Health* **2019**, *16*, 4407. [CrossRef]
9. Esmaeili, Y.; Paidari, S.; Baghbaderani, S.A.; Nateghi, L.; Al-Hassan, A.A.; Ariffin, F. Essential Oils as Natural Antimicrobial Agents in Postharvest Treatments of Fruits and Vegetables: A Review. *Food Meas.* **2022**, *16*, 507–522. [CrossRef]
10. Thielmann, J.; Muranyi, P. Review on the Chemical Composition of *Litsea cubeba* Essential Oils and the Bioactivity of Its Major Constituents Citral and Limonene. *J. Essent. Oil Res.* **2019**, *31*, 361–378. [CrossRef]
11. Wieczyńska, J.; Cavoski, I. Antimicrobial, Antioxidant and Sensory Features of Eugenol, Carvacrol and Trans-Anethole in Active Packaging for Organic Ready-to-Eat Iceberg Lettuce. *Food Chem.* **2018**, *259*, 251–260. [CrossRef] [PubMed]
12. Maleki, G.; Sedaghat, N.; Woltering, E.J.; Farhoodi, M.; Mohebbi, M. Chitosan-Limonene Coating in Combination with Modified Atmosphere Packaging Preserve Postharvest Quality of Cucumber during Storage. *Food Meas.* **2018**, *12*, 1610–1621. [CrossRef]
13. Niu, B.; Yan, Z.; Shao, P.; Kang, J.; Chen, H. Encapsulation of Cinnamon Essential Oil for Active Food Packaging Film with Synergistic Antimicrobial Activity. *Nanomaterials* **2018**, *8*, 598. [CrossRef] [PubMed]
14. Xing, Y.; Lin, H.; Cao, D.; Xu, Q.; Han, W.; Wang, R.; Che, Z.; Li, X. Effect of Chitosan Coating with Cinnamon Oil on the Quality and Physiological Attributes of China Jujube Fruits. *BioMed Res. Int.* **2015**, *2015*, 835151. [CrossRef]
15. Srikandace, Y.; Indrarti, L.; Indriyati; Sancoyorini, M.K. Antibacterial Activity of Bacterial Cellulose-Based Edible Film Incorporated with *Citrus* Spp Essential Oil. *IOP Conf. Ser. Earth Environ. Sci.* **2018**, *160*, 012004. [CrossRef]
16. Ulusoy, S.; Boşgelmez-Tınaz, G.; Seçilmiş-Canbay, H. Tocopherol, Carotene, Phenolic Contents and Antibacterial Properties of Rose Essential Oil, Hydrosol and Absolute. *Curr. Microbiol.* **2009**, *59*, 554–558. [CrossRef]
17. Babaei, A.; Tabaei-Aghdaei, S.R.; Khosh-Khui, M.; Omidbaigi, R.; Naghavi, M.R.; Esselink, G.D.; Smulders, M.J. Microsatellite Analysis of Damask Rose (*Rosa damascena* Mill.) Accessions from Various Regions in Iran Reveals Multiple Genotypes. *BMC Plant Biol.* **2007**, *7*, 12. [CrossRef]
18. Mirjalili, M.H.; Akhoondi, R.; Hadian, J. Chemical Composition of the Aliphatic Compound-Rich Essential Oil of Rosa Foetida from Iran. *Chem. Nat. Compd.* **2015**, *51*, 171–172. [CrossRef]
19. Nedeltcheva-Antonova, D.; Stoicheva, P.; Antonov, L. Chemical Profiling of Bulgarian Rose Absolute (*Rosa damascena* Mill.) Using Gas Chromatography–Mass Spectrometry and Trimethylsilyl Derivatives. *Ind. Crops Prod.* **2017**, *108*, 36–43. [CrossRef]
20. Mohammadhosseini, M.; Sarker, S.D.; Akbarzadeh, A. Chemical Composition of the Essential Oils and Extracts of *Achillea* Species and Their Biological Activities: A Review. *J. Ethnopharmacol.* **2017**, *199*, 257–315. [CrossRef]
21. Nunes, H.; Miguel, M.G. *Rosa damascena* essential oils: A brief review about chemical composition and biological properties. *Trends Phytochem. Res.* **2017**, *3*, 111.
22. Mahboubi, M. *Rosa damascena* as Holy Ancient Herb with Novel Applications. *J. Tradit. Complement. Med.* **2016**, *6*, 10–16. [CrossRef]
23. Najem, W.; El Beyrouthy, M.; Wakim, L.H.; Neema, C.; Ouaini, N. Essential Oil Composition of *Rosa damascena* Mill. from Different Localities in Lebanon. *Acta Bot. Gall.* **2011**, *158*, 365–373. [CrossRef]
24. Bhavaniramya, S.; Vishnupriya, S.; Al-Aboody, M.S.; Vijayakumar, R.; Baskaran, D. Role of Essential Oils in Food Safety: Antimicrobial and Antioxidant Applications. *Grain Oil Sci. Technol.* **2019**, *2*, 49–55. [CrossRef]
25. Dobreva, A.; Kovatcheva, N.; Astatkie, T.; Zheljazkov, V.D. Improvement of Essential Oil Yield of Oil-Bearing (*Rosa damascena* Mill.) Due to Surfactant and Maceration. *Ind. Crops Prod.* **2011**, *34*, 1649–1651. [CrossRef]
26. Lo Fo Wong, D.M.A.; Hald, T.; van der Wolf, P.J.; Swanenburg, M. Epidemiology and Control Measures for *Salmonella* in Pigs and Pork. *Livest. Prod. Sci.* **2002**, *76*, 215–222. [CrossRef]
27. Doulgeraki, A.I.; Papaioannou, M.; Nychas, G.-J.E. Targeted Gene Expression Study of *Salmonella enterica* during Biofilm Formation on Rocket Leaves. *LWT* **2016**, *65*, 254–260. [CrossRef]
28. Silveira, J.B.; Hessel, C.T.; Tondo, E.C. Inactivation of Salmonella Enteritidis on Lettuces Used by Minimally Processed Vegetable Industries. *J. Infect. Dev. Ctries.* **2017**, *11*, 34–41. [CrossRef] [PubMed]
29. Ramos, B.; Brandão, T.R.S.; Teixeira, P.; Silva, C.L.M. Biopreservation Approaches to Reduce *Listeria monocytogenes* in Fresh Vegetables. *Food Microbiol.* **2020**, *85*, 103282. [CrossRef]
30. Gil, M.I.; Selma, M.V.; López-Gálvez, F.; Allende, A. Fresh-Cut Product Sanitation and Wash Water Disinfection: Problems and Solutions. *Int. J. Food Microbiol.* **2009**, *134*, 37–45. [CrossRef]
31. Sharma, S.; Mohler, J.; Mahajan, S.D.; Schwartz, S.A.; Bruggemann, L.; Aalinkeel, R. Microbial Biofilm: A Review on Formation, Infection, Antibiotic Resistance, Control Measures, and Innovative Treatment. *Microorganisms* **2023**, *11*, 1614. [CrossRef] [PubMed]

32. Cylke, C.; Si, F.; Banerjee, S. Effects of Antibiotics on Bacterial Cell Morphology and Their Physiological Origins. *Biochem. Soc. Trans.* **2022**, *50*, 1269–1279. [CrossRef]
33. Paparella, A.; Mazzarrino, G.; Chaves-López, C.; Rossi, C.; Sacchetti, G.; Guerrieri, O.; Serio, A. Chitosan Boosts the Antimicrobial Activity of *Origanum vulgare* Essential Oil in Modified Atmosphere Packaged Pork. *Food Microbiol.* **2016**, *59*, 23–31. [CrossRef]
34. D'Amato, S.; Serio, A.; López, C.C.; Paparella, A. Hydrosols: Biological Activity and Potential as Antimicrobials for Food Applications. *Food Control* **2018**, *86*, 126–137. [CrossRef]
35. Raybaudi-Massilia, R.M.; Mosqueda-Melgar, J.; Martín-Belloso, O. Antimicrobial Activity of Essential Oils on *Salmonella* Enteritidis, *Escherichia coli*, and *Listeria innocua* in Fruit Juices. *J. Food Prot.* **2006**, *69*, 1579–1586. [CrossRef]
36. Gündüz, G.T.; Gönül, Ş.A.; Karapinar, M. Efficacy of Myrtle Oil against Salmonella Typhimurium on Fresh Produce. *Int. J. Food Microbiol.* **2009**, *130*, 147–150. [CrossRef]
37. Silva, J.P.L.D.; Souza, E.F.D.; Modesta, R.C.D.; Gomes, I.A.; Freitas-Silva, O.; Franco, B. Antibacterial Activity of Nisin, Oregano Essential Oil, EDTA, and Their Combination against Salmonella Enteritidis for Application in Mayonnaise. *Vigilância Sanitária Debate* **2016**, *4*, 83. [CrossRef]
38. Rossi, C.; Chaves-López, C.; Možina, S.S.; Di Mattia, C.; Scuota, S.; Luzzi, I.; Jenič, T.; Paparella, A.; Serio, A. *Salmonella enterica* Adhesion: Effect of *Cinnamomum zeylanicum* Essential Oil on Lettuce. *LWT* **2019**, *111*, 16–22. [CrossRef]
39. Srivastava, S.; Bhargava, A. Biofilms and Human Health. *Biotechnol. Lett.* **2016**, *38*, 1–22. [CrossRef]
40. Abdullahi, U.F.; Igwenagu, E.; Mu'azu, A.; Aliyu, S.; Umar, M.I. Intrigues of Biofilm: A Perspective in Veterinary Medicine. *Vet. World* **2016**, *9*, 12–18. [CrossRef]
41. Satpathy, S.; Sen, S.K.; Pattanaik, S.; Raut, S. Review on Bacterial Biofilm: An Universal Cause of Contamination. *Biocatal. Agric. Biotechnol.* **2016**, *7*, 56–66. [CrossRef]
42. Simões, M.; Simões, L.C.; Vieira, M.J. A Review of Current and Emergent Biofilm Control Strategies. *LWT-Food Sci. Technol.* **2010**, *43*, 573–583. [CrossRef]
43. Zhao, X.; Zhao, F.; Wang, J.; Zhong, N. Biofilm Formation and Control Strategies of Foodborne Pathogens: Food Safety Perspectives. *RSC Adv.* **2017**, *7*, 36670–36683. [CrossRef]
44. Milanov, D.; Ljubojević, D.; Čabarkapa, I.; Karabasil, N.; Velhner, M. Biofilm as Risk Factor for Salmonella Contamination in Various Stages of Poultry Production. *Eur. Poult. Sci.* **2017**, *81*, 1–14. [CrossRef]
45. Maifreni, M.; Frigo, F.; Bartolomeoli, I.; Buiatti, S.; Picon, S.; Marino, M. Bacterial Biofilm as a Possible Source of Contamination in the Microbrewery Environment. *Food Control* **2015**, *50*, 809–814. [CrossRef]
46. Kamanula, J.; Sileshi, G.W.; Belmain, S.R.; Sola, P.; Mvumi, B.M.; Nyirenda, G.K.C.; Nyirenda, S.P.; Stevenson, P.C. Farmers' Insect Pest Management Practices and Pesticidal Plant Use in the Protection of Stored Maize and Beans in Southern Africa. *Int. J. Pest Manag.* **2010**, *57*, 41–49. [CrossRef]
47. Mssillou, I.; Saghrouchni, H.; Saber, M.; Zannou, A.J.; Balahbib, A.; Bouyahya, A.; Allali, A.; Lyoussi, B.; Derwich, E. Efficacy and Role of Essential Oils as Bio-Insecticide against the Pulse Beetle *Callosobruchus maculatus* (F.) in Post-Harvest Crops. *Ind. Crops Prod.* **2022**, *189*, 115786. [CrossRef]
48. Ojo, D.O.; Ogunleye, R.F. Comparative Effectiveness of the Powders of Some Underutilized Botanicals for the Control of *Callosobruchus maculatus* (Coleoptera: Bruchidae). *J. Plant Dis. Prot.* **2013**, *120*, 227–232. [CrossRef]
49. Nerio, L.S.; Olivero-Verbel, J.; Stashenko, E.E. Repellent Activity of Essential Oils from Seven Aromatic Plants Grown in Colombia against *Sitophilus zeamais* Motschulsky (Coleoptera). *J. Stored Prod. Res.* **2009**, *45*, 212–214. [CrossRef]
50. Vendan, S.E.; Manivannan, S.; Sunny, A.M.; Murugesan, R. Phytochemical Residue Profiles in Rice Grains Fumigated with Essential Oils for the Control of Rice Weevil. *PLoS ONE* **2017**, *12*, e0186020. [CrossRef]
51. Adams, R.P. *Identification of Essential Oil Components by Gas Chromatography/Mass Spectroscopy*; Allured Publishing Corporation: Carol Stream, IL, USA, 2007.
52. Van Den Dool, H.; Kratz, P.D. A Generalization of the Retention Index System Including Linear Temperature Programmed Gas—Liquid Partition Chromatography. *J. Chromatogr. A* **1963**, *11*, 463–471. [CrossRef]
53. Kačániová, M.; Vukovic, N.L.; Čmiková, N.; Galovičová, L.; Schwarzová, M.; Šimora, V.; Kowalczewski, P.Ł.; Kluz, M.I.; Puchalski, C.; Bakay, L.; et al. Salvia Sclarea Essential Oil Chemical Composition and Biological Activities. *Int. J. Mol. Sci.* **2023**, *24*, 5179. [CrossRef] [PubMed]
54. Kačániová, M.; Čmiková, N.; Vukovic, N.L.; Verešová, A.; Bianchi, A.; Garzoli, S.; Ben Saad, R.; Ben Hsouna, A.; Ban, Z.; Vukic, M.D. Citrus Limon Essential Oil: Chemical Composition and Selected Biological Properties Focusing on the Antimicrobial (In Vitro, In Situ), Antibiofilm, Insecticidal Activity and Preservative Effect against Salmonella Enterica Inoculated in Carrot. *Plants* **2024**, *13*, 524. [CrossRef]
55. Kačániová, M.; Galovičová, L.; Valková, V.; Ďuranová, H.; Borotová, P.; Štefániková, J.; Vukovic, N.L.; Vukic, M.; Kunová, S.; Felsöciová, S.; et al. Chemical Composition and Biological Activity of *Salvia officinalis* Essential Oil. *Acta Hortic. Regiotect.* **2021**, *24*, 81–88. [CrossRef]
56. Kačániová, M.; Galovičová, L.; Ivanišová, E.; Vukovic, N.L.; Štefániková, J.; Valková, V.; Borotová, P.; Žiarovská, J.; Terentjeva, M.; Felšöciová, S.; et al. Antioxidant, Antimicrobial and Antibiofilm Activity of Coriander (*Coriandrum sativum* L.) Essential Oil for Its Application in Foods. *Foods* **2020**, *9*, 282. [CrossRef] [PubMed]

57. Kačániová, M.; Vukovic, N.L.; Čmiková, N.; Bianchi, A.; Garzoli, S.; Ben Saad, R.; Ben Hsouna, A.; Elizondo-Luévano, J.H.; Said-Al Ahl, H.A.H.; Hikal, W.M.; et al. Biological Activity and Phytochemical Characteristics of Star Anise (*Illicium verum*) Essential Oil and Its Anti-*Salmonella* Activity on Sous Vide Pumpkin Model. *Foods* **2024**, *13*, 1505. [CrossRef] [PubMed]
58. Oka, N.; Ikegami, A.; Ohki, M.; Sakata, K.; Yagi, A.; Watanabe, N. Citronellyl Disaccharide Glycoside as an Aroma Precursor from Rose Flowers. *Phytochemistry* **1998**, *47*, 1527–1529. [CrossRef]
59. Knapp, H.; Straubinger, M.; Fornari, S.; Oka, N.; Watanabe, N.; Winterhalter, P. (*S*)-3,7-Dimethyl-5-Octene-1,7-Diol and Related Oxygenated Monoterpenoids from Petals of *Rosa damascena* Mill. *J. Agric. Food Chem.* **1998**, *46*, 1966–1970. [CrossRef]
60. Schieber, A.; Mihalev, K.; Berardini, N.; Mollov, P.; Carle, R. Flavonol Glycosides from Distilled Petals of *Rosa damascena* Mill. *Z. Für Naturforschung C* **2005**, *60*, 379–384. [CrossRef]
61. Andoğan, B.C.; Baydar, H.; Kaya, S.; Demirci, M.; Özbaşar, D.; Mumcu, E. Antimicrobial Activity and Chemical Composition of Some Essential Oils. *Arch. Pharm. Res.* **2002**, *25*, 860–864. [CrossRef]
62. Boskabady, M.H.; Shafei, M.N.; Saberi, Z.; Amini, S. Pharmacological Effects of *Rosa damascena*. *Iran. J. Basic. Med. Sci.* **2011**, *14*, 295–307. [PubMed]
63. Etschmann, M.; Bluemke, W.; Sell, D.; Schrader, J. Biotechnological Production of 2-Phenylethanol. *Appl. Microbiol. Biotechnol.* **2002**, *59*, 1–8. [CrossRef] [PubMed]
64. Aydinli, M.; Tutaş, M. Production of Rose Absolute from Rose Concrete. *Flavour. Fragr. J.* **2003**, *18*, 26–31. [CrossRef]
65. Lei, G.; Wang, L.; Liu, X.; Zhang, A. Fast Quantification of Phenylethyl Alcohol in Rose Water and Chemical Profiles of Rose Water and Oil of *Rosa damascena* and *Rosa rugosa* from Southeast China. *J. Liq. Chromatogr. Relat. Technol.* **2015**, *38*, 823–832. [CrossRef]
66. Yassa, N.; Masoomi, F.; Rohani Rankouhi, S.E.; Hadjiakhoondi, A. Chemical Composition and Antioxidant Activity of the Extract and Essential Oil of *Rosa damascena* from Iran, Population of Guilan. *DARU J. Pharm. Sci.* **2009**, *17*, 175–180.
67. Mahmoodreza Moein, M.M.; Younes Ghasemia, Y.G.; Forough Karami, F.K.; Hossein Tavallali, H.T. Composition of the Essential Oil of *Rosa damascena* Mill. from South of Iran. *Iran. J. Pharm. Sci.* **2010**, *6*, 59–62. [CrossRef]
68. Loghmani-Khouzani, H. Essential Oil Composition of *Rosa damascena* Mill Cultivated in Central Iran. *Sci. Iran.* **2007**, *14*, 316–319.
69. Basim, E.; Basim, H. Antibacterial Activity of *Rosa damascena* Essential Oil. *Fitoterapia* **2003**, *74*, 394–396. [CrossRef]
70. Özkan, G.; Sagdiç, O.; Baydar, N.G.; Baydar, H. Note: Antioxidant and Antibacterial Activities of *Rosa damascena* Flower Extracts. *Food Sci. Technol. Int.* **2004**, *10*, 277–281. [CrossRef]
71. Chouhan, S.; Sharma, K.; Guleria, S. Antimicrobial Activity of Some Essential Oils—Present Status and Future Perspectives. *Medicines* **2017**, *4*, 58. [CrossRef]
72. Ghavam, M.; Afzali, A.; Manca, M.L. Chemotype of Damask Rose with Oleic Acid (9 Octadecenoic Acid) and Its Antimicrobial Effectiveness. *Sci. Rep.* **2021**, *11*, 8027. [CrossRef] [PubMed]
73. Arnold, R.S.; Thom, K.A.; Sharma, S.; Phillips, M.; Kristie Johnson, J.; Morgan, D.J. Emergence of Klebsiella Pneumoniae Carbapenemase-Producing Bacteria. *South Med. J.* **2011**, *104*, 40–45. [CrossRef] [PubMed]
74. Čabarkapa, I.; Čolović, R.; Đuragić, O.; Popović, S.; Kokić, B.; Milanov, D.; Pezo, L. Anti-Biofilm Activities of Essential Oils Rich in Carvacrol and Thymol against *Salmonella* Enteritidis. *Biofouling* **2019**, *35*, 361–375. [CrossRef] [PubMed]
75. Somrani, M.; Debbabi, H.; Palop, A. Antibacterial and Antibiofilm Activity of Essential Oil of Clove against *Listeria monocytogenes* and *Salmonella* Enteritidis. *Food Sci. Technol. Int.* **2022**, *28*, 331–339. [CrossRef] [PubMed]
76. Valeriano, C.; De Oliveira, T.L.C.; De Carvalho, S.M.; Cardoso, M.D.G.; Alves, E.; Piccoli, R.H. The Sanitizing Action of Essential Oil-Based Solutions against Salmonella Enterica Serotype Enteritidis S64 Biofilm Formation on AISI 304 Stainless Steel. *Food Control* **2012**, *25*, 673–677. [CrossRef]
77. Crump, J.A.; Heyderman, R.S. A Perspective on Invasive *Salmonella* Disease in Africa. *Clin. Infect. Dis.* **2015**, *61*, S235–S240. [CrossRef]
78. Mahmoud, A. Effect of Lettuce, Marjoram and Cumin Essential Oils on the Quality and Shelf Life of Minced Meat during Refrigerated Storage. *Zagazig Vet. J.* **2019**, *47*, 288–297. [CrossRef]
79. Hyldgaard, M.; Mygind, T.; Meyer, R.L. Essential Oils in Food Preservation: Mode of Action, Synergies, and Interactions with Food Matrix Components. *Front. Microbio.* **2012**, *3*, 12. [CrossRef]
80. Desai, M.A.; Soni, K.A.; Nannapaneni, R.; Schilling, M.W.; Silva, J.L. Reduction of Listeria Monocytogenes Biofilms on Stainless Steel and Polystyrene Surfaces by Essential Oils. *J. Food Prot.* **2012**, *75*, 1332–1337. [CrossRef]
81. Marinas, I.C.; Oprea, E.; Buleandra, M.; Badea, I.A.; Tihauan, B.M.; Marutescu, L.; Angheloiu, M.; Matei, E.; Chifiriuc, M.C. Chemical Composition, Antipathogenic and Cytotoxic Activity of the Essential Oil Extracted from *Amorpha fruticosa* Fruits. *Molecules* **2021**, *26*, 3146. [CrossRef]
82. Burt, S.A.; Van Der Zee, R.; Koets, A.P.; De Graaff, A.M.; Van Knapen, F.; Gaastra, W.; Haagsman, H.P.; Veldhuizen, E.J.A. Carvacrol Induces Heat Shock Protein 60 and Inhibits Synthesis of Flagellin in *Escherichia coli* O157:H7. *Appl. Environ. Microbiol.* **2007**, *73*, 4484–4490. [CrossRef] [PubMed]
83. Dos Santos Rodrigues, J.B.; De Carvalho, R.J.; De Souza, N.T.; De Sousa Oliveira, K.; Franco, O.L.; Schaffner, D.; De Souza, E.L.; Magnani, M. Effects of Oregano Essential Oil and Carvacrol on Biofilms of *Staphylococcus aureus* from Food-Contact Surfaces. *Food Control* **2017**, *73*, 1237–1246. [CrossRef]
84. Adukwu, E.C.; Allen, S.C.H.; Phillips, C.A. The Anti-Biofilm Activity of Lemongrass (*Cymbopogon flexuosus*) and Grapefruit (*Citrus paradisi*) Essential Oils against Five Strains of *Staphylococcus aureus*. *J. Appl. Microbiol.* **2012**, *113*, 1217–1227. [CrossRef]

85. Almeida, L.D.F.D.D.; Paula, J.F.D.; Almeida, R.V.D.D.; Williams, D.W.; Hebling, J.; Cavalcanti, Y.W. Efficacy of Citronella and Cinnamon Essential Oils on *Candida albicans* Biofilms. *Acta Odontol. Scand.* **2016**, *74*, 393–398. [CrossRef]
86. Karpanen, T.J.; Worthington, T.; Hendry, E.R.; Conway, B.R.; Lambert, P.A. Antimicrobial Efficacy of Chlorhexidine Digluconate Alone and in Combination with Eucalyptus Oil, Tea Tree Oil and Thymol against Planktonic and Biofilm Cultures of Staphylococcus Epidermidis. *J. Antimicrob. Chemother.* **2008**, *62*, 1031–1036. [CrossRef] [PubMed]
87. Kwieciński, J.; Eick, S.; Wójcik, K. Effects of Tea Tree (*Melaleuca alternifolia*) Oil on *Staphylococcus aureus* in Biofilms and Stationary Growth Phase. *Int. J. Antimicrob. Agents* **2009**, *33*, 343–347. [CrossRef]
88. Quendera, A.P.; Barreto, A.S.; Semedo-Lemsaddek, T. Antimicrobial Activity of Essential Oils against Foodborne Multidrug-Resistant Enterococci and Aeromonads in Planktonic and Biofilm State. *Food Sci. Technol. Int.* **2019**, *25*, 101–108. [CrossRef]
89. Sieniawska, E.; Los, R.; Baj, T.; Malm, A.; Glowniak, K. Antimicrobial Efficacy of Mutellina Purpurea Essential Oil and α-Pinene against Staphylococcus Epidermidis Grown in Planktonic and Biofilm Cultures. *Ind. Crops Prod.* **2013**, *51*, 152–157. [CrossRef]
90. El, A.S.; Ibnsouda, K.S.; Latrache, H.; Zineb, G.; Mouradi, H.; Remmal, A. Carvacrol and Thymol Components Inhibiting Pseudomonas Aeruginosa Adherence and Biofilm Formation. *Afr. J. Microbiol. Res.* **2011**, *5*, 3229–3232. [CrossRef]
91. Falcó, I.; Verdeguer, M.; Aznar, R.; Sánchez, G.; Randazzo, W. Sanitizing Food Contact Surfaces by the Use of Essential Oils. *Innov. Food Sci. Emerg. Technol.* **2019**, *51*, 220–228. [CrossRef]
92. Jadhav, S.; Shah, R.; Bhave, M.; Palombo, E.A. Inhibitory Activity of Yarrow Essential Oil on Listeria Planktonic Cells and Biofilms. *Food Control* **2013**, *29*, 125–130. [CrossRef]
93. Sandasi, M.; Leonard, C.M.; Van Vuuren, S.F.; Viljoen, A.M. Peppermint (Mentha Piperita) Inhibits Microbial Biofilms in Vitro. *S. Afr. J. Bot.* **2011**, *77*, 80–85. [CrossRef]
94. Vetas, D.; Dimitropoulou, E.; Mitropoulou, G.; Kourkoutas, Y.; Giaouris, E. Disinfection Efficiencies of Sage and Spearmint Essential Oils against Planktonic and Biofilm *Staphylococcus aureus* Cells in Comparison with Sodium Hypochlorite. *Int. J. Food Microbiol.* **2017**, *257*, 19–25. [CrossRef] [PubMed]
95. Vidács, A.; Kerekes, E.; Rajkó, R.; Petkovits, T.; Alharbi, N.S.; Khaled, J.M.; Vágvölgyi, C.; Krisch, J. Optimization of Essential Oil-Based Natural Disinfectants against Listeria Monocytogenes and Escherichia Coli Biofilms Formed on Polypropylene Surfaces. *J. Mol. Liq.* **2018**, *255*, 257–262. [CrossRef]
96. Amaral, V.C.S.; Santos, P.R.; Da Silva, A.F.; Dos Santos, A.R.; Machinski, M.; Mikcha, J.M.G. Effect of Carvacrol and Thymol on *Salmonella* Spp. Biofilms on Polypropylene. *Int. J. Food Sci. Tech.* **2015**, *50*, 2639–2643. [CrossRef]
97. Braga, P.C.; Culici, M.; Alfieri, M.; Dal Sasso, M. Thymol Inhibits Candida Albicans Biofilm Formation and Mature Biofilm. *Int. J. Antimicrob. Agents* **2008**, *31*, 472–477. [CrossRef]
98. Grant, A.J.; Foster, G.L.; McKinley, T.J.; Brown, S.P.; Clare, S.; Maskell, D.J.; Mastroeni, P. Bacterial Growth Rate and Host Factors as Determinants of Intracellular Bacterial Distributions in Systemic *Salmonella enterica* Infections. *Infect. Immun.* **2009**, *77*, 5608–5611. [CrossRef]
99. Bhunia, A.K. Salmonella Enterica. In *Foodborne Microbial Pathogens: Mechanisms and Pathogenesis*; Bhunia, A.K., Ed.; Springer: New York, NY, USA, 2018; pp. 271–287. ISBN 978-1-4939-7349-1.
100. Zhou, T.; Harrison, A.D.; McKellar, R.; Young, J.C.; Odumeru, J.; Piyasena, P.; Lu, X.; Mercer, D.G.; Karr, S. Determination of Acceptability and Shelf Life of Ready-to-Use Lettuce by Digital Image Analysis. *Food Res. Int.* **2004**, *37*, 875–881. [CrossRef]
101. Nicola, S.; Fontana, E. Fresh-Cut Produce Quality. In *Postharvest Handling*; Elsevier: Amsterdam, The Netherlands, 2014; pp. 217–273. ISBN 978-0-12-408137-6.
102. Cava, R.; Nowak, E.; Taboada, A.; Marin-Iniesta, F. Antimicrobial Activity of Clove and Cinnamon Essential Oils against Listeria Monocytogenes in Pasteurized Milk. *J. Food Prot.* **2007**, *70*, 2757–2763. [CrossRef]
103. Gutierrez, J.; Barry-Ryan, C.; Bourke, P. The Antimicrobial Efficacy of Plant Essential Oil Combinations and Interactions with Food Ingredients. *Int. J. Food Microbiol.* **2008**, *124*, 91–97. [CrossRef]
104. Rattanachaikunsopon, P.; Phumkhachorn, P. Assessment of Factors Influencing Antimicrobial Activity of Carvacrol and Cymene against Vibrio Cholerae in Food. *J. Biosci. Bioeng.* **2010**, *110*, 614–619. [CrossRef]
105. Singh, A.; Singh, R.K.; Bhunia, A.K.; Singh, N. Efficacy of Plant Essential Oils as Antimicrobial Agents against Listeria Monocytogenes in Hotdogs. *LWT-Food Sci. Technol.* **2003**, *36*, 787–794. [CrossRef]
106. Jiménez, M.; Domínguez, J.A.; Pascual-Pineda, L.A.; Azuara, E.; Beristain, C.I. Elaboration and Characterization of O/W Cinnamon (*Cinnamomum zeylanicum*) and Black Pepper (*Piper nigrum*) Emulsions. *Food Hydrocoll.* **2018**, *77*, 902–910. [CrossRef]
107. Avila Gandra, E.; Radünz, M.; Helbig, E.; Dellinghausen Borges, C.; Kuka Valente Gandra, T. A Mini-Review on Encapsulation of Essential Oils. *J. Anal. Pharm. Res.* **2018**, *7*, 00205. [CrossRef]
108. Ali, S.M.; Khan, A.A.; Ahmed, I.; Musaddiq, M.; Ahmed, K.S.; Polasa, H.; Rao, L.V.; Habibullah, C.M.; Sechi, L.A.; Ahmed, N. Antimicrobial Activities of Eugenol and Cinnamaldehyde against the Human Gastric Pathogen Helicobacter Pylori. *Ann. Clin. Microbiol. Antimicrob.* **2005**, *4*, 20. [CrossRef] [PubMed]
109. Nostro, A.; Cellini, L.; Zimbalatti, V.; Blanco, A.R.; Marino, A.; Pizzimenti, F.; Giulio, M.D.; Bisignano, G. Enhanced Activity of Carvacrol against Biofilm of *Staphylococcus aureus* and *Staphylococcus epidermidis* in an Acidic Environment. *APMIS* **2012**, *120*, 967–973. [CrossRef] [PubMed]
110. Skandamis, P.N.; Nychas, G.-J.E. Effect of Oregano Essential Oil on Microbiological and Physico-Chemical Attributes of Minced Meat Stored in Air and Modified Atmospheres. *J. Appl. Microbiol.* **2001**, *91*, 1011–1022. [CrossRef] [PubMed]

111. Tsigarida, E.; Nychas, G.-J.E. Ecophysiological Attributes of a *Lactobacillus* Sp. and a *Pseudomonas* Sp. on Sterile Beef Fillets in Relation to Storage Temperature and Film Permeability. *J. Appl. Microbiol.* **2001**, *90*, 696–705. [CrossRef]
112. Hussain, P.R.; Omeera, A.; Suradkar, P.P.; Dar, M.A. Effect of Combination Treatment of Gamma Irradiation and Ascorbic Acid on Physicochemical and Microbial Quality of Minimally Processed Eggplant (*Solanum melongena* L.). *Radiat. Phys. Chem.* **2014**, *103*, 131–141. [CrossRef]
113. MikeAnosike, E.E.; Braide, W.; Adeleye, S.A.; Oguoma, O.I. Microbiology and Chemical Preservation of Eggplants Sold in Five Popular Markets in Owerri, Imo State, Nigeria. *Biomed J. Sci. Tech. Res.* **2019**, *13*, 10001–10006. [CrossRef]
114. Akhtar, Y.; Isman, M.B. Comparative Growth Inhibitory and Antifeedant Effects of Plant Extracts and Pure Allelochemicals on Four Phytophagous Insect Species. *J. Appl. Entomol.* **2004**, *128*, 32–38. [CrossRef]
115. Isman, M.B.; Miresmailli, S.; Machial, C. Commercial Opportunities for Pesticides Based on Plant Essential Oils in Agriculture, Industry and Consumer Products. *Phytochem. Rev.* **2011**, *10*, 197–204. [CrossRef]
116. Attia, S.; Grissa, K.L.; Ghrabi, Z.G.; Mailleux, A.C.; Lognay, G.; Hance, T. Acaricidal Activity of 31 Essential Oils Extracted from Plants Collected in Tunisia. *J. Essent. Oil Res.* **2012**, *24*, 279–288. [CrossRef]
117. Rasikari, H.L.; Leach, D.N.; Waterman, P.G.; Spooner-hart, R.N.; Basta, A.H.; Banbury, L.K.; Forster, P.I. Acaricidal and Cytotoxic Activities of Extracts from Selected Genera of Australian Lamiaceae. *J. Econ. Entomol.* **2005**, *98*, 1259–1266. [CrossRef] [PubMed]
118. Miresmailli, S.; Bradbury, R.; Isman, M.B. Comparative Toxicity of *Rosmarinus officinalis* L. Essential Oil and Blends of Its Major Constituents against *Tetranychus urticae* Koch (Acari: Tetranychidae) on Two Different Host Plants. *Pest. Manag. Sci.* **2006**, *62*, 366–371. [CrossRef]
119. Badawy, M.E.I.; El-Arami, S.A.A.; Abdelgaleil, S.A.M. Acaricidal and Quantitative Structure Activity Relationship of Monoterpenes against the Two-Spotted Spider Mite, Tetranychus Urticae. *Exp. Appl. Acarol.* **2010**, *52*, 261–274. [CrossRef]
120. Cavalcanti, S.C.H.; Niculau, E.D.S.; Blank, A.F.; Câmara, C.A.G.; Araújo, I.N.; Alves, P.B. Composition and Acaricidal Activity of Lippia Sidoides Essential Oil against Two-Spotted Spider Mite (*Tetranychus urticae* Koch). *Bioresour. Technol.* **2010**, *101*, 829–832. [CrossRef]
121. Roh, H.S.; Lim, E.G.; Kim, J.; Park, C.G. Acaricidal and Oviposition Deterring Effects of Santalol Identified in Sandalwood Oil against Two-Spotted Spider Mite, *Tetranychus urticae* Koch (Acari: Tetranychidae). *J. Pest. Sci.* **2011**, *84*, 495–501. [CrossRef]

foods MDPI

Article

Locust Bean Gum/κ-Carrageenan Film Containing Blueberry or Beetroot Extracts as Intelligent Films to Monitoring Hake (*Merluccius merluccius*) Freshness

Carla S. V. Faria [1], Jorge M. Vieira [1,2], António A. Vicente [1,2] and Joana T. Martins [1,2,*]

1 CEB—Centre of Biological Engineering, University of Minho, 4710-057 Braga, Portugal; jorgevieira@ceb.uminho.pt (J.M.V.); avicente@deb.uminho.pt (A.A.V.)
2 LABBELS—Associate Laboratory, Braga/Guimarães, Portugal
* Correspondence: joanamartins@ceb.uminho.pt

Abstract: The main goal of this work was to develop bio-based and ecofriendly intelligent films as freshness indicators to monitor European hake (*Merluccius merluccius*) quality during storage by using a visual, non-destructive, and real-time technique. Locust bean gum (LBG)/κ-carrageenan (Car) films incorporating blueberry extract (BLE) or beetroot extract (BEE) were developed and their effectiveness to detect hake deterioration during 7 days of storage at 4 °C was evaluated. A visible color response from pink to blue was observed on the BLE films at the end of hake storage, which correlated with the hake deterioration profile, namely an increase in pH values (from 6.60 ± 0.04 to 8.02 ± 0.03), total viable count (TVC, from 4.61 ± 0.36 to 8.61 ± 0.21 log CFU/g), and total volatile basic nitrogen content (TVB-N, from 10.21 ± 1.97 to 66.78 ± 4.81 mg/100 g) beyond the spoilage threshold. The results of this study are very promising, since it was possible to develop a new effective intelligent bio-based responsive indicator film incorporating natural dye BLE, which has the potential to contribute to food waste reduction and improve food safety by detecting the hake freshness status.

Keywords: bio-based film; pH indicator; anthocyanins; betalains; shelf-life

1. Introduction

Food products that exhibit safety limitations due to the presence of microbial agents (bacteria, viruses, and parasites) and/or harmful chemical substances are the cause of more than 200 diseases including diarrhea and cancer, according to the World Health Organization (WHO) [1]. Food deterioration as a result of biological, chemical, or physical phenomena causes changes, in some cases, not directly recognized by the final consumers. Moreover, approximately 30% of all food produced worldwide is lost or wasted every year along the supply chain, according to the Food and Agriculture Organization (FAO). Category wise, food loss is predicted to be 20% in dairy products, 35% in seafood, 45% in fruits and vegetables, and 20% in meat, with the losses predominant at the processing, distribution, and consumption stages [2]. Therefore, it is important to solve the current challenges of food waste, quality, and safety through (i) the detection of changes in food products, in order to avoid product consumption by the final consumer; (ii) the identification of potential health risks; and (iii) the establishment of strategies to reduce or eliminate the occurrence of the previously mentioned events [3].

In this context, the development of innovative packaging with specific functions, namely intelligent packaging, has been observed in recent years. Intelligent packaging emerged with the aim to detect, monitor, and inform the consumer about the quality of food throughout the food chain in a non-destructive way and in real-time [4,5]. Moreover, intelligent packaging has been an important contribution to avoid unnecessary food waste because it is able to monitor and display the food quality status, and consequently avoid food products being thrown away even though they are still suitable for consumption. On the other hand, it has the potential to improve product safety, food logistics, and traceability, increasing the food industries' efficiency [4].

Freshness indicators (such as pH indicators) are devices usually integrated into the packaging that are capable of reacting with metabolites generated by microbial growth and metabolism (e.g., organic acids, ethanol, biogenic amines, and volatile nitrogen compounds) or are able to detect any chemical change, providing visual information about the food product quality [6]. As synthetic dyes are considered carcinogenic or mutagenic agents in humans, their application in food intelligent packaging is not ideal. In an attempt to overcome this limitation, more attention has recently been given to the alternative use of natural pigments. Natural dyes (pigments) extracted from plants are attractive and relevant options in the intelligent packaging field due to their low toxicity, easy preparation, renewable, and non-polluting properties [1,7,8]. Examples of natural dyes that change their color significantly with pH changes are anthocyanins and betalains [9–11].

The increase in consumer and food industry consciousness related to food safety and environmental pollution has guided researchers to focus on the development of safe, environmentally friendly, and biodegradable packaging materials. Thus, our research aim was to replace synthetic, petroleum-based polymers (such as plastics) with eco-friendly materials to be used in food packaging, especially in intelligent packaging [5,12]. The interest in using natural biopolymers such as polysaccharides for the development of bio-based intelligent packaging has increased due to their high abundance and high availability in nature and in low-value agricultural residues/by-products (for example, fruit peels and seeds) [5,10]. For instance, intelligent packaging films have been developed based on locust bean gum (LBG) and polyvinyl alcohol containing betacyanins from the cockscomb flower to indicate shrimp freshness [13]. κ-Carrageenan (Car) is an anionic linear sulfated polysaccharide that consists of the repetition of disaccharide units with 3,6-anhydro-D-galactopyranose residues in its chain [14]. These have been widely used in the food (mainly in dairy products) and pharmaceutical industries due to their biocompatible and biodegradable properties. He et al. [15] developed a film based on Car, gelatin, and curcumin, which work as an intelligent indicator for the freshness detection of grass carp fillets. Thus, polysaccharide-based intelligent films with natural compounds as freshness indicators have the potential to guarantee food quality and safety, reduce environmental impact, and increase the product packaged attractiveness to the retailer and the final consumer [7,12]. However, despite permanent innovations in the food packaging area, research related to this type of bio-based intelligent packaging is still limited.

Fish consumption and popularity have been increasing due to the customers' awareness regarding fish nutritional composition, especially its high protein content [16]. Several studies have been carried out to monitor the freshness of fish products such as Atlantic mackerel (*Scomber scombrus*) and Wuchang bream (*Megalobrama amblycephala*) using bio-based intelligent films containing natural pH dyes [11,17]. However, there is a lack of information regarding hake quality monitoring using bio-based intelligent packaging, which was a research opportunity explored in the present work. Portugal is the largest fish consumer per capita in the European Union due to its geographical location and tradition in fishing and fish consumption, with hake (*Merluccius merluccius*) the third most consumed fish in Portugal [18,19]. Hake is highly perishable due to microbial contamination and enzymatic reactions and can cause high economic losses for the fishery industry as well as a high risk to human health, thus it is crucial to monitor the fish freshness during storage, processing, and transportation.

Thus, to tackle the current concerns related to the consumption of contaminated fish products, food waste, and the environmental impact of petroleum-based packaging, a novel quality indicator sensitive to the changes in hake fish freshness was developed in this work. Therefore, a colorimetric bio-based LBG/Car intelligent film incorporating natural dyes—blueberry extract (BLE) or beetroot extract (BEE)—was developed. Moreover, the effect of BLE or BEE on the LBG/Car films' mechanical, water barrier, and optical properties as well as color response efficiency to different pH conditions and volatile nitrogen compounds were assessed.

2. Materials and Methods

2.1. Materials

Polysaccharides, Car (Gelcarin® DG 5264), and LBG were obtained by FMC Biopolymer (Philadelphia, PA, USA) and Sigma-Aldrich (St. Louis, MO, USA), respectively. The glycerol was purchased from Sigma-Aldrich (St. Louis, MO, USA). Freeze-dried BLE powder (Simplu®, Guimarães, Portugal) and BEE powder (Solgar®, Leonia, NJ, USA) were purchased in Celeiro (Braga, Portugal). Sodium hydroxide (NaOH), ammonia, and plate count agar (PCA) were provided by Merck (Merck SA, Lisbon, Portugal). Boric acid (H_3BO_3), hydrochloric acid (HCl), and sulfuric acid (H_2SO_4) were supplied by Chem-Lab (Chem Lab, Zedelgem, Belgium). Potassium carbonate (K_2CO_3) and methyl red indicator were purchased from Panreac (Panreac AppliChem ITW Companies), Barcelona, Spain). Bromocresol green was supplied by Acros Organic (Acros OrganicsTM, Lisbon, Portugal). The fresh hake was purchased from a local market (Braga, Portugal).

2.2. Production of LBG/Car Film Incorporating BLE or BEE

LBG/Car films were produced based on our previous works [20,21]. Initially, 60% (*w*/*w*) LBG and 40% (*w*/*w*) Car were dissolved in distilled water under stirring for 1 h at 25 °C. Then, 30% (*w*/*w*) of glycerol was added, and the film-forming solution was homogenized at 70 °C under agitation for 30 min. This film-forming solution was used to produce control films (i.e., no-added BLE or BEE) and films incorporating the extracts (1 to 50% (*w*/*w*)). Regarding films incorporating extracts, 30% (*w*/*w*) BLE or 50% (*w*/*w*) BEE was added to the LBG/Car film-forming solution and homogenized at 1200 rpm for 30 min at 40 °C. The selected BLE and BEE concentrations were based on preliminary tests, where the films' visual color uniformity and handling behavior (e.g., brittle) were assessed. All film-forming solutions were added to Petri dishes (90 mm) and placed in an oven with airflow circulation (WTC Binder) at 35 °C for 22 h to produce the films. All films produced were conditioned in a desiccator at 54% relative humidity (RH) until further analysis.

2.3. Physico-Chemical Properties of BLE and BEE Films

2.3.1. Color and Opacity of the Films

The color of films was determined with a Minolta colorimeter (CR 300; Minolta, Japan) [21,22]. To calibrate the instruments, a white standard color plate ($Y = 93.90$; $x = 0.32$; $y = 0.33$) was used as the background for the film color measurements. Lightness (L^*) (white to black) and chromaticity parameters a^* (red to green) and b^* (yellow to blue) were assessed through reflectance measurements using the CIELab scale. The color change of films during storage (ΔE) was determined using Equation (1):

$$\Delta E = \sqrt{\left[(\Delta L*)^2 + (\Delta a*)^2 + (\Delta b*)^2\right]} \quad (1)$$

The opacity of each film sample was obtained according to the Hunterlab method as the relationship between the film opacity on a black standard plate and on a white standard plate. All color and opacity measurements were performed six times for each sample, and three samples of each film were tested.

2.3.2. Evaluation of Color Stability of the Films during Storage

The BLE and BEE films were stored under light or dark conditions for 60 days at 25 °C or 4 °C to evaluate their color stability during storage. The L^*, a^*, and b^* colorimetric parameters were measured in each film using a CR400 colorimeter (Konica, Minolta, Osaka, Japan) according to the methods described in the literature [8,12]. The color parameters of the films were analyzed at 2, 5, 10, 15, and 60 days of storage. The colorimetric measurements were conducted in triplicate on each film sample.

2.3.3. Evaluation of pH-Sensitive Properties of the Films

In order to evaluate the BLE or BEE films' color response to pH changes, a droplet of the HCl (0.1 mol/L) aqueous solution (pH 2, 4, and 6), distilled water (pH 7), or NaOH (0.1 mol/L) aqueous solution (pH 8, 10, 12 and 14) was placed on the surface of the film samples (2 × 2 cm) for 10 min [10]. The color change was evaluated by comparing the color differences between the films before and after contact with the buffer solutions. All measurements were taken at three random locations of each film sample.

2.3.4. BLE and BEE Films Response to Ammonia Vapor

The film's color response to volatile compounds, namely ammonia vapor, was determined as described by Qin et al. [23] and Yun et al. [24]. BLE and BEE film samples (1.5 × 1.5 cm) were placed in the headspace of sealed Petri dishes loaded with 0.4 mol L^{-1} ammonia solution for 60 min at room temperature under no-light conditions. The L^*, a^*, and b^* colorimetric parameters of the films were measured with 10-minute intervals. Color measurements were conducted at three random locations of each film sample.

2.3.5. Water Contact Angle (WCA) Measurements

The WCA on the film surface was measured using an optical contact angle meter (OCA 20, Dataphysics, Filderstadt, Germany) through the sensile drop method [25]. Each film sample was cut into square pieces (1.5 × 1.5 cm) and 5 μL of ultrapure water was dropped onto the film surface using a micro-syringe with a 0.75 mm diameter needle (Hamilton, Bonaduz, Switzerland). Contact angle measurements were performed after 30 s of the droplet placement on the film surface.

2.3.6. Moisture Content (MC) and Water Solubility (WS)

The MC and WS of the films were determined gravimetrically according to the methods reported by other authors [22,26]. The MC was determined by measuring the weight loss of the film samples (2 cm of diameter) after drying at 105 °C in an oven (WTC Binder) until it reached a constant dry sample weight. To determine the WS, dried samples of the film (2 cm of diameter) were weighed and immersed in 50 mL of deionized water for 8 h at room temperature under orbital agitation at 120 rpm (Orbital shaker ES-20/80, BOECO, Hamburg, Germany). The water was removed and the film residue was dried at 105 °C for 24 h to determine the weight of dry matter that was not solubilized in water. For the MC and WC assays, three replicates were performed for each film sample.

2.3.7. Water Vapor Permeability (WVP)

The WVP of the films was determined gravimetrically following the methodology developed by other authors [22,25]. Three samples were cut from each film and sealed on a permeation cup cell containing 50 mL of distilled water (100% RH; 2337 Pa vapor pressure at 20 °C) and placed in a desiccator containing silica gel (0% RH; 0 Pa water vapor pressure at 20 °C). Constant air circulation outside of the test cup was maintained by using a fan inside the desiccator to assure uniform water pressure and steady-state conditions. Cups were weighed every 2 h for 12 h, and the water transferred through the film was determined from the weight loss of the permeation cup cell. All measurements were performed in triplicate. The water vapor transmission rate (WVTR) through the film (g m^{-2} s^{-1}) was calculated through the ratio between the slope of the linear regression of weight loss versus time and the film area. The WVP (g m^{-1} s^{-1} Pa^{-1}) was calculated according to Equation (2):

$$\mathrm{WVP} = \frac{\mathrm{WVTR} \times \mathrm{L}}{\Delta P} \quad (2)$$

where L is the mean film thickness (m) and ΔP is the partial water vapor pressure difference (Pa) across the two sides of the film.

Film thickness was measured using a digital micrometer device (Mitutoyo No. 293-766, Kanagawa, Japan) with 0.001 mm accuracy. Ten thickness measurements were randomly taken from different points of each film sample [27].

2.3.8. Assessment of Mechanical Properties

The tensile strength (TS) and elongation-at-break (EB) of the films were measured using a TA-HD plus Texture Analyzer (Serial RS232, Stable Micro Systems, Surrey, UK), according to Martins et al. [21]. Film specimens (120 mm × 20 mm) from each film sample were cut and placed between the tensile grips. The initial grip separation and crosshead speed were set at 30 mm and at 5 mm min^{-1}, respectively. The TS and EB were determined from the stress–strain curves. At least four measurements were performed for each film.

2.4. Evaluation of BLE and BEE Films as Hake Freshness Colorimetric Indicators

2.4.1. Sample Preparation and Storage Conditions

Fresh hake was purchased from a local market (Continente, Braga, Portugal) and immediately transported to the laboratory. Fishbones were removed, and fish fillet samples (approximately 5 g) were prepared.

To validate the BLE and BEE films as colorimetric indicators sensitive to pH and gas changes, tests with fish samples stored at 4 °C and 25 °C for 7 days were carried out. BLE and BEE film samples (2 cm × 2 cm) were fixed to the top cover of sterile polystyrene 45 mm diameter Petri dishes that were in contact with the dish headspace, as can be seen in Figure 1. Then, the fish samples were placed into Petri dishes and sealed [12,26] (Figure 1). The Petri dish with no-added fish sample was used as the control. All samples were stored at 4 °C and 25 °C and analyzed on days 0, 2, 3, 4, 6, and 7 to detect any physico-chemical or microbial changes. All results obtained over the storage period were compared with the results obtained on day 0. For each sampling day, three replicates were performed for each fish sample.

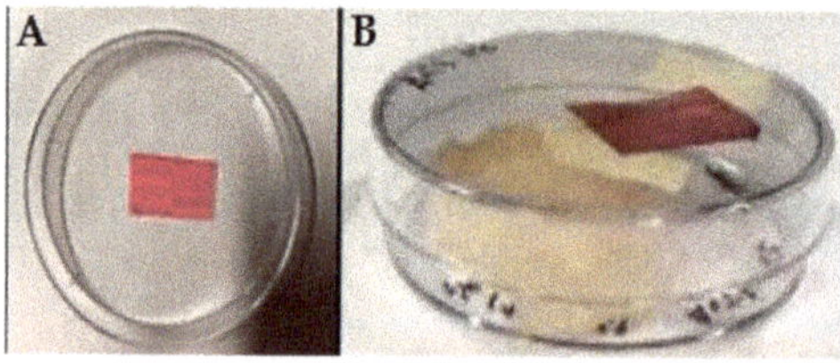

Figure 1. Illustrative example of the experimental setup used to monitor hake freshness during storage at 4 and 25 °C. (**A**) Film sample fixed to the Petri dish cover and (**B**) closed Petri dish system containing the hake sample used to monitor the freshness status.

2.4.2. Film Color Analysis

Film color changes during the loss of hake freshness over the storage time were assessed according to the method described in Section 2.3.1. At least three color measurements were performed on different film samples on each sampling day.

2.4.3. pH Measurement of Fish Samples

The fish samples were blended with 100 mL of distilled water using a stomacher (Stomacher® 3500, Sweard Medical, Worthing, UK) at high speed for 4 min. All pH measurements were performed using a digital pH meter (HI 2210, Hanna Instruments, Porto, Portugal) by immersing the glass electrode in the homogenized samples [12].

2.4.4. Determination of Total Volatile Basic Nitrogen (TVB-N)

TVB-N measurements were conducted according to the Conway micro diffusion method with some modifications [15]. Initially, the hake samples (4 g) were added to 40 mL of distilled water and homogenized using a stomacher (Stomacher® 3500, Sweard Medical, UK) for 8 min. Then, the mixture was filtered using filter paper (type 601A Rotilabo®-Round filters, Carl Roth,

Karlsruhe, Germany). Subsequently, 1 mL of 0.02% boric acid and one drop of pH indicator (methyl red:bromocresol green = 1:5) were added to the inner chamber of the Conway dish. One mL of saturated potassium carbonate solution and 1 mL of the fish filtrate sample were added to the outer dish chamber. A blank sample was also prepared as described, but the filtered sample was not added to the Conway dish. The lid of the Conway dish was closed, and the resulting mixture was allowed to stand at room temperature for 1.5 h. Finally, the solution in the inner chamber was titrated with hydrochloric acid (0.01 mol/L) until a pale pink color appeared. Analyses were conducted in triplicate. The TVB-N content (presented as mg N/100 g of fish sample) was calculated using Equation (3):

$$\mathrm{TVB-N} = \frac{(\mathrm{V1}-\mathrm{V2}) \times \mathrm{c} \times 14 \times 100}{\mathrm{m} \times \frac{\mathrm{V}}{\mathrm{V0}}} \quad (3)$$

where V1 and V2 represent the HCl volume (mL) required for the sample and blank titration, respectively; c is the HCl concentration (mol/L); m is the fish sample weight (g); 14 is the molecular weight of nitrogen; 100 is the conversion factor; V is the volume of filtrate sample (mL); and V0 is the total volume (mL).

2.4.5. Microbiological Analysis

The determination of the total viable count (TVC) was performed in triplicate according to the procedures reported by other authors [15,27]. Hake samples (4 g) were transferred to a stomacher bag. Then, 36 mL of sterile 0.1% peptone water was added to the bag and homogenized for 2 min using a stomacher. Serial decimal dilutions (10^1 to 10^6) were prepared, and 1 mL of the appropriate dilution was placed in a Petri dish and 20 mL of sterile PCA was added to the dish. The dishes were incubated at 35 °C for 72 h. The TVC was determined and expressed as a log colony forming unit per g fish sample (log CFU/g). According to the TVC results , the initial TVC of the hake samples was 4.56 ± 0.36 log CFU.g^{-1}.

2.5. Statistical Analysis

All experiments were performed at least in triplicate. All data were expressed as the mean values ± standard deviation (SD). One-way analysis of variance (ANOVA) was performed, and the significant differences among samples were determined by the Tukey HSD test ($p < 0.05$). Statistics were performed on the SigmaPlot Software 11.0 for Windows.

3. Results and Discussion

3.1. Development of LBG/Car Film Incorporating BLE or BEE

Different BLE or BEE concentrations were tested to select the films' formulations for use as potential freshness colorimetric indicators. LBG/Car films presented a soft yellow color and were transparent, while films incorporating BLE or BEE showed a pink color (Figure 2). Based on visual examination of the film surface, namely the color uniformity throughout the film and handling properties, films containing 30% (*w*/*w*) of BLE or 50% (*w*/*w*) of BEE concentrations were selected for further analysis. Films containing BLE and BEE concentrations below 30% and 50%, respectively, presented very light and non-homogenous coloration and were very brittle when handled.

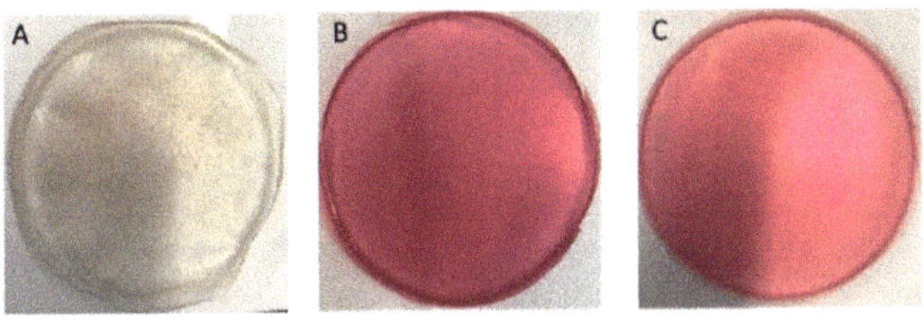

Figure 2. Photographs of the (**A**) control film (no-added extracts), (**B**) film incorporating 30% (*w*/*w*) BLE, and (**C**) film incorporating 50% (% *w*/*w*) BEE.

3.2. *Physicochemical Properties of the Developed Films*

3.2.1. Optical Properties: Color and Opacity

The color parameters (L^*, a^*, and b^*) and opacity of the various films developed are shown in Table 1. The control film showed a higher transparency and luminosity (i.e., higher L^* value) than the BLE and BEE films. The BLE addition to the film led to a decrease in L^* and b^* parameters ($p < 0.05$), which demonstrated a decrease in film brightness and an increase in the blue color of the film (Table 1). On the other hand, there was a significant increase in a^* value ($p < 0.05$) compared to the control film, which represented an increment in red coloration of the film. This result was in line with the reddish/purple color of the BLE, contributing to the red/purple color of the BLE film. Qin et al. [5] also concluded that the rise in the anthocyanin-rich extract from *Lycium ruthenicum* Murr content within starch films significantly increased the a^* value and decreased the b^* value, indicating a change in film color to red. Merz et al. [28] observed that the color of the chitosan and polyvinyl alcohol film changed to red when anthocyanins from the jambolana fruit were added. The addition of BEE also caused a decrease in the L^* and b^* values ($p < 0.05$) of the control films (Table 1). Furthermore, the a^* value of the BEE film had a higher increase when compared to the BLE film ($p < 0.05$). For this reason, the pink/red color was more visible on the BEE films (Figure 2). These changes in the color parameters after adding red pigments such as betalains from beetroot have previously been reported. For instance, Wu et al. [13] showed that films based on LBG/polyvinyl alcohol with betacyanins from the cockscomb flower were reddish-purple, which caused an increase in the a^* value. A similar color was also observed in pectin-based films incorporating the beetroot extract [29].

Table 1. Color parameters (L^*, a^* and b^*) and opacity of the control, BLE, and BEE films.

Films	L^*	a^*	b^*	Opacity (%)
Control	89.13 ± 2.51 ^{a}	0.14 ± 0.05 ^{a}	5.18 ± 0.82 ^{a}	13.55 ± 1.81 ^{a}
BLE	41.01 ± 1.46 ^{b}	40.76 ± 3.20 ^{b}	4.74 ± 1.90 ^{b}	37.41 ± 2.73 ^{b}
BEE	56.80 ± 0.84 ^{b}	51.58 ± 0.96 ^{c}	-4.35 ± 0.23 ^{c}	24.86 ± 1.18 ^{c}

$^{a–c}$ Different superscript letters in the same column indicate significant differences ($p < 0.05$).

Film opacity refers to the film's ability to act as a barrier to light. The higher the opacity value, the lower the film transparency. The control film showed the lowest opacity value compared to the BLE and BEE films, confirming its high transparency (Table 1). The BLE film presented the highest opacity value compared to the BEE films. Thus, the BLE film reduced the film transparency and hindered the passage of light through it, which has great potential to prevent oxidative rancidity in foods caused by exposure to light [30]. Other works have reported similar film opacity results where the increase in film opacity was due to the incorporation of natural extracts in bio-based intelligent films [31,32].

3.2.2. Color Stability of the BLE and BEE Films during Storage

It is necessary to evaluate the influence of environment conditions (namely, temperature and light) on the produced films in order to confirm that the intelligent film can provide reliable visual feedback to consumers. Thus, the BLE and BEE films' color stability was tested at different temperatures (4 °C and 25 °C) and under light/no light exposure to simulate possible food storage conditions.

As shown in Figure 3, the ΔE values of the BLE and BEE films were influenced by the incidence of light and temperature after 2 days of storage. The untrained human eye detects a color change when ΔE values are higher than 5 [33,34]. In all of the tested conditions, the BLE and BEE films presented ΔE values higher than 5 (Figure 3), suggesting that the films' color change can be detected by the consumer [35]. These changes may be the result of the photodegradation of natural pigments (such as anthocyanins and betalains) in the films. However, the ΔE values of the BLE and BEE films remained stable from day 2 until day 60 ($p > 0.05$) when they were stored at 4 °C under the presence or absence of light. It can be concluded that if the BLE and BEE films are stored under refrigeration conditions, the films color changes will not be directly associated with the photodegradation due to light exposure, but rather to changes to the food product quality. This result is in accordance to Etxabide et al. [36], who studied the color stability of anthocyanins and betalains containing colorants under different storage conditions for the development of intelligent packaging. They concluded that the color remained stable under 4 °C and no light incidence storage conditions for 28 days, although a small color fading was observed during the final days. On the other hand, the ΔE results related to storage at 25 °C under light storage conditions showed an evident increase in ΔE values ($p < 0.05$), possibly due to BLE and BEE degradation (Figure 3). However, the BLE and BEE films showed no significant differences ($p > 0.05$) in ΔE values after 60 days of storage at 25 °C under no light conditions. Thus, an increase in temperature and exposure to light (e.g., during food products transport chain), which could compromise the food quality, could be detected by both films studied. Gao et al. [37] also concluded that the ΔE of intelligent films based on Car incorporated with anthocyanins and betalains increased gradually with the extension of storage time at 25 °C. The authors stated that the color stability decreased progressively, which may be related to the oxidation and decomposition of anthocyanins and betalains. The results of film stability as an intelligent indicator for monitoring hake spoilage during specific storage conditions were very promising. However, it should be explored in more detail in future work.

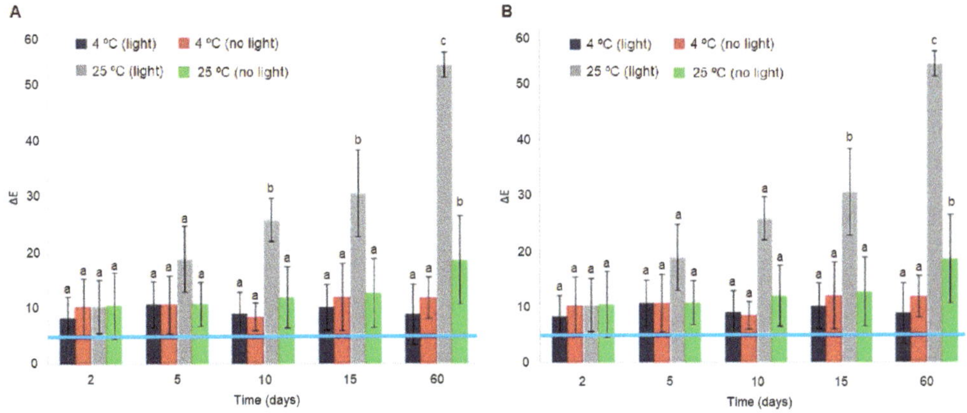

Figure 3. Total color difference (ΔE) of (**A**) the BLE and (**B**) BEE films during 60 days of storage under different temperature and light conditions. Values are given as the mean ± SD (n = 9). The blue line represents a ΔE equal to 5, a value at which the consumer can visually detect colorimetric changes. [a–c] Different letters at the same storage condition over time represent statistically significant differences ($p < 0.05$).

3.2.3. pH-Sensitive Properties of the BLE and BEE Films

Figure 4 illustrates the visible color response (ranging from red to green color) of each film sample when exposed to different buffer solutions (pH 2 to 14). The BLE film presented a red color at pH = 2; a light purple color at a pH range of 4 to 7; a dark purple color at pH = 10; a blue/green color at pH = 12, and yellow color at pH = 14 [5,12,38]. The color parameter analysis of the BLE films showed that the L^* value increased (39.13 ± 4.04 to 44.63 ± 5.87), and the a^* (40.61 ± 3.02 to 26.00 ± 2.23) and b^* (26.27 ± 4.16 to 10.36 ± 1.32) values significantly decreased ($p < 0.05$) between pH 2 and 4. Therefore, in acidic solutions, the BLE film showed a decrease in its red color and an increase in its blue color. The values of the L^*, a^*, and b^* parameters remained similar ($p > 0.05$) between pH 4 and 10, which showed that the BLE film did not change its color under this pH range. As the pH value increased from 10 to 12, the L^* values increased (45.05 ± 2.29 to 51.05 ± 3.35), and the a^* (28.66 ± 2.21 to 7.60 ± 5.81) and b^* (9.66 ± 1.99 to 3.85 ± 2.39) values dropped sharply. Thus, it can be concluded that the BLE film showed a more intense blue color, as can be seen in Figure 4. During the pH transition from 12 to 14, the L^* value increased (51.05 ± 3.35 to 62.18 ± 5.03), the a^* value decreased (7.60 ± 5.81 to 5.03 ± 1.54), and the b^* value increased (3.85 ± 2.39 to 26.98 ± 3.68). These results demonstrate that the film became more yellow. Research has shown that the color stability of the anthocyanins present in blueberry is influenced by external conditions such as pH, storage temperature, and incidence of light [7]. Indeed, anthocyanins are more stable in acidic than in alkaline solutions. The flavylium cation form of anthocyanins is stable under acidic solutions, with its double bonds prone to conjugation (e.g., cation hydration), forming the carbinol basic form (pink-light violet color). At pH 10, anthocyanins are in the quinoidal base form due to cation deprotonation, leading to a color change to violet. At pH 12, a color change to blue-green is observed, since the quinoidal anhydro base form occurs by deprotonation from the oxygen and carbon ring. Finally, the retro-chalcone formation occurs at pH 14 through carbinol tautomerization, which is responsible for the appearance of a yellow-green color [39]. Ma et al. [40] reported similar color changes of polyvinyl alcohol-chitosan films containing anthocyanins from mulberry, while Choi et al. [12] observed color changes of agar/potato starch films incorporating anthocyanin extracts (from purple sweet potato) from bright red at pH 2, to pink at pH 3–4, to brown at pH 5–6, and finally to greenish-brown at pH 9–10.

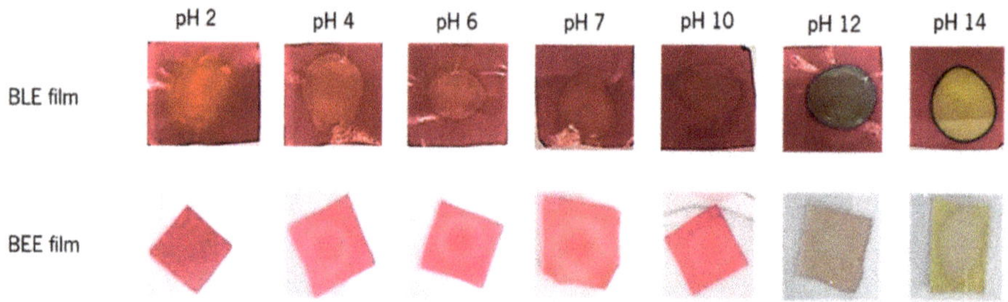

Figure 4. Color changes of the BLE and BEE films as a result of the contact of different buffer solutions (pH 2 to 14) to the films.

Regarding the BEE films, they exhibited a pink-purple color when the films were in contact with buffer solutions ranging from pH 2 to 10 (Figure 4). At this pH range, the L^*, a^*, and b^* values did not present statistically significant differences ($p > 0.05$), concluding that the BEE film did not present visible colorimetric changes under these conditions. When the pH values changed from 10 to 12, the L^* value increased (55.08 ± 1.51 to 60.17 ± 6.27), and the a^* (44.55 ± 3.37 to 21.97 ± 3.54) and b^* (4.94 ± 1.23 to 3.37 ± 1.72) values decreased, which means that the BEE film had a higher lightness and showed a green-blue color. At pH 12 to 14, the color film changed to yellow. In this case, the L^* value (60.17 ± 6.27 to 74.50 ± 2.22) continued to increase, the a^* decreased significantly (21.97 ± 3.54 to 1.07 ± 3.26), and b^* increased significantly ($p < 0.05$) (3.37 ± 1.72 to 23.06 ± 1.94). Beetroot contains two groups of betalain pigments: red-violet betacyanins (at pH 4 to 10) and yellow betaxanthins (at pH > 10). Betalains are relatively stable at a pH range of 2 to 10, unlike anthocyanins [28]. For this reason, it is not easy to visually distinguish the film color change over a wide range of pH values, with the most visible change between pH 10 and 12. Similar to our results, Wu et al. [13] described the color changes of LGB/polyvinyl alcohol films incorporating betacyanins (from cockscomb flower) at different pH. The films exhibited stable reddish-purple colors at pH 3 to 8 and turned light brown or light purple at pH 9 to 12. These color changes were related to the structural degradation of betacyanins in alkaline solutions. Thus, it can be concluded that the color of all films tested were strongly pH-dependent, showing their potential to be applied in the intelligent packaging field [5,28].

3.2.4. Films Response to Ammonia Vapor

Generally, the spoilage of many protein-rich animal foods can produce a high amount of volatile nitrogenous compounds (e.g., ammonia, dimethylamine, and trimethylamine) due to the deterioration of proteins, which directly affect the food products' pH status [9]. Thus, all films were exposed to an ammonia solution to simulate the volatile nitrogenous compounds released from food products presenting a high protein content, and the colorimetric film response was evaluated.

As shown in Table 2, the control film's color did not change over time, and the a^* and b^* values remained nearly constant ($p > 0.05$). As expected, the BLE film showed a significant color change from pink/violet to blue after 10 min, and pale green/yellow after 60 min when exposed to ammonia vapor. Accordingly, the a^* and b^* values significantly decreased ($p < 0.05$) after 30 min of exposure time, which means that the BLE film became blue/green. After 30 min, the blue/green color of the film started to fade, and some grey color portions began to appear (Table 2). The a^* and b^* values increased, which consequently led to a decrease in the green and blue color. After 60 min, there was a decrease in the a^* and b^* values again, but at 120 min, they increased again, causing another considerable color change in the film (blue to pale green/yellow). The mechanism that possibly explains these color changes is the fact that ammonia can diffuse through the films, combining with water molecules and subsequently hydrolyzing into hydroxyl ions, producing an alkaline environment in films. As previously discussed, BLE films, as exposed to increasingly alkaline conditions, change their color from pink/violet to blue, and finally, to green/yellow [41]. Jiang et al. [33] also reported that the film color from carboxymethylcellulose/starch/purple sweet potato anthocyanins changed from red to green after 60 min of exposure time to ammonia. Other research has shown a significant color change from reddish/pink to pale green of methylcellulose/chitosan nanofiber/barberry anthocyanin films after 15 min, and to yellow after 30 min when exposed to ammonia vapor [42].

Table 2. Color response of the control, BLE, and BEE films to ammonia vapor.

	Control Film			BLE Film			BEE Film		
Time (min)	Film Photo	a^*	b^*	Film Photo	a^*	b^*	Film Photo	a^*	b^*
0	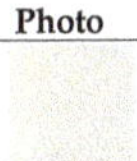	0.25 ± 0.07 [a]	4.32 ± 0.50 [a]		43.49 ± 3.44 [a]	5.42 ± 2.10 [a]	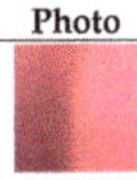	46.09 ± 4.53 [a]	-4.96 ± 0.52 [ab]
10		0.13 ± 0.05 [a]	4.00 ± 0.39 [a]		3.34 ± 2.31 [b]	-11.59 ± 0.93 [b]		33.61 ± 3.77 [a]	-5.70 ± 1.01 [b]
20		0.17 ± 0.04 [a]	3.71 ± 0.19 [a]		0.25 ± 0.94 [c]	-6.35 ± 1.21 [b]	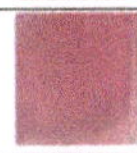	27.35 ± 4.51 [b]	-4.45 ± 1.21 [abc]
30	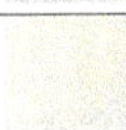	0.16 ± 0.07 [a]	3.67 ± 0.59 [a]		0.62 0.65 [bc]	-3.41 ± 1.79 [bc]	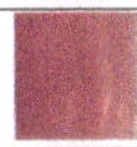	30.50 ± 3.27 [b]	-3.83 ± 0.78 [ab]
40		0.16 ± 0.06 [a]	4.00 ± 0.68 [a]		1.54 ± 1.77 [bcd]	2.22 ± 5.38 [bc]		28.27 ± 4.12 [b]	-2.24 ± 1.38 [ad]
50		0.17 ± 0.06 [a]	4.15 ± 0.46 [a]		1.52 ± 1.32 [bcd]	5.45 ± 4.33 [c]	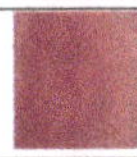	29.91 ± 2.61 [b]	-1.82 ± 1.07 [cd]
60		0.17 ± 0.05 [a]	4.13 ± 0.31 [a]		0.83 ± 1.07 [bcd]	5.02 ± 3.88 [c]		26.85 ± 7.00 [b]	-1.18 ± 0.82 [d]
120		0.09 ± 0.10 [a]	3.99 ± 0.36 [a]		4.21 ± 1.27 [d]	13.83 ± 4.10 [d]		25.98 ± 5.22 [b]	3.40 ± 2.02 [d]

[a–d] Different superscript letters in the same column indicate significant differences between samples ($p < 0.05$).

Regarding the BEE film, the colorimetric changes were not visually significant ($p > 0.05$), as shown in Table 2. The films showed a pink/violet color during 120 min when exposed to ammonia. Overall, the a^* value decreased and the b^* value increased during the exposure time, which allowed for an increase in the green and yellow color of the BEE film, despite not being easily visible. The ammonia diffusion on the film surface caused alkalinization of the environment, which led to the degradation of betacyanins, changing the red color into colorless cyclo-dopa-5-O-(malonyl)-β-gucoside and yellow betalamic acid, consequently inducing minor colorimetric changes in the film [23,43]. Naghdi et al. [44] developed a starch film containing paper flower betacyanin and reported the film's sensitivity to ammonia gas as the color changed from pink to yellow after 30 min of exposure. Qin et al. [23] observed that the pink color of the polyvinyl alcohol film/starch with red pitaya betalains faded after 10 min of exposure, and the yellowness degree gradually intensified after 20 min. Despite these results being obtained by other researchers, the BEE film developed did not significantly display color changes compared to the BLE film. This can be due to several factors: the amount of betalains incorporated in the film, the polymeric base used in the bio-based film development, and the time of exposure to ammonia or the ammonia solution concentration used.

The real-time, immediate, and sensitive reactions of colorimetric films are important properties of the color and pH-sensing intelligent packaging applications. As a result, the BLE film showed the most evident colorimetric differences in a short time of exposure to ammonia compared to the BEE film. Thus, BLE could be applied to protein-rich animal food to monitor its freshness.

3.2.5. Water Contact Angle (WCA), Moisture Content (MC), Water Solubility (WS), and Water Vapor Permeability (WVP)

The films' hydrophobicity was assessed through the measurement of the contact angle of water drop on the surface of the films. All WCAs measured were less than 90 °C, which indicated that the surface of all films had a hydrophilic character [37,45]. This result could be related to the presence of free hydrophilic groups on the surface of the LBG/Car film, which make the film more hydrophilic [37]. Moreover, no significant differences were observed between the control film (34.50 ± 5.51°) and the other films (i.e., BLE (41.43 ± 3.32°) and BEE (30.62 ± 3.62°) film samples ($p > 0.05$)). However, the BLE film presented a higher hydrophobicity (i.e., higher WCA value) than the BEE film ($p < 0.05$). Wu et al. [46] reported similar results and concluded that the addition of polyphenols (such as anthocyanins) into polysaccharide-based films led to the formation of hydrogen bonds and non-covalent hydrophobic interactions. Luchese et al. [47] also described that adding blueberry residue to starch films increased the WCA, and consequently, the film hydrophobicity.

As shown in Table 3, the MC values of the BLE and BEE films were not significantly different from the MC values of the control film ($p > 0.05$). These results were different from the ones reported by Zhai et al. [48]. The authors showed a significant decrease in the MC of the starch/polyvinyl alcohol–roselle anthocyanins films, which was attributed to the interactions between the film matrix and anthocyanins that decreased the availability of hydroxyl groups in the matrix to interact with the water molecules. The MC of the BEE films were lower than the MC of the BLE films ($p < 0.05$) (Table 3). This is possibly because less hydroxyl group interactions between the water molecules and BEE films occurred compared to BLE [5,49]. Similarly, Jamróz et al. [17] verified that the MC significantly decreased in furcellaran films by adding BEE.

Table 3. Moisture content (MC), water solubility (WS), water vapor permeability (WVP), tensile strength (TS), and elongation-at-break (EB) of the control, BLE, and BEE films.

Films	MC (%)	WS (%)	WVP × 10^{-10} (g m^{-1} s^{-1} Pa^{-1})	TS (MPa)	EB (%)
Control	21.05 ± 1.30 ^{ab}	94.41 ± 0.01 ^{b}	2.09 ± 0.29 ^{a}	11.70 ± 3.38 ^{a}	16.50 ± 1.48 ^{a}
BLE	23.54 ± 0.98 ^{a}	39.66 ± 0.11 ^{a}	2.30 ± 0.58 ^{a}	40.26 ± 5.79 ^{b}	28.39 ± 1.19 ^{b}
BEE	18.23 ± 1.47 ^{b}	92.90 ± 0.03 ^{b}	3.13 ± 0.46 ^{b}	22.83 ± 2.75 ^{c}	27.76 ± 0.90 ^{b}

$^{a–c}$ Different superscript letters in the same column indicate significant differences ($p < 0.05$).

The film solubilization starts with water access into the biopolymeric matrix, followed by a breakdown of hydrogen bonds and Van der Waals forces between the chains of biopolymers [50]. The WS results are shown in Table 3, and it is possible to observe that the control and BEE films lost their integrity after 8 h. This may be due to the high number of hydrophilic groups, which can form strong hydrogen bonds with water molecules, leaving the films very hydrated. On the other hand, the incorporation of BLE on the film led to a significant reduction in the WS values obtained ($p < 0.05$) compared to the other films (Table 3). This could possibly be explained by hydrophobic groups present in BLE, which decreased the film's ability to form bonds with water molecules. Thus, interaction between the phenolic compounds in BLE with the hydroxyl groups of the LBG and Car matrix led to a more compact inner-structure formation, with stronger bonds. Huang et al. [27] reported a similar WS behavior of an agar-based film incorporating *Arnebia euchroma* root extracts rich in anthocyanins, concluding that the WS was reduced significantly due to the addition of water-insoluble extracts.

The WVP is an important feature in food packaging, which assesses the water barrier properties of the film. Many films used in the food industry must have reduced low WVP values to delay food deterioration and to extend its shelf life and ensure its quality [44]. Furthermore, it has been described that the thickness, film matrix integrity, and interactions between the functional groups of film components can influence the films' WVP [23,44]. The results obtained for the control, BLE, and BEE films are shown in Table 3. The addition of BLE to the biopolymer-based matrix did not cause a significant effect on the WVP values of the films ($p > 0.05$). As previously discussed in the WCA, WS, and MC results section, the BLE can possibly form intermolecular interactions with LBG and Car, which limited the interactions between water vapor and film matrix [49]. Gao et al. [37] also reported that the addition of an anthocyanin-rich purple sweet potato extract had no significant effect on the WVP of the Car-based film. On the other hand, in our study, a significant increase in the WVP values were observed ($p < 0.05$) when the BEE was incorporated into the LBG/Car film matrix. It is possible that the increase in the WVP can be attributed to the BEE hydrophilicity (in agreement with WCA, WS, and MC results) and the increase in micro-paths in the network microstructure, which facilitated the diffusion of water molecules through the matrix. Zamudio-Flores et al. [51] determined that the addition of a beet extract significantly increased the WVP of oxidized starch films, which was also attributed to the hydrophilic property of the betalains present in the extract composition.

3.2.6. Mechanical Properties

The intelligent film must have enough strength to withstand the mechanical stress to suit its application and retain its integrity during transport and storage [21].

The effect of the incorporation of the extracts on the mechanical properties (TS and EB) of the biopolymeric matrix are shown in Table 3. The incorporation of the BLE and BEE extracts into film significantly increased the TS and EB values compared with the control film ($p < 0.05$). This result may be related to the reduction in intermolecular interactions between the LBG and Car, improving the chain mobility and the overall film elasticity. Additionally, the observed increase in TS and EB values may be associated with the interactions between the phenolic compounds and LBG/Car groups. Accordingly, the internal structure of the film matrix with the addition of natural extracts changed, becoming more compact and able to withstand higher mechanical forces [17,37]. Zhang et al. [52] showed a significant increase in the TS and EB values in starch/polyvinyl alcohol incorporating anthocyanin-rich purple potato. Qin et al. [23] found a similar trend when betalains from red pitaya peel were incorporated in a starch-polyvinyl alcohol matrix. The authors stated that the increase in the film mechanical strength film was due to a weakening of intermolecular forces between adjacent macromolecules, which facilitated the motion of polymer chains.

3.3. *Monitoring Hake Fish Freshness Using the Developed Colorimetric Films*

3.3.1. Selection of Colorimetric Film to Be Applied in Fish Freshness Study

The BLE and BEE films were tested as freshness indicators during hake storage at 4 °C. Overall, it was possible to observe gradual colorimetric changes of both films at 4 °C (Figure 5).

According to previous reports, a ΔE value higher than 5 can be detected by the human eye. However, only values above 12 represent an absolute difference in color, which is very noticeable, even by untrained panelists/consumers [27,30]. The BLE film showed more distinguishable color variations to the naked eye (i.e., $\Delta E > 12$) on day 2 of storage when compared to the BEE film (ΔE around 5.4). At day 6, a significant increase in ΔE values ($p < 0.05$) occurred in both films, with colorimetric changes easily detectable. Similarly, Zhai et al. [48] reported that films based on starch/polyvinyl alcohol incorporated with roselle anthocyanins changed their color from the initial pink to purple in 3–4 days and to green after 6 days of fish storage at 4 °C. Qin et al. [23] also described that the color of starch/polyvinyl alcohol films incorporating betalains from red pitaya changed from purple to yellow after fish storage at 4 °C for 8 days.

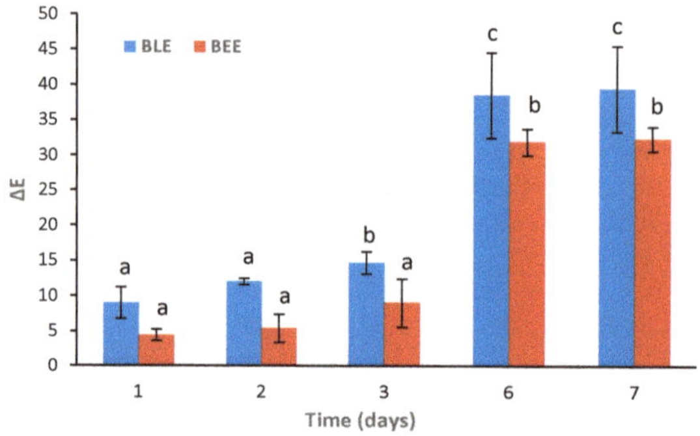

Figure 5. Color changes (ΔE) of the BLE and BEE films during fish storage during 7 days at 4 °C. Values are given as the mean ± SD ($n = 3$). [a–c] Different letters indicate significant differences between the same film sample over time ($p < 0.05$).

Both colorimetric films displayed continuous color changes within the fish shelf life, suggesting that they are capable of indicating the real-time fish freshness [48]. However, the BLE film color change was more easily detectable early on (day 1) than the BEE film (day 2). Moreover, it was possible to observe a significant perceptible color change ($p < 0.05$) of the BLE film on day 3. These results could be due to the anthocyanin-rich BLE's ability to detect more easily and react faster with volatile compounds than betalain-rich BEE. Thus, the BLE film would be more suitable for monitoring real-time fish freshness in the early days of storage than the BEE film. Hence, we decided to only apply the BLE film as an intelligent freshness indicator for the hake samples to evaluate other parameters (i.e., microbiological, pH, and TVB-N) during storage.

3.3.2. Relationship between Color Changes of Freshness Indicator Film and Microbiological and Physico-Chemical Characteristics of Fish during Storage

Microbiological spoilage is one of the main factors in which to monitor fish quality during cold storage since seafood is very perishable and susceptible to microbial deterioration. This is mainly due to its higher water activity (fundamental factor for the growth and survival of microorganisms), high fat content (favorable to oxidation), and neutral pH values (useful to the development and growth of most microorganisms) when compared to other food products [30,44].

Therefore, the microbiological quality of the hake samples was assessed as a function of storage time. During hake storage, the microbial load increased ($p < 0.05$), being equal to 8.61 ± 0.21 log CFU g^{-1} on the last day of storage (day 7) (Figure 6A). As the storage time increases, microorganisms generally grow faster and generate various metabolites responsible for off-odors, off-flavors, texture, and color changes, resulting in sensorial rejection by the consumer [44,52]. Fish is considered tobe acceptable for human consumption if the samples' microbial load does not exceed 7.0 log CFU g^{-1}, the maximum accepted value for marine species [53]. As can be observed in Figure 6A, this limit was exceeded between days 3 and 4 of storage at 4 °C, pointing out that the hake samples had lost their quality/freshness and were no longer suitable for consumption. Huang et al. [27] described a novel colorimetric indicator based on agar incorporated with anthocyanin extracts for monitoring fish freshness at 4 °C. They concluded that the fish samples were not suitable for consumption (7.11 log CFU g^{-1}) after 5–6 days.

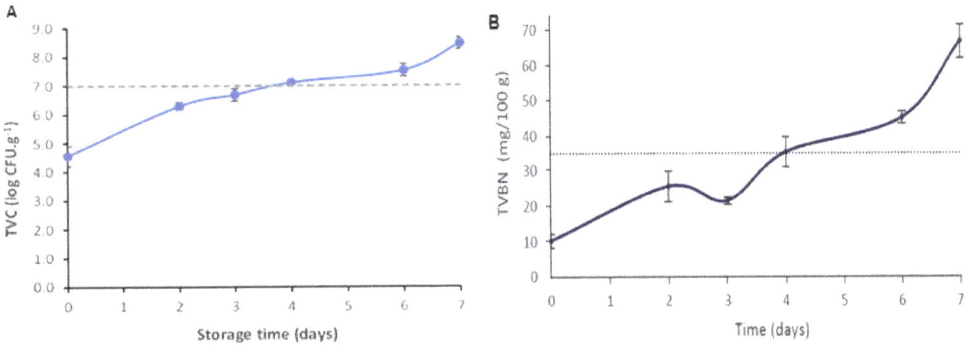

Figure 6. (**A**) Total viable count (TVC) of the hake samples stored for 7 days at 4 °C. The dashed horizontal line represents the maximum acceptable intake limit value (7 log CFU g^{-1}. Values are the average ± SD (n = 6). (**B**) TVB-N levels of the hake samples during storage at 4 °C. The acceptable TVB-N limit for fish consumption (35 mg TVB-N/100 g) is represented by the dashed horizontal line. Values are the average ± SD (n = 3).

Volatile amines such as ammonia, di-, and trimethyl amine, collectively recognized as TVB-N, are produced due to fish proteolysis during storage. For this reason, it is considered an important parameter indicative of muscle tissue degradation and the degree of fish freshness [30,44]. Usually, microbiological spoilage leads to an increase in TVB-N content because numerous enzymes produced by microorganism in the fish post-mortem phase will degrade the protein present in fish muscle tissue, and consequently, an unpleasant fish taste and odor will develop [5]. The average TVB-N values obtained throughout the storage of the hake fish samples at 4 °C are shown in Figure 6B. The TVB-N content increased significantly throughout the storage period ($p < 0.05$). On the last day of storage, the TVB-N values reached 66.78 ± 4.81 mg/100 g. This result agrees with the increase in the microbial growth on the fish samples (Figure 6A), which is related to the microbial activity and action of endogenous enzymes [23,27,30,44]. According to Regulation (EU) No. 2019/627, the TVB-N limit value for species belonging to the *Merlucciidae* family to be considered fit for human consumption is 35 mg/100 g [54]. Thus, hake samples exceeded the maximum TVB-N limit of acceptability after 4 days of storage (Figure 6B). Furthermore, it was also noticed that the lapsed time to reach the maximum TVB-N limit of acceptability (day 4) (Figure 6B) and TVC threshold (between day 3 and 4) (Figure 6A) was very similar. This phenomenon has previously been observed by other researchers who stated that the generation of TVB-N inherently follows the increase in microbial population [27].

One of the parameters associated with fish freshness and quality is the pH value. The initial pH value of the fresh hake was 6.60 ± 0.04, which gradually increased to 8.02 ± 0.03, on day 7 (Table 4). The pH increase may be correlated to compounds (such as ammonia and amines) generated from protein degradation in hake muscle, resulting from endogenous and microbial proteolytic enzyme activity [30,37,52]. The hake samples presented a higher pH value on day 4 compared to previous days ($p < 0.05$) (Table 4). The obtained pH result showed a good connection with the microbiological analyses and TVB-N results, since the threshold of the recommended TVC and TVB-N limits for fish freshness were also reached at day 4 of storage (Figure 6).

Table 4. pH values of hake and the color parameters (L^*, a^* and b^*) values of the BLE film during storage at 4 °C.

Storage Time (Days)	pH	L^*	a^*	b^*	Film Color	Freshness Stage
0	6.60 ± 0.04^{a}	62.86 ± 3.00^{a}	27.91 ± 1.98^{a}	0.58 ± 0.97^{a}		Fresh
2	6.87 ± 0.03^{b}	50.89 ± 4.04^{a}	20.56 ± 1.37^{a}	3.09 ± 0.59^{a}		Medium fresh
3	6.68 ± 0.06^{a}	46.47 ± 1.41^{a}	21.59 ± 4.31^{ab}	2.88 ± 0.55^{a}		Medium fresh
4	7.04 ± 0.03^{c}	55.41 ± 10.75^{a}	19.25 ± 6.03^{b}	2.69 ± 0.23^{a}		Spoiled
6	7.92 ± 0.04^{d}	57.40 ± 9.34^{a}	1.04 ± 0.44^{c}	2.86 ± 2.71^{a}		Spoiled
7	8.02 ± 0.03^{d}	57.14 ± 5.59^{a}	1.11 ± 0.16^{c}	3.12 ± 2.93^{a}		Spoiled

$^{a\text{–}d}$ Different superscript letters in the same column indicate significant differences ($p < 0.05$).

Regarding the BLE film colorimetric performance, the results showed remarkable color changes during the storage period (Table 4). The film color change from pink/purple (days 1 to 3) to blue/green (day 7) may be due to the pH sensitive characteristic of BLE films. No significant differences in the film brightness (L^*) values were observed ($p > 0.05$) during storage (Table 4). It was also possible to observe that the a^* value decreased significantly ($p < 0.05$) over 7 days, indicating a considerable loss of the film's red color. It is possible that an increase in TVB-N released from the hake samples led to an increase in pH values, resulting in film color changes. According to the Table 4 results, the color of the film changed from pink to light purple from 0 to 4 days of storage, with the a* values significantly decreasing ($p < 0.05$) during this time period. Meanwhile, the TVB-N levels of hake increased from 10 to 36 mg/100 g, suggesting that the hake was spoiled and not safe for consumption (Figure 6B). Thus, it was possible to observe a correlation between the changes in the indicator film color and changes in TVB-N in hake. The reduction in hake freshness due to the microbial growth in muscle tissue (i.e., TVC increased) led to the release of increasing amounts of TVB-N (such as ammonia, trimethylamine, and dimethylacetamide) [50]. It is possible that these basic compounds were detected by anthocyanins present in BLE, leading to changes to the anthocyanins' chemical structure (i.e., flavylium cation converted into quinoidal bases), and consequently decreasing the a^* values of the films. Regarding the b^* parameter, there was no significant differences ($p > 0.05$) during storage (Table 4). Similarly, Qin et al. [5] also described a color variation from purple to green/yellow of cassava starch with anthocyanins, in accordance with the change in pH and TVB-N values of pork during storage. Furthermore, a starch/polyvinyl alcohol film incorporated with roselle anthocyanins that changed color from pink/purple to green when in contact with spoiled fish has been reported [48]. The most noticeable color change for the human eye to recognize occurred on day 4, corresponding to the period when the fish was considered to be unsuitable for consumption (Table 4).

Although intelligent bio-based films have been developed in recent years for monitoring fish freshness, to the best of our knowledge, hake freshness monitoring films that are fully bio-based have not been developed yet. Moreover, studies related to the effects of light and temperature on the stability of film indicators, especially those developed with BLE, are limited. Hence, we report in this work an alternative and stable intelligent film for hake pH monitoring based on Car, LBG, and BLE, which has been successfully applied as a natural pH indicator.

The BLE-based film indicator can be integrated into packaging as a sticky label fixed to the headspace of a transparent package to monitor seafood freshness. For instance, a disc of BLE-based film with a printed ring presenting a reference color scale around it should be used for purposes of comparison.

4. Conclusions

Novel colorimetric LBG/Car-based films incorporating BLE or BEE as freshness indicators were successfully developed in the present work. The developed bio-based films demonstrated a high sensitivity to pH and TVB-N content changes with a visible transformation from a pink to red or yellow color under acidic or alkaline environments, respectively. Moreover, the color stability test showed that the colorimetric films were stable within 60 days at 4 °C. Based on the physicochemical properties of the developed films, the BLE film was chosen in order to evaluate its ability to monitor the hake quality during storage. The mechanical properties of the control films were enhanced when BLE was incorporated into the films as the EB and TS of the films increased 58% and 29%, respectively. Moreover, the results showed that films containing BLE became more hydrophobic (WCA = $41.43 \pm 3.32°$) than the control film (WCA = $34.50 \pm 5.51°$), which could be advantageous since water is an accelerating agent in the food deterioration process. The hake application trial suggests that the visible color transition point of the colorimetric BLE film is correlated with the loss of fish freshness due to the spoilage process over the storage time as verified by the TVC, TVB-N, and pH results. The BLE film color changed from pink to light purple after 4 days of storage, which could be observed with the naked eye, showing the shift in hake from fresh to spoiled.

In summary, this work reveals the great potential of LBG/Car-based films incorporating BLE as a dynamic colorimetric indicator suitable for sensing the actual condition of packaged and refrigerated hake, which might to some extent contribute to a reduction in food waste. Future research on the industrial use of the developed film freshness indicator including scale-up processing methods and validation tests under different storage conditions to avoid any false negative should be conducted. Furthermore, assessment of the film's ability to act as indicators of freshness when applied to other protein-based food products such as meat should be explored.

Author Contributions: Conceptualization, C.S.V.F. and J.T.M.; Methodology, C.S.V.F., J.M.V., and J.T.M.; Software, C.S.V.F.; Validation, C.S.V.F., J.M.V., A.A.V., and J.T.M.; Formal analysis, C.S.V.F., J.M.V., A.A.V., and J.T.M.; Investigation, C.S.V.F., J.M.V., and J.T.M.; Resources, A.A.V.; Data curation, J.T.M.; Writing—original draft preparation, C.S.V.F. and J.T.M.; Writing—review and editing, C.S.V.F., J.M.V., A.A.V., and J.T.M.; Visualization, C.S.V.F. and J.T.M.; Supervision, A.A.V. and J.T.M.; Project administration, J.T.M.; Funding acquisition, A.A.V. All authors have read and agreed to the published version of the manuscript.

Funding: This research was funded by the Portuguese Foundation for Science and Technology (FCT) under the scope of the strategic funding of the UIDB/04469/2020 unit with DOI 10.54499/UIDB/04469/2020 (accessed on 1 July 2024) and by LABBELS—Associate Laboratory in Biotechnology, Bioengineering and Microelectromechanical Systems, LA/P/0029/2020.

Institutional Review Board Statement: Not applicable.

Informed Consent Statement: Not applicable.

Data Availability Statement: The original contributions presented in the study are included in the article, further inquiries can be directed to the corresponding author.

Acknowledgments: Joana T. Martins acknowledges FCT for her Assistant Research contract obtained under the scope of Scientific Stimulus Employment with reference 2022.00788.CEECIND/CP1718/CT0024 (https://doi.org/10.54499/2022.00788.CEECIND/CP1718/CT0024, accessed on 1 July 2024).

Conflicts of Interest: The authors declare no conflicts of interest.

References

1. Zhai, X.; Li, Z.; Zhang, J.; Shi, J.; Zou, X.; Huang, X.; Zhang, D.; Sun, Y.; Yang, Z.; Holmes, M.; et al. Natural Biomaterial-Based Edible and pH-Sensitive Films Combined with Electrochemical Writing for Intelligent Food Packaging. *J. Agric. Food Chem.* **2018**, *66*, 12836–12846. [CrossRef] [PubMed]
2. Kalpana, S.; Priyadarshini, S.R.; Maria Leena, M.; Moses, J.A.; Anandharamakrishnan, C. Intelligent packaging: Trends and applications in food systems. *Trends Food Sci. Technol.* **2019**, *93*, 145–157. [CrossRef]
3. Biji, K.B.; Ravishankar, C.N.; Mohan, C.O.; Srinivasa Gopal, T.K. Smart packaging systems for food applications: A review. *J. Food Sci. Technol.* **2015**, *52*, 6125–6135. [CrossRef] [PubMed]
4. Muller, P.; Schmid, M. Intelligent Packaging in the Food Sector: A Brief Overview. *Foods* **2019**, *8*, 16. [CrossRef]
5. Qin, Y.; Liu, Y.; Yong, H.; Liu, J.; Zhang, X.; Liu, J. Preparation and characterization of active and intelligent packaging films based on cassava starch and anthocyanins from Lycium ruthenicum Murr. *Int. J. Biol. Macromol.* **2019**, *134*, 80–90. [CrossRef]
6. Poyatos-Racionero, E.; Ros-Lis, J.V.; Vivancos, J.-L.; Martínez-Máñez, R. Recent advances on intelligent packaging as tools to reduce food waste. *J. Clean. Prod.* **2018**, *172*, 3398–3409. [CrossRef]
7. Balbinot-Alfaro, E.; Craveiro, D.V.; Lima, K.O.; Costa, H.L.G.; Lopes, D.R.; Prentice, C. Intelligent Packaging with pH Indicator Potential. *Food Eng. Rev.* **2019**, *11*, 235–244. [CrossRef]
8. Luchese, C.L.; Abdalla, V.F.; Spada, J.C.; Tessaro, I.C. Evaluation of blueberry residue incorporated cassava starch film as pH indicator in different simulants and foodstuffs. *Food Hydrocoll.* **2018**, *82*, 209–218. [CrossRef]
9. Liu, J.; Wang, H.; Wang, P.; Guo, M.; Jiang, S.; Li, X.; Jiang, S. Films based on κ-carrageenan incorporated with curcumin for freshness monitoring. *Food Hydrocoll.* **2018**, *83*, 134–142. [CrossRef]
10. Prietto, L.; Mirapalhete, T.C.; Pinto, V.Z.; Hoffmann, J.F.; Vanier, N.L.; Lim, L.-T.; Guerra Dias, A.R.; da Rosa Zavareze, E. pH-sensitive films containing anthocyanins extracted from black bean seed coat and red cabbage. *LWT* **2017**, *80*, 492–500. [CrossRef]
11. Silva-Pereira, M.C.; Teixeira, J.A.; Pereira-Júnior, V.A.; Stefani, R. Chitosan/corn starch blend films with extract from Brassica oleraceae (red cabbage) as a visual indicator of fish deterioration. *LWT—Food Sci. Technol.* **2015**, *61*, 258–262. [CrossRef]
12. Choi, I.; Lee, J.Y.; Lacroix, M.; Han, J. Intelligent pH indicator film composed of agar/potato starch and anthocyanin extracts from purple sweet potato. *Food Chem.* **2017**, *218*, 122–128. [CrossRef]
13. Wu, Y.; Tang, P.; Quan, S.; Zhang, H.; Wang, K.; Liu, J. Preparation, characterization and application of smart packaging films based on locust bean gum/polyvinyl alcohol blend and betacyanins from cockscomb (*Celosia cristata* L.) flower. *Int. J. Biol. Macromol.* **2021**, *191*, 679–688. [CrossRef]
14. Shahbazi, M.; Rajabzadeh, G.; Ettelaie, R.; Rafe, A. Kinetic study of kappa-carrageenan degradation and its impact on mechanical and structural properties of chitosan/kappa-carrageenan film. *Carbohydr. Polym.* **2016**, *142*, 167–176. [CrossRef]
15. He, F.; Kong, Q.; Jin, Z.; Mou, H. Developing a unidirectionally permeable edible film based on k-carrageenan and gelatin for visually detecting the freshness of grass carp fillets. *Carbohydr. Polym.* **2020**, *241*, 116336. [CrossRef]
16. Souza, B.W.; Cerqueira, M.A.; Ruiz, H.A.; Martins, J.T.; Casariego, A.; Teixeira, J.A.; Vicente, A.A. Effect of chitosan-based coatings on the shelf life of salmon (*Salmo salar*). *J. Agric. Food Chem.* **2010**, *58*, 11456–11462. [CrossRef]
17. Jamróz, E.; Kulawik, P.; Guzik, P.; Duda, I. The verification of intelligent properties of furcellaran films with plant extracts on the stored fresh Atlantic mackerel during storage at 2 °C. *Food Hydrocoll.* **2019**, *97*, 105211. [CrossRef]
18. Almeida, C.; Karadzic, V.; Vaz, S. The seafood market in Portugal: Driving forces and consequences. *Mar. Policy* **2015**, *61*, 87–94. [CrossRef]
19. Mendes, R.; Silva, H.; Oliveira, P.; Oliveira, L.; Teixeira, B. Quality of Frozen Hake Fillets in the Portuguese Retail Market: A Case Study of Inadequate Practices in the European Frozen White Fish Market. *Foods* **2021**, *10*, 848. [CrossRef]
20. Martins, J.T.; Bourbon, A.I.; Pinheiro, A.C.; Souza, B.W.S.; Cerqueira, M.A.; Vicente, A.A. Biocomposite Films Based on κ-Carrageenan/Locust Bean Gum Blends and Clays: Physical and Antimicrobial Properties. *Food Bioprocess Technol.* **2012**, *6*, 2081–2092. [CrossRef]
21. Martins, J.T.; Cerqueira, M.A.; Bourbon, A.I.; Pinheiro, A.C.; Souza, B.W.S.; Vicente, A.A. Synergistic effects between κ-carrageenan and locust bean gum on physicochemical properties of edible films made thereof. *Food Hydrocoll.* **2012**, *29*, 280–289. [CrossRef]
22. Casariego, A.; Souza, B.W.S.; Cerqueira, M.A.; Teixeira, J.A.; Cruz, L.; Díaz, R.; Vicente, A.A. Chitosan/clay films' properties as affected by biopolymer and clay micro/nanoparticles' concentrations. *Food Hydrocoll.* **2009**, *23*, 1895–1902. [CrossRef]
23. Qin, Y.; Liu, Y.; Zhang, X.; Liu, J. Development of active and intelligent packaging by incorporating betalains from red pitaya (*Hylocereus polyrhizus*) peel into starch/polyvinyl alcohol films. *Food Hydrocoll.* **2020**, *100*, 105410. [CrossRef]
24. Yun, D.; Cai, H.; Liu, Y.; Xiao, L.; Song, J.; Liu, J. Development of active and intelligent films based on cassava starch and Chinese bayberry (*Myrica rubra* Sieb. et Zucc.) anthocyanins. *RSC Adv.* **2019**, *9*, 30905–30916. [CrossRef]
25. Michelin, M.; Marques, A.M.; Pastrana, L.M.; Teixeira, J.A.; Cerqueira, M.A. Carboxymethyl cellulose-based films: Effect of organosolv lignin incorporation on physicochemical and antioxidant properties. *J. Food Eng.* **2020**, *285*, 110107. [CrossRef]
26. Zhou, X.; Yu, X.; Xie, F.; Fan, Y.; Xu, X.; Qi, J.; Xiong, G.; Gao, X.; Zhang, F. pH-responsive double-layer indicator films based on konjac glucomannan/camellia oil and carrageenan/anthocyanin/curcumin for monitoring meat freshness. *Food Hydrocoll.* **2021**, *118*, 106695. [CrossRef]
27. Huang, S.; Xiong, Y.; Zou, Y.; Dong, Q.; Ding, F.; Liu, X.; Li, H. A novel colorimetric indicator based on agar incorporated with Arnebia euchroma root extracts for monitoring fish freshness. *Food Hydrocoll.* **2019**, *90*, 198–205. [CrossRef]

28. Merz, B.; Capello, C.; Leandro, G.C.; Moritz, D.E.; Monteiro, A.R.; Valencia, G.A. A novel colorimetric indicator film based on chitosan, polyvinyl alcohol and anthocyanins from jambolan (*Syzygium cumini*) fruit for monitoring shrimp freshness. *Int. J. Biol. Macromol.* **2020**, *153*, 625–632. [CrossRef]
29. Sucheta; Rai, S.K.; Chaturvedi, K.; Yadav, S.K. Evaluation of structural integrity and functionality of commercial pectin based edible films incorporated with corn flour, beetroot, orange peel, muesli and rice flour. *Food Hydrocoll.* **2019**, *91*, 127–135. [CrossRef]
30. Kanatt, S.R. Development of active/intelligent food packaging film containing Amaranthus leaf extract for shelf life extension of chicken/fish during chilled storage. *Food Packag. Shelf Life* **2020**, *24*, 100506. [CrossRef]
31. Kanatt, S.R.; Chawla, S.P. Shelf life extension of chicken packed in active film developed with mango peel extract. *J. Food Saf.* **2017**, *38*, e12385. [CrossRef]
32. Kanatt, S.R.; Jethwa, T.; Sawant, K.; Chawla, S.P. PVA-Gelatin Films Incorporated with Tomato Pulp: A Potential Primary Food Packaging Film. *Int. J. Curr. Microbiol. Appl. Sci.* **2017**, *6*, 1428–1441.
33. Jiang, G.; Hou, X.; Zeng, X.; Zhang, C.; Wu, H.; Shen, G.; Li, S.; Luo, Q.; Li, M.; Liu, X.; et al. Preparation and characterization of indicator films from carboxymethyl-cellulose/starch and purple sweet potato (*Ipomoea batatas* (L.) lam) anthocyanins for monitoring fish freshness. *Int. J. Biol. Macromol.* **2020**, *143*, 359–372. [CrossRef]
34. Roy, S.; Rhim, J.W. Anthocyanin food colorant and its application in pH-responsive color change indicator films. *Crit. Rev. Food Sci. Nutr.* **2021**, *61*, 2297–2325. [CrossRef]
35. Mokrzycki, W.S.; Tatol, M. Colour difference ΔE—A survey. *Mach. Graph. Vis. Int. J.* **2011**, *20*, 383–411.
36. Etxabide, A.; Kilmartin, P.A.; Maté, J.I. Color stability and pH-indicator ability of curcumin, anthocyanin and betanin containing colorants under different storage conditions for intelligent packaging development. *Food Control* **2021**, *121*, 107645. [CrossRef]
37. Gao, L.; Liu, P.; Liu, L.; Li, S.; Zhao, Y.; Xie, J.; Xu, H. κ-carrageenan-based pH-sensing films incorporated with anthocyanins or/and betacyanins extracted from purple sweet potatoes and peels of dragon fruits. *Process Biochem.* **2022**, *121*, 463–480. [CrossRef]
38. Krga, I.; Milenkovic, D. Anthocyanins: From Sources and Bioavailability to Cardiovascular-Health Benefits and Molecular Mechanisms of Action. *J. Agric. Food Chem.* **2019**, *67*, 1771–1783. [CrossRef]
39. Khan, P.; Farooqui, M. Analytical Applications of Plant Extract as Natural pH Indicator: A Review. *J. Adv. Sci. Res.* **2011**, *2*, 20–27.
40. Ma, Q.; Lu, X.; Wang, W.; Hubbe, M.A.; Liu, Y.; Mu, J.; Wang, J.; Sun, J.; Rojas, O.J. Recent developments in colorimetric and optical indicators stimulated by volatile base nitrogen to monitor seafood freshness. *Food Packag. Shelf Life* **2021**, *28*, 100634. [CrossRef]
41. Agunos, R.I.F.; Mendoza, D.V.M.; Rivera, M.A.S. Anthocyanin Colorimetric Strip for Volatile Amine Determination. *Int. J. Food Sci.* **2020**, *2020*, 1672851. [CrossRef]
42. Alizadeh-Sani, M.; Tavassoli, M.; Mohammadian, E.; Ehsani, A.; Khaniki, G.J.; Priyadarshi, R.; Rhim, J.W. pH-responsive color indicator films based on methylcellulose/chitosan nanofiber and barberry anthocyanins for real-time monitoring of meat freshness. *Int. J. Biol. Macromol.* **2021**, *166*, 741–750. [CrossRef]
43. Jiang, H.; Zhang, W.; Pu, Y.; Chen, L.; Cao, J.; Jiang, W. Development and characterization of a novel active and intelligent film based on pectin and betacyanins from peel waste of pitaya (*Hylocereus undatus*). *Food Chem.* **2023**, *404*, 134444. [CrossRef]
44. Naghdi, S.; Rezaei, M.; Abdollahi, M. A starch-based pH-sensing and ammonia detector film containing betacyanin of paperflower for application in intelligent packaging of fish. *Int. J. Biol. Macromol.* **2021**, *191*, 161–170. [CrossRef]
45. Rodríguez-Félix, F.; Corte-Tarazón, J.A.; Rochín-Wong, S.; Fernández-Quiroz, J.D.; Garzón-García, A.M.; Santos-Sauceda, I.; Plascencia-Martínez, D.F.; Chan-Chan, L.H.; Vásquez-López, C.; Barreras-Urbina, C.G.; et al. Physicochemical, structural, mechanical and antioxidant properties of zein films incorporated with no-ultrafiltered and ultrafiltered betalains extract from the beetroot (*Beta vulgaris*) bagasse with potential application as active food packaging. *J. Food Eng.* **2022**, *334*, 111153. [CrossRef]
46. Wu, L.T.; Tsai, I.L.; Ho, Y.C.; Hang, Y.H.; Lin, C.; Tsai, M.L.; Mi, F.L. Active and intelligent gellan gum-based packaging films for controlling anthocyanins release and monitoring food freshness. *Carbohydr. Polym.* **2021**, *254*, 117410. [CrossRef]
47. Luchese, C.L.; Sperotto, N.; Spada, J.C.; Tessaro, I.C. Effect of blueberry agro-industrial waste addition to corn starch-based films for the production of a pH-indicator film. *Int. J. Biol. Macromol.* **2017**, *104*, 11–18. [CrossRef]
48. Zhai, X.; Shi, J.; Zou, X.; Wang, S.; Jiang, C.; Zhang, J.; Huang, X.; Zhang, W.; Holmes, M. Novel colorimetric films based on starch/polyvinyl alcohol incorporated with roselle anthocyanins for fish freshness monitoring. *Food Hydrocoll.* **2017**, *69*, 308–317. [CrossRef]
49. Hu, H.; Yao, X.; Qin, Y.; Yong, H.; Liu, J. Development of multifunctional food packaging by incorporating betalains from vegetable amaranth (*Amaranthus tricolor* L.) into quaternary ammonium chitosan/fish gelatin blend films. *Int. J. Biol. Macromol.* **2020**, *159*, 675–684. [CrossRef]
50. Yoshida, C.M.P.; Maciel, V.B.V.; Mendonça, M.E.D.; Franco, T.T. Chitosan biobased and intelligent films: Monitoring pH variations. *LWT—Food Sci. Technol.* **2014**, *55*, 83–89. [CrossRef]
51. Zamudio-Flores, P.; Ochoa Reyes, E.; Ornelas-Paz, J.; Tirado-Gallegos, J.; Bello-Perez, L.; Rubio Ríos, A.; Cardenas-Felix, R. Physicochemical, mechanical, and structural features of oxidized oat and banana starch films enriched with betalains. *Agrociencia* **2015**, *49*, 483–498. Available online: http://www.scielo.org.mx/scielo.php?script=sci_arttext&pid=S1405-31952015000500002&lng=es&nrm=iso (accessed on 23 June 2024).
52. Zhang, K.; Huang, T.S.; Yan, H.; Hu, X.; Ren, T. Novel pH-sensitive films based on starch/polyvinyl alcohol and food anthocyanins as a visual indicator of shrimp deterioration. *Int. J. Biol. Macromol.* **2020**, *145*, 768–776. [CrossRef] [PubMed]

53. Horwitz, W. Microorganisms in Foods. 2. Sampling for Microbiological Analysis: Principles and Specific Applications. *J. AOAC Int.* **1975**, *58*, 1308. [CrossRef]
54. Commission, E. Regulation (EU) 2019/627 of the European Commission of 5 March 2019 Laying down Uniform Practical Arrangements for the Performance of Official Controls on Products of Animal Origin Intended for Human Consumption in Accordance with Regulation (EU) 2017/625 of the European Parliament and of the Council and Amending Commission Regulation (EC) No 2074/2005 as Regards Official Controls. 2019. Available online: https://eur-lex.europa.eu/legal-content/EN/TXT/PDF/?uri=CELEX:32019R0627 (accessed on 25 June 2024).

foods

MDPI

Article

Postharvest Quality of Arugula (*Eruca sativa*) Microgreens Determined by Microbiological, Physico-Chemical, and Sensory Parameters

Marina R. Komeroski [1,*], Thais Beninca [1], Keyla A. Portal [2], Patrícia S. Malheiros [1], Tâmmila V. Klug [3], Simone H. Flores [1] and Alessandro O. Rios [1]

1 Postgraduate Program in Food Science, Institute of Food Science and Technology, Federal University of Rio Grande do Sul, Porto Alegre 90010-150, Brazil; thais_beninca@hotmail.com (T.B.); patricia.malheiros@ufrgs.br (P.S.M.); simone.flores@ufrgs.br (S.H.F.); alessandro.rios@ufrgs.br (A.O.R.)
2 Department of Nutrition, School of Medicine, Federal University of Rio Grande do Sul, Porto Alegre 90010-150, Brazil; keyla200215@gmail.com
3 Postgraduate Program in Science and Food Technology, Department of Food Science, Farroupilha Federal Institute, Santa Maria 97050-685, Brazil; tammilaklug@gmail.com
* Correspondence: marina_rochak@hotmail.com; Tel.: +55-51-33089937

Abstract: (1) Background: Cultivating microgreens is emerging as an excellent market opportunity. Their easy, short, and sustainable production methods are the main reasons they are approved by growers. However, a feature that still prevents its further spread is the microbiological risk and their rapid senescence. The present study was conducted to evaluate the post-harvest storage and shelf life of arugula microgreens in different packaging through microbiological, physico-chemical, and sensory parameters; (2) Methods: Plants were stored at 5 °C in open air, vacuum sealed, and under modified atmosphere bags and tested at 0, 3, 5, 7, and 10 days; (3) Results: Microgreens stored in all packaging were safe for consumption within ten days. Regarding physical and chemical parameters, open packaging proved to be promising, with less weight loss and slower chlorophyll degradation. The sensory analysis demonstrated that the microgreens stored in the vacuum-sealed packaging showed a decrease in quality from the fifth day onwards for all attributes. However, the MAP presented good scores with a better visual quality, similar to the fresh microgreens.

Keywords: Brassicaceae microgreens; packaging; shelf-life assessment; microbiological analysis; physico-chemical parameters

1. Introduction

Arugula is one of the most consumed vegetables from the Brassicaceae family in the world [1,2] and the intensity of its aroma, pungency, and crunchiness seem to be decisive in consumer acceptance [3]. Microgreens of this vegetable are consumed fresh in salads but are also served as a garnish in other dishes, such as soups and sandwiches [4]. These plants contain significant amounts of important bioactive compounds and minerals [5], often being higher than adult plants of the same species [6], since they receive only light treatments [7] and are preceded by the germination stage [8]. This crop has a fast production cycle [9] and can be produced in greenhouses, in soil, or, more commonly, in soilless systems using solid organic or inorganic growing media or hydroponics [10], demonstrating the potential of these products to adapt the production of leafy vegetables to different scales [11]. If they are produced hydroponically, the soil is replaced by a substrate and seedlings are fed with a solution containing all the essential elements for their growth [12], allowing them to be grown organically [13]. Komeroski et al. [14] showed a high 45 protein, total fiber, and soluble fiber content of arugula microgreens grown using this system.

Choosing a culture medium with adequate microbiological characteristics is extremely important to ensure the safe consumption of microgreens, since the chosen medium may represent a contamination source [15]. For example, heat and humidity are the same ideal growth conditions for microgreens and pathogens such as *Salmonella, Listeria*, and *E. coli* O157:H7. These bacteria can infect seeds through small cracks and multiply to high levels during sprouting [16].

Before harvest, the greatest risks of contamination are related to irrigation water, substrate, and other factors [15]. In this regard, a delicate balance is required to maintain lower temperature (0–5 °C) and humidity conditions (50–85%) that optimizes the quality retention and microgreens shelf life while discouraging the growth of microbes and human pathogens [17]. However, data on potential microbiological hazards post-harvest are lacking. Currently, this type of product is packaged for subsequent sale without any disinfectant treatment, increasing consumer health risks.

According to Mir et al. [18], advances in packaging technology will help maintain the quality of microgreens for longer periods and extend their shelf life. In addition to quality parameters, functional information from these plants will help select the specific crop for a particular type of storage.

Modified atmosphere packaging (MAP) extends shelf life by ensuring persistent storage temperature and limited oxygen or moisture flows [19] by changing the gas composition, creating an appropriate atmosphere inside the packaging film, and effecting the integrity of the leaf tissue membrane [20]. Nowadays, it is one of the most effective technologies in maintaining the quality and extending the shelf life of fresh produce [21].

This work goal was to evaluate the post-harvest storage and shelf life of arugula microgreens in open, vacuum-sealed, and modified atmosphere packaging through microbiological, sensorial, and physico-chemical analysis.

2. Materials and Methods

2.1. Growing Conditions

Arugula (*Eruca sativa* L.) seeds were purchased from Isla Sementes® Ltda. (Rio Grande do Sul, Brazil). The cultivation was carried out at the Laboratory of Bioactive Compounds of the Institute of Food Science and Technology of the Federal University of Rio Grande do Sul (ICTA/UFRGS) according to the methodology proposed by Di Gioia et al. [10], with some modifications:

Seeds were placed in Green-up® phenolic foams with central holes and soaked in water for 48 h to promote germination according to the supplier's recommendation. The cultivation system used was floating hydroponic. In this case, these sown foams were placed on top of perforated polystyrene boards and irrigated using a nutrient solution prepared with potable water containing (mg L^{-1}) 6% nitrogen, 9% phosphorus, 29% potassium, 2.7% magnesium, 5% sulfur, 0.2% iron, 0.05% manganese, 0.02% zinc, 0.05% boron, 0.03% copper, 0.002% cobalt, 0.006% nickel, and 0.01% molybdenum. An air pump was used to oxygenate the nutrient solution. Seeded trays were kept at 25 °C in a controlled environment and were exposed to sunlight for a 12 h photoperiod.

Microgreens were harvested after 14 days of germination, cutting the plants manually with scissors a few millimeters above the phenolic foam.

The microgreens were packed (30 g) in polypropylene bags ((15 × 20 × 1.2 cm; OTR 50 $cm^3/m^2/24$ h/bar at 23 °C and 75% RH; residual pressure in the vacuum package was <100 Pa; 12 mm of thickness)-TecMaq®, São Paulo, Brazil) in different forms: open packaging (Open), the plastic bag remains with the upper side open; vacuum sealed packaging (Sealed), using the Fastvac® sealer (model TM 250, TecMaq®, São Paulo, Brazil), and modified atmosphere packaging (MAP), in the same equipment, with the following gas composition: 5% carbon dioxide, 2.1% oxygen, and 92.9% nitrogen, as recommended by Chitarra and Chitarra [22]. They were then stored at 5 °C in a climate chamber (model NL-41-ALT, New Lab®, São Paulo, Brazil) under a 12-h photoperiod (light/dark cycle) for ten days. Sampling was performed after 3, 5, 7, and 10 days of storage. Three packs were sampled at each time and analyzed as independent replicates for shelf-life evaluation.

2.2. Microbiological Analysis

2.2.1. *Salmonella* spp. and *Listeria monocytogenes* Analysis

The isolation of *Salmonella* spp. and *Listeria monocytogenes* in samples of microgreens stored in open, sealed, and modification atmosphere bags for 0, 5, and 10 days was carried out according to the ISO 6579-1:2017/Amd 1:2020 [23] and ISO 11290-1:2017 [24] methodology, respectively.

For *Salmonella* pre-enrichment, 10 g of the sample was added to 90 mL of buffered peptone water (BPW; HIMEDIA®) and incubated at 37 °C for 24 h. Subsequently, selective enrichment was carried out in Muller Kauffmann Tetrationate Broth (MKTTn; HIMEDIA®, Kelton, PA, USA) at 37 °C for 24 h and Rappaport–Vassilidis Soy Broth (RVS; HIMEDIA®) at 37 °C and 43 °C for 24 h. Then, seeding was carried out on Xylose Lysine Desoxycholate Agar (XLD; KASVI®, Paraná, Brazil) and Hektoen Enteric Agar (HE; HIMEDIA®) plates, both being incubated at 37 °C for 24 h [25]. Typical colonies were selected and subjected to identification through proteomic analysis using matrix-assisted laser desorption/ionization mass spectrometry and time of flight (MALDI-TOF/MS), model Autoflex Speed (Bruker Corporation, Bremen, Germany), following the method of Barbosa et al. [26].

For *Listeria* pre-enrichment, 10 g of the sample was added to 90 mL of Half-Fraser broth (HIMEDIA®) and incubated at 30 °C for 25 h. Subsequently, sowing was carried out on Listeria Ottaviani & Agosti (ALOA®) and OXFORD® agar plates, incubating at 37 °C for 24 h to 48 h. Furthermore, 100 µL of the contents of the pre-enrichment tube were transferred to 10 mL of Fraser broth and incubated at 37 °C for 24 h. Afterward, seeding was carried out on ALOA® and OXFORD® plates incubated at 37 °C for 24 to 48 h [25]. All experiments were performed in duplicate.

2.2.2. Total Enterobacteriaceae and *Escherichia coli* Count

Total Enterobacteriaceae and *Escherichia coli* in samples of microgreens stored in open air, sealed packaging, and under a modified atmosphere for 0 and 10 days was determined according to the Petrifilm^{TM} method [18]. Subsequently, the plates were incubated at 35 °C for 24 h to count total Enterobacteriaceae and 48 h to count *Escherichia coli* [25]. Experiments were performed in duplicate.

2.2.3. Mesophilic, Psychrophilic, and Molds and Yeasts Count

The total count of mesophilic, psychrophilic, and molds and yeasts in samples of microgreens stored in open air, sealed packages, and under a modified atmosphere for 0, 5, and 10 days was determined according to the plating methods APHA 08:2015, APHA 13.61:2015, and APHA 21:2015, respectively. The following media and incubation conditions were used: Plate Count Agar (PCA) for mesophilic (35 °C for 48 h) and psychrotrophilic (7 °C for 10 days) and Dichloran Rose Bengal Chloramphenicol (DRBC) agar plates for molds and yeasts (25 °C for 5 days). The results were expressed as log CFU/g [25]. Experiments were performed in duplicate.

2.3. Physicochemical Analysis

2.3.1. Color

Microgreen samples were measured using a colorimeter from Konica Minolta®-Osaka/Japan (model Chrona Meter CR400). This equipment identifies the color spectrum in a three-dimensional system (CIE standard illuminant D65 and standard observer 10), with the vertical axis, "L", referring to the color of the sample from black to white; axis "a", from green to red; and axis "b" from blue to yellow. The L-axis ranges from 0 to 100, with values over 50 being the lighter samples and below 50 the darker samples. All readings were performed in triplicate.

2.3.2. pH, Soluble Solid Content and Titratable Acidity

Microgreens were ground in a tissue homogenizer Ultra Turrax IKA® (model T25, IKA, São Paulo, Brazil) in a ratio of 1:5 (leaves:water) and then filtered, and its pH was measured at 25 °C using a Satra® pH meter (model pHS-3E, Pró-Análise, Porto Alegre, Brazil). The Brix° were measured using an Atago® refractometer (model PAL-3, Atago, São Paulo, Brazil) according to the AOAC methodology [27]. The titratable acidity (TA) was measured by titrating a 10 mL aliquot of microgreen juice extracts with 0.1 mol/L NaOH to the end point of pH 8.2. The results were expressed as the concentrations of (H^+) in the unit of mol/L based on the fresh weight mass of the sample [28].

2.3.3. Weight Loss

Weight loss was performed by scaling the microgreens before and after storage. Weight loss was computed according to the following formula [29]:

$$\text{Weight loss (\%)} = (m_0 - m_1)/m_0 \times 100\% \quad (1)$$

where:

m_0 = the weight (g) of microgreens before storage
m_1 = the weight (g) of microgreens after storage

2.3.4. Chlorophylls

The samples (0.4 g) were macerated with 10% (*w*/*v*) of 80% (*v*/*v*) acetone–water and then vortexed for 10 min. The homogenized tissue was centrifuged (5000× *g* for 15 min) and the supernatants were filtered. Samples were analyzed in UV–VIS spectrophotometer with the absorbance of the solution mixture at 647 nm for chlorophyll "a" and 663 nm for chlorophyll "b", and the concentration was determined using the CHLs equations, according to method of Lichtenthaler and Wellburn [30]. The results were expressed on a dry basis.

2.4. Sensory Evaluation

Fresh (control, harvested on each day of analysis) and stored microgreens (3, 5, and 7 days) were submitted to sensory analysis at the Laboratory of Sensory Evaluation of the Food Science and Technology Institute (ICTA) of UFRGS to evaluate the following attributes: appearance, texture, taste, odor, color, and overall acceptance.

40 Adult judges of both genders (the average age was ±34 years, ranging from 18–49; 72.5% of the judges (n = 29) were female while 27.5% (n = 11) were male) were randomly and voluntarily recruited to receive a slice of approximately 1 g of each formulation, with three-digit random codes, a glass of water to clean the taste buds, and an evaluation sheet for sensory analysis, which contained a hedonic scale of nine points (1 = Dislike extremely, 9 = Like extremely). The analysis was carried out over a period of 20 min and the judges signed the Term of Informed Consent in accordance with the rules of research involving human beings.

This procedure was approved by the ethics and research committee of the UFRGS. Number of the Certificate of Presentation of Ethical Review: 65303222.5.0000.5347.

To calculate the acceptability index (AI) of each treatment (>70% = well accept), the following expression was used, as described by Viana [31]:

$$\text{AI (\%)} = A \times 100/B\ (1) \quad (2)$$

where:

A = average grade obtained for each treatment
B = maximum grade given for each treatment

2.5. Statistical Analysis

Data regarding shelf life in different packaging were subjected to two-way analysis of variance ANOVA. The mean values were compared using Tukey's test and separated by a least significant difference test ($p < 0.05$). Statistical analysis was performed using the software Statistica 14.0.1.

3. Results and Discussion

Microbiological analyses were performed to evaluate the contamination of the growing media and microbial growth levels on microgreens stored in different packaging. *Salmonella* spp. and *Listeria monocytogenes* were not observed in arugula microgreens, regardless of the day of storage and packaging used (Table 1). The same was observed by Priti et al. [32], who worked with mung bean (*Vigna radiata* L.), lentil (*Lens culinaris* subsp. culinaris), and Indian mustard (*Brassica juncea* L.) microgreens and did not detect *Salmonella* or *Listeria* in any of the samples tested.

Table 1. Presence or absence of *Salmonella* spp. and *Listeria monocytogenes* in arugula microgreens stored in different packages for 0, 5, and 10 days.

Samples/Time	0 Days	5 Days	10 Days
	Salmonella spp.		
Open	Absent	Absent	Absent
Sealed	Absent	Absent	Absent
MAP	Absent	Absent	Absent
	Listeria monocytogenes		
Open	Absent	Absent	Absent
Sealed	Absent	Absent	Absent
MAP	Absent	Absent	Absent

Open: open packaging, the plastic bag remains with the upper side open; Sealed: vacuum sealed packaging; MAP: modified atmosphere packaging.

Generally, mesophilic and psychrotrophic bacteria counts and enumeration of total Enterobacteriaceae are useful for indicating the shelf-life duration and microbial quality of foods [33]. Regarding the Enterobacteriaceae count (Table 2), the presence of these microorganisms was verified. Despite this, and in accordance with resolution (RDC 724/2022) and normative instruction (IN 161/2022) of the Brazil National Health Surveillance Agency [34,35] for microbiological food standards for fresh and prepared vegetable products, with no specific standard for microgreens, it proves good hygiene conditions and correct handling of samples during harvest and storage, regardless of the packaging used. We also observed the absence of *E. coli* on all samples.

Table 2. Total Enterobacteriaceae and *Escherichia coli* (log CFU/g) in arugula microgreens stored in different packages for 0 and 10 days.

Samples (S)	Time (T)		*p* Value		
	0 Days	10 Days	S	T	S × T
	Total Enterobacteriaceae				
Open	7.61 ± 0.60^{aA}	6.95 ± 0.44^{bA}			
Sealed	7.61 ± 0.60^{aA}	7.22 ± 0.54^{bA}	0.862	0.039	0.862
MAP	7.61 ± 0.60^{aA}	6.91 ± 0.38^{bA}			
	Escherichia coli				
Open	Absent	Absent	-	-	-
Sealed	Absent	Absent	-	-	-
MAP	Absent	Absent	-	-	-

Open: open packaging, the plastic bag remains with the upper side open; Sealed: vacuum sealed packaging; MAP: modified atmosphere packaging. Different lowercase superscript in the same line indicates statistically significant difference for the time in the same package type by Tukey's test ($p < 0.05$). Different uppercase superscript in the same column indicates statistically significant difference for the package in the same time by Tukey's test ($p < 0.05$).

Jablasone, Warriner, and Griffiths [36] found *E. coli* O157:H7 in the internal tissues of watercress, lettuce, radish, and spinach seedlings, but not in the mature plants' tissues. The pathogen preferentially colonized the epidermal root junctions since, during seed germination, the seed releases a mixture of carbohydrates and peptides that can attract neighboring bacteria in the rhizosphere.

Chandra et al. [37] suggested that bacterial populations can easily grow on microgreens' delicate and immature tissue structure and may be stimulated by sugars and other organic molecules derived from the endosperm breakdown during germination. Postharvest, respiration becomes the primary physiological process of the plant, which uses its own previously accumulated metabolic reserves. However, depending on the intensity of biochemical reactions, tissues can reach senescence faster, becoming more susceptible to moisture loss and the development of microorganisms [22].

Table 3 shows the microbial population of arugula microgreens after 0, 5, and 10 days of storage. The total aerobic mesophilic bacteria during the initial phase of storage was 7.3 ± 0.22 log CFU/g, with a statistical difference ($p < 0.05$) for subsequent storage days. On day 10 of storage, we can observe a statistically significant difference ($p < 0.05$) in relation to the packaging used. This fact can be explained by fermentative activity, which is a characteristic of biological oxidations in an oxygen-free environment, such as vacuum-sealed packaging. In this environment, pyruvic acid is converted into carbon dioxide and acetaldehyde [15].

Table 3. Total count (log CFU/g) of mesophilic, psychrotrophilic, and molds and yeasts in arugula microgreens stored in different packages for 0, 5, and 10 days.

Samples (S)	Time (T)			p Value		
	0 Days	5 Days	10 Days	S	T	S × T
		Mesophilics				
Open	7.30 ± 0.22 ^{aA}	7.40 ± 0.40 ^{aA}	7.90 ± 0.26 ^{aB}			
Sealed	7.30 ± 0.22 ^{bA}	7.90 ± 0.23 ^{bA}	8.80 ± 0.42 ^{aA}	0.008	<0.000	0.013
MAP	7.30 ± 0.22 ^{aA}	7.90 ± 0.37 ^{aA}	7.70 ± 0.15 ^{aB}			
		Psychrotrophs				
Open	5.70 ± 0.10 ^{bA}	7.30 ± 0.16 ^{aB}	7.90 ± 0.58 ^{aA}			
Sealed	5.70 ± 0.10 ^{cA}	7.60 ± 0.05 ^{bAB}	8.60 ± 0.48 ^{aA}	0.087	<0.000	0.003
MAP	5.70 ± 0.10 ^{bA}	8.30 ± 0.50 ^{aA}	7.70 ± 0.15 ^{aA}			
		Yeasts and Molds				
Open	5.70 ± 0.06 ^{aA}	6.50 ± 0.28 ^{aB}	7.00 ± 1.09 ^{aB}			
Sealed	5.70 ± 0.06 ^{cA}	7.60 ± 0.09 ^{bAB}	8.90 ± 0.53 ^{aA}	0.002	<0.000	0.001
MAP	5.70 ± 0.06 ^{cA}	8.30 ± 0.83 ^{aA}	7.00 ± 0.01 ^{bB}			

Open: open packaging, the plastic bag remains with the upper side open; Sealed: vacuum sealed packaging; MAP: modified atmosphere packaging. Different lowercase superscript in the same line indicates statistically significant difference for the time in the same package type by Tukey's test ($p < 0.05$). Different uppercase superscript in the same column indicates statistically significant difference for the package in the same time by Tukey's test ($p < 0.05$).

Similar values of total aerobic mesophilic bacteria were found in the work of Paradiso et al. [5] with chicory microgreens. During the initial storage phase, the number was 6.69 ± 0.07 log CFU per g and 8.19 ± 0.07 log CFU/g after 10 days of storage. In their work, psychrotrophic microorganism counts were very similar to those of mesophilic microorganisms, which matches this study at 5 and 10 days of storage.

Authors such as Chandra et al. [37] and Xiao et al. [38] identified the presence of mesophilic aerobic bacteria and molds and yeasts in these vegetables, with up to 10^7 CFU/g^{-1} and 10^5 CFU/g^{-1}, respectively. These levels can be considered potentially dangerous, both for the food's safety and its sensory quality and preservation capacity. It is worth mentioning that the International Commission on Microbiological Specifications for Foods (ICMSF) [33] states that foods with aerobic microorganism counts above 10^6 CFU/g typically show noticeable signs of spoilage, such as off-smells, off-tastes, and changes in appearance.

The growth of yeasts and molds on buckwheat microgreens was relatively slow during the initial 8 days of storage (about 5.5 log CFU/g) and increased obviously from 8 to 12 days (up to 8.2 log CFU/g) storage in the study of Yan et al. [29]. Similar behavior was observed in the microgreens in this study, except for those in the open packaging, which did not differ statistically ($p < 0.05$).

According to Kyriacou et al. [39], the optimal conditions for microgreen growth are a pH between 6.56 and 7.54. Unfortunately, this is the same range for the development of neutrophil bacteria. Data in the study of Huang, Luo, and Nou [40] showed that the proliferation of *Salmonella* and *L. monocytogenes* was more significantly impacted by long-term suboptimal refrigeration or frequent temperature fluctuation than short-term terminal exposure to higher temperatures in fresh-cut cantaloupe.

The environmental conditions that influence the development of plants in a hydroponic system are light, temperature, and humidity [41]. The FDA [42] requires that all foods in the "Time and Temperature Control for Safety" be maintained at a temperature not exceeding 5 °C. Fresh-cut leafy green vegetables all belong to this category.

However, controlling climatic factors seems to be challenging for microgreens growers, who occasionally treated seeds with hydrogen peroxide before planting to mitigate the potential proliferation of mold and pathogens [43]. However, as reported by these authors, there is little evidence regarding the effectiveness of H_2O_2 for controlling microorganisms, including pathogens, on nonfood-contact and food-contact surfaces.

In Table 4 we show the physico-chemical evaluation of microgreens up to the tenth day of storage. In this work, we observed a drop in pH within seven days of storage, with a statistical difference ($p < 0.05$) between open and modified atmosphere packaging. This can be explained by the degradation of nitrogenous compounds present in the leaves of microgreens, releasing ammonia. Ammonia, when combined with water, resulting from cellular respiration, forms ammonium hydroxide, a weak base. However, the production of acids during cellular respiration often exceeds the production of bases, resulting in tissue acidification [22].

Table 4. Physicochemical analysis of arugula microgreens stored in different packages for 0, 3, 5, and 10 days.

Samples (S)	Time (T)					*p* Value		
	0 Days	3 Days	5 Days	7 Days	10 Days	S	T	S × T
			pH					
Open	5.34 ± 0.10 ^{bA}	6.39 ± 0.04 ^{aA}	6.05 ± 0.53 ^{aA}	4.54 ± 0.16 ^{cA}	5.93 ± 0.10 ^{aA}			
Sealed	5.34 ± 0.10 ^{bA}	6.35 ± 0.11 ^{aA}	6.24 ± 0.82 ^{aA}	4.85 ± 0.11 ^{cA}	6.29 ± 0.06 ^{aA}	0.432	<0.000	0.908
MAP	5.34 ± 0.10 ^{bA}	6.20 ± 0.07 ^{aA}	6.12 ± 0.87 ^{aA}	4.95 ± 0.12 ^{cA}	6.26 ± 0.23 ^{aA}			
			Soluble solids					
Open	9.43 ± 0.05 ^{bA}	9.26 ± 0.05 ^{cB}	9.70 ± 0.00 ^{aA}	9.66 ± 0.05 ^{aA}	9.50 ± 0.00 ^{bA}			
Sealed	9.43 ± 0.05 ^{bA}	9.43 ± 0.05 ^{bA}	9.70 ± 0.00 ^{aA}	9.66 ± 0.05 ^{aA}	9.46 ± 0.05 ^{bA}	0.114	<0.000	0.017
MAP	9.43 ± 0.05 ^{bA}	9.43 ± 0.05 ^{bA}	9.66 ± 0.05 ^{aA}	9.66 ± 0.05 ^{aA}	9.56 ± 0.05 ^{abA}			
			Titratable acidity					
Open	2.46 ± 0.92 ^{aA}	0.90 ± 0.17 ^{bcA}	1.43 ± 0.66 ^{bA}	0.63 ± 0.15 ^{cA}	0.53 ± 0.32 ^{cA}			
Sealed	2.46 ± 0.92 ^{aA}	0.86 ± 0.30 ^{bcA}	1.46 ± 0.89 ^{bA}	0.66 ± 0.05 ^{cA}	0.40 ± 0.10 ^{cA}	0.988	<0.000	0.999
MAP	2.46 ± 0.92 ^{aA}	1.03 ± 0.25 ^{bcA}	1.50 ± 0.70 ^{bA}	0.76 ± 0.15 ^{cA}	0.23 ± 0.05 ^{cA}			
			Weight loss					
Open	-	8.93 ± 2.02 ^{bA}	9.05 ± 2.09 ^{bC}	15.03 ± 1.36 ^{aB}	16.85 ± 3.63 ^{aB}			
Sealed	-	7.15 ± 2.43 ^{cA}	13.10 ± 1.00 ^{bB}	15.02 ± 2.38 ^{abB}	17.79 ± 1.35 ^{aB}	<0.000	<0.000	<0.000
MAP	-	9.86 ± 0.34 ^{dA}	16.44 ± 1.76 ^{cA}	20.32 ± 1.90 ^{bA}	24.14 ± 1.27 ^{aA}			

Open: open packaging, the plastic bag remains with the upper side open; Sealed: vacuum sealed packaging; MAP: modified atmosphere packaging. Different lowercase superscript in the same line indicates statistically significant difference for the time in the same package type by Tukey's test ($p < 0.05$). Different uppercase superscript in the same column indicates statistically significant difference for the package in the same time by Tukey's test ($p < 0.05$).

The increasing acidity at 5 days, which happened in this study, may be attributed to the biochemical conversion of fatty acids to acids over time [44]. After that, it is expected to decrease over time due to the plant's physiological processes.

In this study, the effect of time on the increase in soluble solids (SS) occurred on days 5 and 7, whereas the interaction with the MAP did not differ significantly in 10 days of storage. This can be attributed to differential utilization of metabolites in by respiration getting influenced by the permeabilities of the packaging material to gasses in this package [44].

Leafy vegetables are highly susceptible to water loss after harvest, and respiration and other senescence-related metabolic processes are the primary cause of postharvest loss [45]. In this work, the weight loss increased with increasing storage time, with no statistical difference ($p < 0.05$) for 7 and 10 days, but with a difference for MAP. It was observed that although the weight loss was greater in the MAP, the microgreens presented a better visual quality compared to those in the open packaging, since contact with the environment caused dehydration in the microgreens closest to the opening.

In the work of Patil et al. [21], the post-harvest treatment of ascorbic acid + citric acid along with MAP of broccoli microgreens significantly ($p < 0.05$) suppressed the weight loss and helped to retain better firmness. According to Khan and Mittal [46], the efficiency of MAP depends upon multiple factors like appropriate gas composition, oxygen transmission rate (OTR), freshness, degree of processing of the product, product surface area, metabolism, respiration rate, microbial quality of produce, storage temperature, and relative humidity.

As Kou et al. [19] and Xiao et al. [38] explained, a favorable O_2/CO_2 balance and absence of anaerobic conditions that cause physiological damage to leaf tissue is required. When controlling these factors, the effect of storage temperature on the shelf life of microgreens appears to be more critical than the gas permeability of the packaging. At a storage temperature of 5 °C, the microgreens remained good for consumption for up to ten days of analysis.

Chlorophyll content is important for the health benefits it offers and has an effect on the appearance of the microgreens. Various shades of greenness in microgreens add to their aesthetic appeal [47]. In this work, the microgreens' total chlorophyll content (Table 5) ranged from 31.03 ± 1.38 mg/100 g to 15.04 ± 2.43 mg/100 g at 10 days in MAP. In general, the decline in chlorophylls can be observed from the third day onwards, with a statistical difference ($p < 0.05$). Chandra et al. [37] observed that polypropylene films, due to the more significant accumulation of CO_2, caused faster and more irreversible damage to the membrane compared to polyethylene, generating unpleasant odors. Furthermore, electrolyte leakage may contribute to faster yellowing, tissue senescence, and chlorophyll degradation of vegetables [38].

Similar values were found for total chlorophyll in the work of Kowitcharoen et al. [48], who analyzed some Brassicaceae microgreens such as Rat-tailed radish and red cabbage (36.61 and 39.79 mg/100 g, respectively). In the study of Ghoora, Hanldipur, and Srividya [49], the total chlorophyll of radish microgreens was 50.9 mg/100 g, but with a similar proportion of chlorophyll a and b (chlorophyll a approximately 2× higher) in relation to this work.

Color parameters are presented in Table 5. The L coordinate indicates lightness, which decreased significantly over the storage period after 3 days. The a* coordinate, denoting greenness, becomes less negative, signifying a decrease in ing samples. The b* value, representing yellowing, also showed a regressive incline over greenness over the storage period. However, no significant difference ($p < 0.05$) was observed in open packaging over the same storage period for the a* coordinate.

Katsenios et al. [50] reported that a bright green color corresponds to the high-quality index for microgreens of green vegetable species, and yellowing suggests the product's quality deterioration. Green basil microgreens showed similar behavior in the work of Ciriello et al. [51]: luminosity (L; 48.06), greenness (a*; −17.30), and yellowness (b*; 28.92). El-Nakhel et al. [52] and Petropoulos et al. [53] reported that nutrient availability may affect the color of microgreens' leaves and, thus, improve the visual quality of the final product.

Table 5. Chlorophyll and color analysis of arugula microgreens stored in different packages for 0, 3, 5, 7, and 10 days.

Samples (S)	Time (T)					p Value		
	0 Days	3 Days	5 Days	7 Days	10 Days	S	T	S × T
			Chlorophyll a					
Open	19.67 ± 0.64 ^{aA}	18.69 ± 0.84 ^{aA}	17.47 ± 1.22 ^{aA}	14.80 ± 1.26 ^{bA}	12.38 ± 0.73 ^{bA}			
Sealed	19.67 ± 0.64 ^{aA}	13.96 ± 0.30 ^{cB}	16.04 ± 1.22 ^{bA}	14.08 ± 0.03 ^{cA}	11.42 ± 0.68 ^{dA}	<0.000	<0.000	<0.000
MAP	19.67 ± 0.64 ^{aA}	17.62 ± 0.39 ^{bA}	17.16 ± 0.24 ^{bA}	10.88 ± 0.22 ^{cB}	8.70 ± 0.18 ^{dB}			
			Chlorophyll b					
Open	11.35 ± 2.02 ^{aA}	9.37 ± 1.60 ^{bA}	8.59 ± 1.32 ^{bA}	11.31 ± 2.24 ^{bA}	10.00 ± 0.83 ^{bA}			
Sealed	11.35 ± 2.02 ^{aA}	6.77 ± 1.07 ^{bA}	6.33 ± 1.37 ^{bA}	5.91 ± 0.07 ^{bA}	6.88 ± 1.94 ^{bA}	<0.000	<0.000	0.720
MAP	11.35 ± 2.02 ^{aA}	8.71 ± 0.42 ^{bA}	6.17 ± 1.51 ^{bA}	5.51 ± 0.37 ^{bA}	6.34 ± 2.32 ^{bA}			
			Total chlorophyll					
Open	31.03 ± 1.38 ^{aA}	28.07 ± 0.93 ^{abA}	26.07 ± 0.15 ^{bcA}	26.11 ± 3.50 ^{bcA}	22.39 ± 0.52 ^{cA}			
Sealed	31.03 ± 1.38 ^{aA}	20.73 ± 1.21 ^{bcB}	22.38 ± 0.91 ^{bB}	19.99 ± 0.10 ^{bcB}	18.31 ± 1.26 ^{cB}	<0.000	<0.000	<0.000
MAP	31.03 ± 1.38 ^{aA}	26.33 ± 0.32 ^{bA}	25.79 ± 1.72 ^{bA}	16.39 ± 0.37 ^{cB}	15.04 ± 2.43 ^{cB}			
			Color L					
Open	53.41 ± 6.56 ^{aA}	51.64 ± 1.17 ^{bA}	44.30 ± 1.96 ^{bcA}	44.62 ± 0.03 ^{cA}	39.08 ± 0.02 ^{cA}			
Sealed	53.41 ± 6.56 ^{aA}	44.58 ± 5.67 ^{bA}	41.50 ± 1.14 ^{bcA}	41.90 ± 0.03 ^{cA}	41.39 ± 0.03 ^{cA}	0.222	<0.000	0.377
MAP	53.41 ± 6.56 ^{aA}	45.35 ± 3.17 ^{bA}	44.94 ± 0.09 ^{bcA}	40.53 ± 0.01 ^{cA}	39.83 ± 0.29 ^{cA}			
			Color a*					
Open	−14.82 ± 1.50 ^{aA}	−11.23 ± 3.82 ^{aA}	−12.39 ± 0.01 ^{aA}	−11.12 ± 0.03 ^{aB}	−9.93 ± 0.02 ^{aC}	0.596	<0.000	0.029
Sealed	−14.82 ± 1.50 ^{bA}	−11.82 ± 2.99 ^{bA}	−13.86 ± 1.69 ^{bA}	−11.08 ± 0.01 ^{bB}	−6.43 ± 0.01 ^{aA}			
MAP	−14.82 ± 1.50 ^{cA}	−13.01 ± 2.10 ^{cA}	−11.79 ± 0.05 ^{bcA}	−7.75 ± 0.01 ^{aA}	−9.15 ± 0.03 ^{abB}			
			Color b*					
Open	25.15 ± 1.20 ^{aA}	22.35 ± 3.84 ^{abA}	26.81 ± 0.45 ^{aA}	18.66 ± 0.25 ^{bcA}	15.52 ± 0.02 ^{cA}	0.002	<0.000	<0.000
Sealed	25.15 ± 1.20 ^{aA}	20.60 ± 2.84 ^{abA}	22.49 ± 3.25 ^{abA}	17.88 ± 0.03 ^{bcB}	14.57 ± 0.02 ^{cB}			
MAP	25.15 ± 1.20 ^{aA}	24.18 ± 1.54 ^{aA}	17.19 ± 0.05 ^{bB}	15.56 ± 0.02 ^{bcC}	14.71 ± 0.21 ^{cB}			

Open: open packaging, the plastic bag remains with the upper side open; Sealed: vacuum sealed packaging; MAP: modified atmosphere packaging. Different lowercase superscript in the same line indicates statistically significant difference for the time in the same package type by Tukey's test ($p < 0.05$). Different uppercase superscript in the same column indicates statistically significant difference for the package in the same time by Tukey's test ($p < 0.05$).

According to sensory analysis results (Table 6), the sealed package showed a significant difference ($p < 0.05$) at 5 and 7 days in all attributes when compared to the third day. This result can be proven by the visual quality of the microgreens in this packaging, as shown in Supplementary Figure S1 (FS1). In contrast, we can observe a similarity between the open and MAP treatments (Supplementary Figure S2 (FS2)). In the study of Dhaka et al. [54], mustard microgreen was the first to undergo deterioration, and its visual quality showed some changes after four days.

In the descriptive analysis conducted in the work of Bafumo et al. [55], the evaluators identified rocket and watercress microgreens as the most astringent, standing out for pronounced bitterness and sourness, distinguishing them from the rest. Assessed microgreens evaluated by Caracciolo et al. [56] had consistently greater appearance scores than those concerning texture and flavor. Their statistical analysis indicated that the observed differences in microgreens acceptability depended on two main sensory dimensions experienced through the consumer test: astringency/sourness and bitterness.

In relation to the acceptability index (Figure 1), as expected, the fresh sample obtained the best result (91.38%) for color, followed by appearance (91.11%) and texture (86.94%). It is noteworthy that regardless of the sealed packaging, at 5 and 7 days, all acceptability indexes of all treatments reached values above 60%, which is considered satisfactory.

Table 6. Sensory evaluation of arugula microgreens stored in different packages for 3, 5, and 7 days.

Samples (S)	Time (T)			*p* Value		
	3 Days	5 Days	7 Days	S	T	S × T
		Appearance				
Fresh	7.55 ± 1.63 ^{abA}	7.43 ± 1.50 ^{bA}	8.20 ± 0.96 ^{aA}	<0.00	<0.00	<0.00
Open	7.66 ± 1.54 ^{aA}	7.46 ± 1.37 ^{aA}	7.37 ± 1.39 ^{aA}			
Sealed	7.63 ± 1.57 ^{aA}	3.07 ± 1.72 ^{bB}	3.27 ± 1.89 ^{bB}			
MAP	7.60 ± 1.28 ^{aA}	7.35 ± 1.42 ^{aA}	7.42 ± 1.52 ^{aA}			
		Odor				
Fresh	6.63 ± 1.63 ^{aAB}	5.89 ± 1.55 ^{aA}	6.57 ± 1.66 ^{aA}	<0.000	<0.000	0.009
Open	6.50 ± 1.90 ^{aAB}	5.84 ± 1.49 ^{aA}	6.00 ± 1.63 ^{aA}			
Sealed	6.33 ± 1.86 ^{aB}	4.33 ± 1.82 ^{bB}	4.57 ± 2.12 ^{bB}			
MAP	7.25 ± 1.31 ^{aA}	5.74 ± 1.75 ^{bA}	6.40 ± 1.56 ^{bA}			
		Color				
Fresh	7.80 ± 1.41 ^{abA}	7.38 ± 1.42 ^{bA}	8.22 ± 0.97 ^{aA}	<0.000	<0.000	<0.000
Open	7.88 ± 1.48 ^{aA}	7.43 ± 1.41 ^{aA}	7.62 ± 1.35 ^{aA}			
Sealed	7.80 ± 1.45 ^{aA}	4.07 ± 1.97 ^{bB}	4.50 ± 2.18 ^{bB}			
MAP	7.62 ± 1.06 ^{aA}	7.41 ± 1.39 ^{aA}	7.32 ± 1.54 ^{aA}			
		Texture				
Fresh	7.50 ± 1.39 ^{aA}	7.74 ± 1.31 ^{aA}	7.82 ± 1.15 ^{aA}	<0.000	<0.000	<0.000
Open	7.45 ± 1.56 ^{aA}	7.61 ± 1.33 ^{aA}	7.62 ± 1.35 ^{aA}			
Sealed	7.26 ± 1.84 ^{aA}	4.33 ± 2.46 ^{bB}	3.80 ± 2.15 ^{bB}			
MAP	7.58 ± 1.27 ^{aA}	7.41 ± 1.44 ^{aA}	7.60 ± 1.48 ^{aA}			
		Taste				
Fresh	6.32 ± 2.35 ^{aB}	6.64 ± 2.06 ^{aA}	7.07 ± 1.99 ^{aA}	<0.000	0.156	<0.000
Open	6.03 ± 2.29 ^{aB}	6.64 ± 1.85 ^{aA}	6.82 ± 2.06 ^{aA}			
Sealed	6.25 ± 2.48 ^{aB}	4.74 ± 2.33 ^{bB}	3.77 ± 2.42 ^{bB}			
MAP	7.37 ± 1.46 ^{aA}	6.66 ± 1.72 ^{aA}	6.80 ± 1.84 ^{aA}			
		Global acceptance				
Fresh	6.76 ± 1.88 ^{bAB}	6.79 ± 1.55 ^{bA}	7.67 ± 1.26 ^{aA}	<0.000	0.002	<0.000
Open	6.55 ± 2.00 ^{aB}	6.84 ± 1.64 ^{aA}	7.20 ± 1.34 ^{aA}			
Sealed	6.56 ± 1.98 ^{aB}	4.53 ± 2.24 ^{bB}	3.80 ± 2.17 ^{bB}			
MAP	7.60 ± 1.21 ^{aA}	6.82 ± 1.47 ^{bA}	7.25 ± 1.37 ^{abA}			

Open: open packaging, the plastic bag remains with the upper side open; Sealed: vacuum sealed packaging; MAP: modified atmosphere packaging. Different lowercase superscript in the same line indicates statistically significant difference for the time in the same package type by Tukey's test ($p < 0.05$). Different uppercase superscript in the same column indicates statistically significant difference for the package in the same time by Tukey's test ($p < 0.05$).

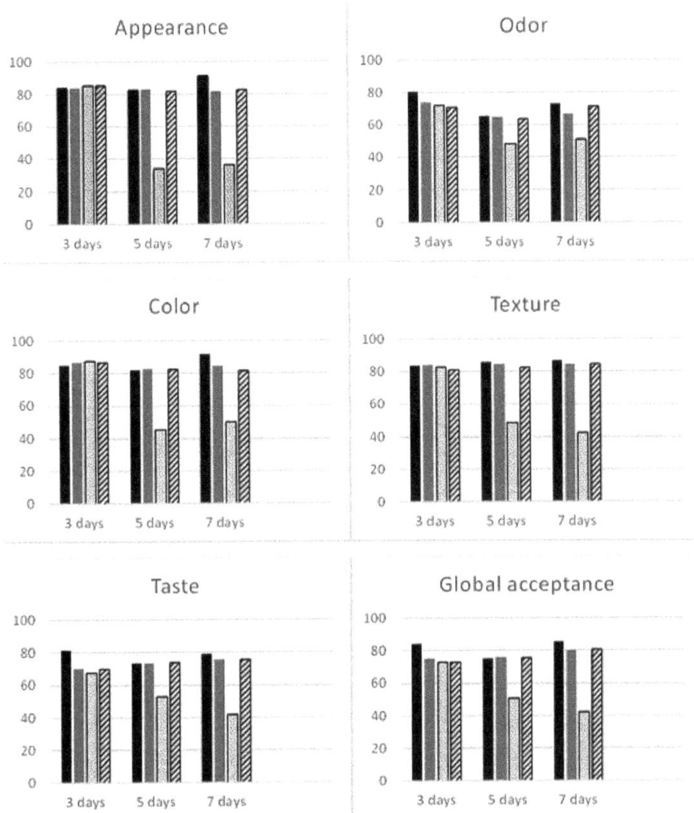

Figure 1. Acceptability index of sensory attributes of fresh (in black), open (in gray), sealed (with dots), and MAP (with stripes) arugula microgreens.

4. Conclusions

Microgreens stored in all packaging types were safe for consumption over ten days. Regarding physico-chemical parameters, open packaging proved to be promising, with less weight loss and slower chlorophyll degradation, with maintenance of the greenness. The sensory analysis demonstrated that the microgreens stored in the vacuum-sealed packaging showed a worsening in quality from the fifth day onwards for all attributes. However, the MAP presented good scores, with a better visual quality, similar to fresh microgreens.

Further analysis can be conducted to enhance microgreens' shelf life. We hope that the findings of the study will contribute to this field and inspire investigation into effective methods for food packaging and preservation.

Supplementary Materials: The following supporting information can be downloaded at: https://www.mdpi.com/article/10.3390/foods13193020/s1, Figure S1: Visual quality of fresh (without packaging), open (open packaging, the plastic bag remains with the upper side open), sealed (vacuum sealed packaging) and MAP (modified atmosphere packaging) arugula microgreens in seven days of storage; Figure S2: Sensory analysis radar chart of fresh (without packaging), open (open packaging, the plastic bag remains with the upper side open), sealed (vacuum sealed packaging) and MAP (modified atmosphere packaging) arugula microgreens.

Author Contributions: Conceptualization, M.R.K. and A.O.R.; methodology, M.R.K., A.O.R., S.H.F., P.S.M. and T.B.; software, M.R.K. and S.H.F.; validation, M.R.K., A.O.R., S.H.F., T.V.K. and P.S.M.; formal analysis, M.R.K., K.A.P. and T.B.; investigation, M.R.K., K.A.P. and T.B.; resources, M.R.K.,

A.O.R. and S.H.F.; data curation, M.R.K., A.O.R. and S.H.F.; writing—original draft preparation, M.R.K.; writing—review and editing, M.R.K., A.O.R., S.H.F., T.V.K. and P.S.M.; visualization, M.R.K., A.O.R., S.H.F. and T.V.K.; supervision, A.O.R., S.H.F., T.V.K. and P.S.M.; project administration, M.R.K. and A.O.R.; funding acquisition, M.R.K. and A.O.R. All authors have read and agreed to the published version of the manuscript.

Funding: This research received no external funding.

Institutional Review Board Statement: The study was conducted in accordance with the Declaration of Helsinki, and approved by the Ethics Committee of Federal University of Rio Grande do Sul (protocol 65303222.5.0000.5347 approved in February 2023).

Informed Consent Statement: Informed consent was obtained from all subjects involved in the study.

Data Availability Statement: The original contributions presented in the study are included in the article/Supplementary Materials, further inquiries can be directed to the corresponding author.

Acknowledgments: "Capes" for the study fellowship.

Conflicts of Interest: The authors declare no conflicts of interest.

References

1. Alruwaih, N.A.; Yaylayan, V.A. Comparative evaluation of bioactive compounds in lyophilized and tray-dried rocket (*Eruca sativa*). *J. Food Process. Preserv.* **2017**, *41*, e13205. [CrossRef]
2. Xiao, Z.; Rausch, S.; Luo, Y.; Sun, J.; Yu, L.; Wang, Q.; Chein, P.; Stommel, J. Microgreens of Brassicaceae: Genetic diversity of phytochemical concentrations and antioxidant capacity. *LWT* **2019**, *101*, 731–737. [CrossRef]
3. Zhang, Y.; Xiao, Z.; Ager, E.; Kong, L.; Tan, L. Nutritional quality and health benefits of microgreens, a crop of modern agriculture. *J. Future Foods* **2021**, *1*, 58–66. [CrossRef]
4. Turner, E.R.; Luo, Y.; Buchanan, R.L. Microgreen nutrition, food safety, and shelf life: A review. *J. Food Sci.* **2020**, *85*, 870–882. [CrossRef] [PubMed]
5. Paradiso, V.M.; Castellino, M.; Renna, M.; Gattullo, C.E.; Calasso, M.; Terzano, R.; Santamaria, P. Nutritional characterization and shelf-life of packaged microgreens. *Food Funct.* **2018**, *9*, 5629–5641. [CrossRef]
6. Butkutė, B.; Taujenis, L.; Norkevičienė, E. Small-seeded legumes as a novel food source. Variation of nutritional, mineral and phytochemical profiles in the chain: Raw seeds-sprouted seeds-microgreens. *Molecules* **2018**, *24*, 133. [CrossRef]
7. Di Gioia, F.; Renna, M.; Santamaria, P. Sprouts, microgreens and "Baby Leaf" vegetables. In *Minimally Processed Refrigerated Fruits and Vegetables*; Yildiz, F., Wiley, R.C., Eds.; Springer: Boston, MA, USA, 2017; pp. 403–432.
8. Johnson, S.; Prenni, J.; Heuberger, A.; Isweiri, H.; Chaparro, J.; Newman, S.; Uchanski, M.; Omerigic, H.; Michell, K.; Bunning, M.; et al. Comprehensive evaluation of metabolites and minerals in 6 microgreen species and the influence of maturity. *Curr. Dev. Nutr.* **2021**, *5*, nzaa180. [CrossRef]
9. Kopsell, D.; Pantanizopoulos, N.; Sams, C.; Kopsell, D. Shoot tissue pigment levels increase in "Florida Broadleaf" mustard (*Brassica juncea* L.) microgreens following high light treatment. *Sci. Hortic.* **2012**, *140*, 96–99. [CrossRef]
10. Di Gioia, F.; Mininni, C.; Santamaria, P. Come coltivare i micro-ortaggi—How to grow microgreens—Cómo cultivar micro-hortalizas. In *Microgreens: Novel Fresh and Functional Food to Explore All the Value of Biodiversity*; Di Gioia, F., Santamaria, P., Eds.; Eco-logica: Bari, Italy, 2015; pp. 51–79.
11. Kyriacou, M.; Rouphael, Y.; Di Gioia, F.; Kyratzis, A.; Serio, F.; Renna, M.; de Pascale, S. Micro-scale vegetable production and the rise of microgreens. *Trends Food Sci. Technol.* **2016**, *57*, 103–115. [CrossRef]
12. Renna, M.; Castellino, M.; Leoni, B.; Paradiso, V.M.; Santamaria, P. Microgreens production with low potassium content for patients with impaired kidney function. *Nutrients* **2018**, *10*, 675. [CrossRef]
13. Ebert, A.; Wu, T.; Yang, R.Y. Amaranth sprouts and microgreens—A homestead vegetable production option to enhance food and nutrition security in the rural–urban continuum. In *Proceedings of the Regional Symposium on Sustaining Small-Scale Vegetable Production and Marketing Systems for Food and Nutrition Security*; AVRDC Publication: Taiwan, China, 2014; pp. 233–244.
14. Komeroski, M.R.; Portal, K.A.; Comiotto, J.; Klug, T.V.; Flores, S.H.; Rios, A.D.O. Nutritional quality and bioactive compounds of arugula (*Eruca sativa* L.) sprouts and microgreens. *Int. J. Food Sci. Technol.* **2023**, *58*, 5089–5096. [CrossRef]
15. Riggio, G.; Wang, Q.; Kniel, K.; Gibson, K. Microgreens—A review of food safety considerations along the farm to fork continuum. *Int. J. Food Microbiol.* **2019**, *290*, 76–85. [CrossRef] [PubMed]
16. Wright, K.M.; Holden, N.J. Quantification and colonisation dynamics of *Escherichia coli* O157: H7 inoculation of microgreens species and plant growth substrates. *Int. J. Food Microbiol.* **2018**, *273*, 1–10. [CrossRef] [PubMed]
17. Di Gioia, F.; De Bellis, P.; Mininni, C.; Santamaria, P.; Serio, F. Physicochemical, agronomical and microbiological evaluation of alternative growing media for the production of rapini (*Brassica rapa* L.) microgreens. *J. Sci. Food Agric.* **2016**, *97*, 1212–1219. [CrossRef]
18. Mir, S.A.; Shah, M.A.; Mir, M.M. Microgreens: Production, shelf life, and bioactive components. *Crit. Rev. Food Sci. Nutr.* **2016**, *57*, 2730–2736. [CrossRef]

19. Sharma, A.; Hazarika, M.; Heisnam, P.; Pandey, H.; Devadas, V.S.; Singh, D.; Kartha, B.D. Influence of storage conditions, packaging, post-harvest technology, nanotechnology and molecular approaches on shelf life of microgreens. *J. Agric. Food Res.* **2023**, *14*, 100835. [CrossRef]
20. Kou, L.; Luo, Y.; Yang, T.; Xiao, Z.; Turner, E.R.; Lester, G.E.; Camp, M.J. Postharvest biology, quality and shelf life of buckwheat microgreens. *LWT-Food Sci. Technol.* **2013**, *51*, 73–78. [CrossRef]
21. Patil, M.; Sharma, S.; Sridhar, K.; Anurag, R.K.; Grover, K.; Dharni, K.; Sharma, M. Effect of postharvest treatments and storage temperature on the physiological, nutritional, and shelf-life of broccoli (*Brassica oleracea*) microgreens. *Sci. Hortic.* **2024**, *327*, 112805. [CrossRef]
22. Chitarra, M.I.F.; Chitararra, A.B. *Pós-colheita de frutas e hortaliças: Fisiologia e manuseio*, 2nd ed.; rev. amp.; Universidade Federal de Lavras: Lavras, Brazil, 2005; 783p.
23. *ISO 6579-1:2017/Amd 1:2020*; Microbiology of the Food Chain—Horizontal Method for the Detection, Enumeration and Serotyping of Salmonella—Part 1: Detection of *Salmonella* spp. Amendment 1: Broader Range of Incubation Temperatures, Amendment to the Status of Annex D, and Correction of the Composition of MSRV and SC. International Organisation for Standardization: Geneva, Switzerland, 2020.
24. *ISO 11290-1:2017*; Microbiology of the Food Chain—Horizontal Method for the Detection and Enumeration of *Listeria monocytogenes* and of *Listeria* spp. Part 1: Detection Method. International Organisation for Standardization: Geneva, Switzerland, 2017.
25. Silva, N.; Junqueira, V.; Silveira, N.; Taniwaki, M.; Gomes, R.; Okazaki, M.; Iamanaka, B. *Manual de Métodos de Análise Microbiológica de Alimentos e Água*, 6th ed.; Blucher: São Paulo, Brazil, 2021.
26. Barbosa, A.D.; Alexandre, B.; Tondo, E.C.; Malheiros, P.S. Microbial survival in gourmet hamburger thermally processed by different degrees of doneness. *Int. J. Gastron. Food Sci.* **2022**, *28*, 100501. [CrossRef]
27. AOAC. *Official Methods of Analysis of AOAC International*, 16th ed.; AOAC International: Rockville, MD, USA, 1995; Volume 1.
28. Xiao, Z.; Lester, G.E.; Park, E.; Saftner, R.A.; Luo, Y.; Wang, Q. Evaluation and correlation of sensory attributes and chemical compositions of emerging fresh produce: Microgreens. *Postharvest Biol. Technol.* **2015**, *110*, 140–148. [CrossRef]
29. Yan, H.; Li, W.; Chen, H.; Liao, Q.; Xia, M.; Wu, D.; Zhao, J. Effects of Storage Temperature, Packaging Material and Wash Treatment on Quality and Shelf Life of Tartary Buckwheat Microgreens. *Foods* **2022**, *11*, 3630. [CrossRef] [PubMed]
30. Wellburn, A.R.; Lichtenthaler, H. Formulae and program to determine total carotenoids and chlorophylls a and b of leaf extracts in different solvents. In *Advances in Photosynthesis Research: Proceedings of the VIth International Congress on Photosynthesis, Brussels, Belgium, August 1–6, 1983 Volume 2*; Springer: Dordrecht, The Netherlands, 1984; pp. 9–12.
31. Viana, L.T. Sensory analysis in the food industry. *Rev. Inst. Laticínios Cândido Tostes* **2009**, *64*, 12–21.
32. Priti; Sangwan, S.; Kukreja, B.; Mishra, G.P.; Dikshit, H.K.; Singh, A.; Aski, M.; Kumar, A.; Taak, V.; Stobdan, T.; et al. Yield optimization, microbial load analysis, and sensory evaluation of mungbean (*Vigna radiata* L.), lentil (*Lens culinaris* subsp. *culinaris*), and Indian mustard (*Brassica juncea* L.) microgreens grown under greenhouse conditions. *PLoS ONE* **2011**, *17*, e0268085. [CrossRef] [PubMed]
33. International Commission on Microbiological Specifications for Foods—ICMSF. *Microorganisms in Foods 8: Microbiological Testing in Food Safety Management*; Kluwer Academic: New York, NY, USA, 2011.
34. BRASIL. RESOLUÇÃO—RDC N° 724 2011, DE 01 DE JULHO DE. 2022. Available online: https://antigo.anvisa.gov.br/documents/10181/2718376/RDC_724_2022_.pdf/33c61081-4f32-43c2-9105-c318fa6069ce (accessed on 22 July 2024).
35. BRASIL. INSTRUÇÃO NORMATIVA N° 161, DE 01 DE JULHO DE. 2022. Available online: https://antigo.anvisa.gov.br/documents/10181/2718376/IN_161_2022_.pdf/b08d70cb-add6-47e3-a5d3-fa317c2d54b2 (accessed on 22 July 2024).
36. Jablasone, J.; Warriner, K.; Griffiths, M. Interactions of *Escherichia coli* O157: H7, *Salmonella typhimurium* and *Listeria monocytogenes* plants cultivated in a gnotobiotic system. *Int. J. Food Microbiol.* **2005**, *99*, 7–18. [CrossRef] [PubMed]
37. Chandra, D.; Kim, J.G.; Kim, Y.P. Changes in microbial population and quality of microgreens treated with different sanitizers and packaging films. *Hortic. Environ. Biotechnol.* **2012**, *53*, 32–40. [CrossRef]
38. Xiao, Z.; Luo, Y.; Lester, G.E.; Kou, L.; Yang, T.; Wang, Q. Postharvest quality and shelf life of radish microgreens as impacted by storage temperature, packaging film, and chlorine wash treatment. *LWT—Food Sci. Technol.* **2014**, *55*, 551–558. [CrossRef]
39. Kyriacou, M.C.; El-Nakhel, C.; Pannico, A.; Graziani, G.; Soteriou, G.A.; Giordano, M.; Rouphael, Y. Phenolic constitution, phytochemical and macronutrient content in three species of microgreens as modulated by natural fiber and synthetic substrates. *Antioxidants* **2020**, *9*, 252. [CrossRef]
40. Huang, J.; Luo, Y.; Nou, X. Growth of *Salmonella enterica* and *Listeria monocytogenes* on fresh-cut cantaloupe under different temperature abuse scenarios. *J. Food Prot.* **2015**, *78*, 1125–1131. [CrossRef]
41. Rusu, T.; Moraru, P.I.; Mintas, O.S. Influence of environmental and nutritional factors on the development of lettuce (*Lactuca sativa* L.) microgreens grown in a hydroponic system: A review. *Not. Bot. Horti Agrobot. Cluj-Napoca* **2021**, *49*, 12427. [CrossRef]
42. FDA. Recommendations for the Temperature Control of Cut Leafy Greens during Storage and Display in Retail Food Establishments. 2009. Available online: https://www.fda.gov/media/78982/download (accessed on 14 August 2024).
43. Hamilton, A.N.; Fraser, A.M.; Gibson, K.E. Barriers to implementing risk management practices in microgreens growing operations in the United States: Thematic analysis of interviews and survey data. *Food Control* **2023**, *152*, 109836. [CrossRef]
44. Dalal, N.; Siddiqui, S. Effect of chemical treatment, storage and packaging on physico-chemical properties of sunflower microgreens. *Int. J. Chem. Stud.* **2019**, *7*, 1046–1050.

45. Gómez, P.A.; Egea-Gilabert, C.; Giménez, A.; Benaissa, R.R.; Amoruso, F.; Signore, A.; Gallegos-Cedillo, V.M.; Ochoa, J.; Fernández, J.A. Biodegradable Food Packaging of Wild Rocket (*Diplotaxis tenuifolia* L. [DC.]) and Sea Fennel (*Crithmum maritimum* L.) Grown in a Cascade Cropping System for Short Food Supply Chain. *Horticulturae* **2023**, *9*, 621. [CrossRef]
46. Khan, M.A.M.; Mittal, A. Modified atmosphere packaging technique: An overview. In Proceedings of the International Conference on Recent Advances in Engineering and Science (ICRAES-2020), Aligarh, India, 11–12 January 2020.
47. De La Fuente, B.; López-García, G.; Máñez, V.; Alegría, A.; Barberá, R.; Cilla, A. Evaluation of the bioaccessibility of antioxidant bioactive compounds and minerals of four genotypes of Brassicaceae microgreens. *Foods* **2019**, *8*, 250. [CrossRef] [PubMed]
48. Kowitcharoen, L.; Phornvillay, S.; Lekkham, P.; Pongprasert, N.; Srilaong, V. Bioactive composition and nutritional profile of microgreens cultivated in Thailand. *Appl. Sci.* **2021**, *11*, 7981. [CrossRef]
49. Ghoora, M.D.; Haldipur, A.C.; Srividya, N. Comparative evaluation of phytochemical content, antioxidant capacities and overall antioxidant potential of select culinary microgreens. *J. Agric. Food Res.* **2020**, *2*, 100046. [CrossRef]
50. Katsenios, N.; Christopoulos, M.V.; Kakabouki, I.; Vlachakis, D.; Kavvadias, V.; Efthimiadou, A. Effect of pulsed electromagnetic field on growth, physiology and postharvest quality of kale (*Brassica oleracea*), wheat (*Triticum durum*) and spinach (*Spinacia oleracea*) microgreens. *Agronomy* **2021**, *11*, 1364. [CrossRef]
51. Ciriello, M.; Formisano, L.; El-Nakhel, C.; Zarrelli, A.; Giordano, M.; De Pascale, S.; Rouphael, Y. Iodine biofortification of four microgreens species and its implications for mineral composition and potential contribution to the recommended dietary intake of iodine. *Sci. Hortic.* **2023**, *320*, 112229. [CrossRef]
52. El-Nakhel, C.; Pannico, A.; Graziani, G.; Kyriacou, M.C.; Gaspari, A.; Ritieni, A.; De Pascale, S.; Rouphael, Y. Nutrient supplementation configures the bioactive profile and production characteristics of three *Brassica* L. microgreens species grown in peat-based media. *Agronomy* **2021**, *11*, 346. [CrossRef]
53. Petropoulos, S.A.; El-Nakhel, C.; Graziani, G.; Kyriacou, M.C.; Rouphael, Y. The effects of nutrient solution feeding regime on yield, mineral profile, and phytochemical composition of spinach microgreens. *Horticulturae* **2021**, *7*, 162. [CrossRef]
54. Dhaka, A.S.; Dikshit, H.K.; Mishra, G.P.; Tontang, M.T.; Meena, N.L.; Kumar, R.R.; Praveen, S. Evaluation of growth conditions, antioxidant potential, and sensory attributes of six diverse microgreens species. *Agriculture* **2023**, *13*, 676. [CrossRef]
55. Bafumo, R.F.; Alloggia, F.P.; Ramirez, D.A.; Maza, M.A.; Fontana, A.; Moreno, D.A.; Camargo, A.B. Optimal brassicacea family microgreens from a phytochemical and sensory perspective. *Food Res. Int.* **2024**, *193*, 114812. [CrossRef] [PubMed]
56. Caracciolo, F.; El-Nakhel, C.; Raimondo, M.; Kyriacou, M.C.; Cembalo, L.; De Pascale, S.; Rouphael, Y. Sensory attributes and consumer acceptability of 12 microgreens species. *Agronomy* **2020**, *10*, 1043. [CrossRef]

foods

MDPI

Article

Osmodehydrofreezing of Tomatoes: Optimization of Osmotic Dehydration and Shelf Life Modeling

Efimia Dermesonlouoglou, Lefteris Pittas, Petros Taoukis and Maria Giannakourou *

Laboratory of Food Chemistry and Technology, School of Chemical Engineering, National Technical University of Athens, 15780 Athens, Greece; efider@chemeng.ntua.gr (E.D.); lefteris2352000@gmail.com (L.P.); taoukis@chemeng.ntua.gr (P.T.)

* Correspondence: mgian@chemeng.ntua.gr; Tel.: +30-210-7723166

Abstract: The objective was to review, using an integrated approach, all parameters related to osmotic dehydration, freezing, and frozen storage when assessing the advantages of the osmodehydrofreezing-ODF process. Peeled cherry tomatoes were treated at (T) 25, 35, and 45 °C (t) up to 180 min in glycerol-based OD-solution (50, 60, 70% *w*/*w*). OD was studied and optimized by applying the Response Surface Methodology, combined with selected desirability criteria to define the optimum process parameters. Water loss-WL, solid gain-SG, water activity reduction-a_w, texture and color changes were monitored during the process. Untreated and OD-treated at optimal OD conditions (C = 61.5%, T = 36 °C; t = 72 min) samples were frozen and stored at isothermal (T, −5, −8, −14, −23 °C) and non-isothermal temperature conditions (T_{eff}, −7.3 °C). OD samples presented acceptable color, increased firmness, low drip loss and high vitamin C/lycopene retention during frozen storage. OD increased the shelf life of frozen cherry tomato (up to 3.5 times based on sensory quality loss). The kinetic models obtained for vitamin and lycopene degradation and sensory quality loss were validated at non-isothermal conditions.

Keywords: cherry tomato; osmodehydrofreezing; process optimization; frozen storage; quality loss; shelf-life

1. Introduction

Osmotic dehydration (OD) entails submerging food products in hypertonic aqueous solutions that contain carbohydrates and/or salts to partially remove water without causing a phase change [1]. Mass transfer occurring during the OD process involves the water transfer from the product matrix to the osmotic solution, the solute transfer (taking place in opposite directions), and, to a lesser extent, the leaching of the product's solutes [2]. The goal of osmotic processing is to reduce the water activity of the product without significantly altering its quality, unlike other dehydration methods (such as air-drying) [2,3]. Food material characteristics, molecular weight and concentration of the solutes in the osmotic solution, the temperature of the process, the length of immersion, and the agitation level affect the rate of mass transfer [4–10]. Furthermore, studies have been carried out to better understand mass transfer and model the mechanisms of the OD process [11,12]. Especially for plant-based food, such as fruits and vegetables, that are susceptible to significant tissue damage when high temperatures are applied, OD is found to lead to improved quality retention, particularly when compared to other, more intense dehydration techniques like air-drying [13,14]. However, osmotically dehydrated fruits and vegetables do not reach a water activity level to render them microbiologically stable and a complementary preservation method such as freezing or air-drying is required to obtain an extended shelf life. Besides lowering food water activity and thus improving its stability, OD could be also used as a tool in matrix engineering since it typically results in the selective uptake of solutes, such as bioactive components, components targeting texture, flavor, or nutritional value enhancement [4,15,16].

Osmodehydrofreezing (ODF) is the application of partial dehydration in a hypertonic solution prior to a subsequent traditional freezing procedure [17]. The application of osmotic pre-treatment before conventional freezing lowers the freezing point and significantly reduces the water amount to be frozen thereafter. When a solute is impregnated and the moisture content decreases, the glass transition temperature of the modified tissue rises, greatly increasing its stability in the subsequent frozen storage [11,12]. Osmodehydrofreezing has several positive effects on the freezing process itself, such as better preservation of structure and other quality attributes, lower packaging, distribution, and storage costs, lower energy to freeze the remaining water and a decrease in the refrigeration load [18]. Despite the above-mentioned advantages, ODF's wider exploitation and industrial application are practically limited [2]. This is primarily because it is challenging to model and control mass transport phenomena, additionally due to issues related to osmotic solution management [19,20]. In this context, there is a need to implement an integrated kinetic approach aiming at optimizing the ODF, using criteria such as lowering the product water activity as well as the quality loss reaction rates. Response surface methodology (RSM) has been frequently proposed as a means of characterizing and pinpointing the primary elements impacting processes. Moreover, the desirability approach is used, as a complementary RSM tool, to optimize multiple responses, when there is a need to apply multiple optimization criteria [21] and this strategy has been recently implemented for optimizing the osmotic dehydration process [6,7,16,22–24].

Cherry tomato is abundant in nutrients and bioactive compounds, and recently, an economically important commodity in multiple recipes [25]. Its small size, tomato-like flavor, and variety of colors make it a favorite option. Cherry tomatoes in different colors and tastes are bought and eaten raw; however, they can also be found in processed foods [26]. Its thin skin and juicy characteristics make it perishable during storage and transportation, [27,28]. Therefore, it is necessary to find suitable processing and/or packaging methods to improve the storage life of cherry tomatoes. Freezing, as a single preservation method, gives poor results, especially when frozen tomatoes are used for direct consumption [29].

In this study, the quality of osmo-dehydrofrozen (produced at optimized OD conditions using RSM, coupled with selected desirability functions) cherry tomatoes was kinetically studied during frozen storage. More specifically, mass exchange, color and texture changes were monitored during the OD process. After processing tomato pieces at the optimal OD conditions, samples were frozen and packed. Vitamin C and lycopene, color, texture, drip loss and overall acceptability of OD pre-treated as well as conventionally (untreated) frozen samples were measured during storage at isothermal and non-isothermal temperatures at a wide range of sub-zero temperatures and product shelf life was calculated. Applying this integrated process (ODF), involving three steps (OD-freezing-frozen storage), one can better assess its advantages, compared to the conventional freezing preservation, in terms of the stability and the improved shelf life of the ODF tomato samples, obtained.

2. Materials and Methods

2.1. Sample Preparation

Cherry tomatoes (*Lycopersicon esculentum*) (diameter: 30 ± 10 mm) were purchased at the local market and kept in temperature-controlled cabinets for a maximum of 1 day (approximately 4.0 °C). Before osmotic treatment, tomatoes were washed with tap water, and then, blanched in water at 80 °C for 30 s, for both peel removal and enzyme inactivation. Water activity (a_w) was measured at 0.9505 ± 0.0076, total soluble solids (TSS) were 8.8 ± 0.6 °Brix and pH was 4.108 ± 0.012 (mean ± standard deviation; five samples). According to Gould (2013), all measured values fall within the acceptable limits for ripeness [30]. Striking is the high content of total soluble solids (approximately 7.0 °Brix) which is commonly perceived as a good predictor for both ripeness and high sensory quality [24,25].

2.2. Osmotic Dehydration

Osmotic dehydration (OD) was carried out in solutions by dissolving glycerol (Glycerine EP 212, OLEOGEN, Genova, Italy), (C) 50, 60, and 70% (w/w), 3.5% NaCl and 1.5% $CaCl_2$, for (t) up to 180 min at (T) 25, 35, and 45 °C [8]. to yield a solid to liquid ratio of 1:5 (w/w). Glycerol was used as the main lowering a_w agent sugar alcohol with low sweetness (Regulation EC No. 1333/2008) [31]. Additionally, it acts as a cryoprotectant and it improves the texture. Sodium chloride was included in the osmotic solution to balance the slight sweetness obtained during osmotic pretreatment as well as to increase the driving force of the process [32]. Calcium chloride was used to retain or even to improve the texture properties of the osmodehydrated product. Calcium may interact with cellular pectins and other wall components modifying the structural and mechanical properties of plant tissue [33]. Tomatoes were pre-weighed, put in cylindrical glass containers and placed in water baths of controlled temperature, under constant agitation (240 rpm). At times (t) 15, 30, 60, 90, 120, 150, and 180 min samples (triplicates) were removed from jars and weighed after carefully blotting the excess coating solution.

2.2.1. Water, Total Solids and Water Activity

Water content and total solids were monitored during the process, measured by drying at 105 °C for 24 h according to the AOAC method (1990). Water activity (a_w) was monitored using Aqua LAB 4TEV (Decagon Devices, Inc., Pullman, WA, USA).

2.2.2. Mass Transfer

Water loss (*WL*) and solid gain (*SG*), which describe the mass transfer phenomena during the OD process, were calculated (Equations (1) and (2)) [34]:

$$WL = \frac{(M_0 - m_0) - (M - m)}{m_0} \quad (1)$$

$$SG = \frac{(m - m_0)}{m_0} \quad (2)$$

where M_0 and m_0 are the initial total mass and dry mass of the tomato sample at OD time zero, respectively, and M and m are the total mass and dry mass of the tomato sample at OD time t, respectively.

2.2.3. Experimental Design and RSM Implementation

The optimization of the OD process was performed by applying a symmetrical 15-run three-level Box–Behnken design (BBD). Temperature (x1), 25, 35, 45 °C; time of osmotic treatment (x2), 30, 60, 90 min and glycerol concentration (x3), 50, 60, 70% were the factors under optimization or independent variables, while mass exchange (*WL*, *SG*), water activity (a_w), texture (firmness), and color (ΔE) changes were the response of the model or dependent variable, which were adequately modeled using a second-order polynomial model describing the effect of the most important OD processing factors. Since the examined factors have different natural units, their actual values should be transformed into dimensionless normalized coded units (−1, 0, +1) to directly compare and evaluate their effect and the effect of their interactions on the dependent variables. The actual and coded values of the investigated extraction factors are presented in Table 1.

Table 1. Actual and coded values of the investigated factors.

Independent OD Variable	OD Temperature (°C)	OD Time (min)	Glycerol Concentration (% *w*/*w*)	X_1	X_2	X_3
High	25	30	50	+1	+1	+1
Center	35	60	60	0	0	0
Low	45	90	70	−1	−1	−1
Standard order	X_1	X_2	X_3			
1	1	0	1			
2	−1	1	0			
3	0	−1	1			
4	0	1	−1			
5	1	−1	0			
6	1	0	−1			
7	−1	0	1			
8	−1	0	−1			
9	−1	−1	0			
10	0	0	0			
11	0	0	0			
12	0	1	1			
13	1	1	0			
14	0	−1	−1			
15	0	0	0			

2.2.4. Factor Interactions during Osmotic Dehydration

A second-order polynomial was used to describe the interactions between response variables Y (water loss WL and solid gain SG, water activity a_w) and the factor variables X_i (concentration of glycerol co C) 50, 60, 70%, process temperature (T) 25, 35, 45 °C, and process time (t) 5–180 min (Equation (3)) [35].

$$y = a_0 + \sum a_i x_i + \sum a_{ii} x_i^2 + \sum a_{ij} x_i x_j \tag{3}$$

where a_0 is the constant, a_i is the linear, a_{ii} is the quadratic, and a_{ij} is the interaction effects of the factors. The quadratic variables indicate the surface curvature, the linear variables the coordinates of the greatest value anticipated, and the bi-factorial cross-variables the directions of the axes of the geometric figure formed by sectioning the surface area [35]. Positive values of the coefficient indicate an effect that favours the response, whereas negative values show an opposite effect of the factor on the response. After assessing the polynomial equations' coefficients and calculating the model, analysis of variance (ANOVA) was applied to investigate how well the model describes the data. The corresponding p value served as a statistical significance indicator for this purpose.

2.3. Freezing Process

Cherry tomatoes, osmodehydrated and non-osmodehydrated (untreated), were quick frozen at −40 °C (forced air convection characteristics, h = 11 W/m^{2}K) (Sanyo MIR 553, Sanyo Electric Co., Gunma, Japan), packed in pouches used for commercial frozen vegetable products [laminate film characteristics: 20 lm bio-oriented polypropylene (BOPP)-48 lm polyethylene (PE), water-vapor transmission rate (WVTR) < 6 g/m^2 d], and kept at −40 °C.

2.4. Storage at Sub-Freezing Temperatures

Osmodehydrated and non-osmodehydrated (untreated) cherry tomatoes were stored in temperature-controlled cabinets (Sanyo MIR 153, 253, Sanyo Electric Co., Gunma, Japan) at isothermal temperatures (T) −5, −8, −14, −23 °C (±0.5 °C), monitored by temperature dataloggers (COX TRACER®, Belmont, NC, USA). The experimental set-up was according to ASLT (Accelerated Shelf-Life Testing) principles [36]. Quality and sensory parameters (color, texture, drip loss, vitamin C, lycopene) of osmodehydrofrozen (OD) and non-osmodehydrated (untreated) frozen cherry tomato samples were determined at predetermined time intervals during storage based on a rough estimate of an expected temperature-dependence of quality loss [4].

The applicability of the developed (at isothermal conditions) kinetic models was tested by an independent experiment (at non-isothermal conditions). A temperature-time scenario was applied, by repeating three successive cycles of −12 °C–24 h, −5 °C–24 h and −8 °C–24 h (Sanyo MIR 153, Sanyo Electric Co., Gunma, Japan) equal to an effective temperature (T_{eff}) of −7.3 °C [37].

2.4.1. Color

Tomato color, expressed in CIELab scale, was measured by a chromameter (CR-200, which uses the D65 illuminant and a 10-degree observer for its measurements, Minolta Co., Tokyo, Japan). The total color change during the OD process as well as during the frozen storage, was mathematically described by Equation (4) [38].

$$\Delta E = \sqrt{(L - L_0)^2 + (a - a_0)^2 + (b - b_0)^2} \quad (4)$$

where L (lightness), a (from +: red to −: green), b (from +: yellow to −: blue), and L_0, a_0, b_0 values at time t, and at time zero (0), either of the OD process or of the frozen storage, respectively.

2.4.2. Texture

Texture measurements of cherry tomato samples (OD and untreated), equilibrated to room temperature for 30 min, were measured by a texture analyzer (TA-XT2i of Stable Micro Systems, England). The test took place on a non-lubricated flat platform with a knife probe, used for cutting samples at a test speed of 1 mm/s and a depth of 1/3 of the initial (approximately 5 mm). The maximum peak force (F_{max}, g) was reported as hardness/firmness. Eight to ten samples (3 replicates) were used.

2.4.3. Drip Loss

Osmodehydrated and non-osmodehydrated (untreated) cherry tomatoes were put on a sieve to allow their exuded water to drip off and then left to defrost at room temperature for 60 min. Samples were weighed at time intervals of 10 min. The difference between the weight at any time t and the initial weight referred to as the initial value was defined as the drip loss of the sample [39] (Equation (5)).

$$Drip\ loss\ (\%) = \frac{W_{before\ thawing} - W_{after\ thawing}}{W_{before\ thawing}} * 100 \quad (5)$$

where Wi, (g) the weight of tomato before freezing and $W_{after\ thawing}$ the weight of the tomato after freezing.

2.4.4. Vitamin C

Vitamin C (mg L-ascorbic acid/100 g fresh/OD material) was determined by a high-performance liquid chromatography method (HPLC) [37]. Two replicates were used. Five g of osmodehydrofrozen or untreated sample homogenates were mechanically stirred in 15 mL of a 4.5% (w/v) solution of metaphosphoric acid (Merck, Darmstadt, Germany) for

15 min (two replicates for each measurement). The mixture was filtered under vacuum and diluted using HPLC grade water (Merck, Darmstadt, Germany); the total final volume was recorded and an aliquot was filtered through a 0.45 m filter (Chromafil PVDF-45/25, Macherey-Nagel, Germany) prior to injection into the chromatographic column. The details of the HPLC instrumentation were HP Series 1100 (quarternary pump, vacuum degasser, a Rheo dyne 20 μL injection loop and a Diode-Array Detector, controlled by HPChemStation 1100 series software); Hypersil ODS column (250 × 4.6 mm) of particle size 5 μm; mobile phase: HPLC grade water with metaphosphoric acid to pH 2.2; detection at 245 nm; calibrated by external standard method.

2.4.5. Lycopene

Lycopene (mg total lycopene/100 g fresh/OD material) was determined using a high-performance liquid chromatography method [40]. Two and a half g of osmodehydrofrozen or untreated sample homogenates were added to a mixture consisting of 50 mL of hexane, 25 mL of ethanol, 25 mL of acetone and 0.05% (*w*/*v*) butylated hydroxytoluene, then stoppered and mechanically agitated for 5 min (two replicates for each measurement). Cold, doubly distilled water (10 mL) was added and the suspension was agitated for an additional 5 min. The solution was allowed to stand for 15 min for separation of polar and non-polar layers. The upper hexane layer, containing the lycopene, was filtered through a 0.22-lm filter (Chromafil PET-20/25; Macherey-Nagel, Duren, Germany) and then injected into the chromatographic column. The instrumentation was the same as in Section 2.4.4. The mobile phase was methanol:tetrahydrofuran:water at a ratio of 67:27:6. The detection was performed at 472 nm. The method was calibrated by an external standard method.

2.4.6. Sensory Parameters

Sensory evaluation was carried out by a panel of eight to ten trained assessors [70% female and 30% male, 60% young (18–35 years old) and 40% middle-aged (36–55 years old)] of the sensory laboratory of NTUA, according to ISO standards [41,42] (ISO 6658:2017, ISO 20613:2019), on the following attributes: appearance for red color; firmness and juiciness for texture, sweetness, sourness for flavor and aftertaste. The intensity was evaluated using a 9-point scale (1 = low intensity and 9 = high intensity). The panel also performed an overall impression test using a 9-point hedonic scale (1 = extremely dislike to 9 = extremely like).

2.4.7. Data Analysis

The evolution of the selected quality and sensory parameters of osmodehydrofrozen (OD) and non-osmodehydrofrozen (untreated) cherry tomato samples were plotted and mathematically modeled, at all studied temperatures; The quality loss rates k were calculated. The vitamin C (ascorbic acid) and lycopene loss were described by an apparent first-order reaction (Equation (6)) [43], whereas the sensory quality loss was described by an apparent zero-order reaction (Equation (7)) [6].

$$C_{\frac{asc}{lyc}} = C_{\frac{asc}{lyc}0} \cdot exp\left(-k_{\frac{asc}{lyc}} \cdot t\right) \tag{6}$$

$$S_{overall} = S_{overall\ 0} - k_{sens} \cdot t \tag{7}$$

where $C_{asc/lyc}$ and $Casc_{/lyc0}$ are the concentrations of vitamin C or lycopene at time t and zero, respectively, $S_{overall}$ and $S_{overall,0}$ are the sensory scores at time t and zero, respectively, and $k_{asc/lyc}$, $k_{sens,overall}$ is the apparent reaction rate of vitamin C loss, lycopene loss or overall sensory quality loss.

The temperature dependence of the deterioration rates $k_{asc/lyc}/k_{sens,overall}$ was modelled by the Arrhenius equation (Equation (8)) [36]:

$$k_{\frac{asc}{lyc},overall} = k_{ref} \cdot \exp\left[-\frac{E_a}{R}\left(\frac{1}{T_{ref}} - \frac{1}{T}\right)\right] \tag{8}$$

The shelf life, determined (based on the vitamin C or lycopene loss), was calculated using Equations (6) and (8) (Equation (9)). The shelf life (SL), determined by the overall sensory quality, was calculated using Equations (7) and (8) (Equation (10)):

$$SL_{asc/lyc} = \frac{lnC_{\frac{asc,final}{lyc,\ final}} - lnC_{\frac{asc}{lyc},0}}{k_{asc/lyc_{Tref}} \exp\left(-\frac{E_{a_{asc/lyc}}}{R}\left(\frac{1}{T} - \frac{1}{T_{ref}}\right)\right)} \quad (9)$$

$$SL_s = \frac{S_0 - S_L}{k_{s_{Tref}} \exp\left(-\frac{E_{a_S}}{R}\left(\frac{1}{T} - \frac{1}{T_{ref}}\right)\right)} \quad (10)$$

where $E_{a\ asc/lyc}$ is the activation energy of the parameter k, k_{Tref} is the deterioration rate at the reference temperature T_{ref} (−18 °C, in this study), R is the universal gas constant, $C_{\frac{asc,final}{lyc,final}}$ is the acceptance limit of vitamin C/lycopene content (50% loss) [37,44], S_L is the acceptance limit of the sensory parameter S (overall impression score = 5) [8]. Using these equations, the shelf life can be predicted at different storage temperatures (in the same temperature range). In the case of variable temperature profiles, common for temperature abuse during real conditions, where T becomes a function of time (T(t)), Equations (7) and (8) are transformed to integrals, and accordingly solved to obtain SL decrease at each iso-thermal time step.

Variable temperature profiles were expressed by the effective temperature concept [37]. Effective temperature T_{eff}, is defined as the constant temperature leading to the same quality parameter value as the variable temperature distribution (that can be discretized into small isothermal time steps, t_i) over the same time period, is based on the Arrhenius model and integrates into a single value the effect of the variable temperature profile as given in Equation (11),

$$k_{ref}\sum_i\left(exp\left[-\frac{E_A}{R}\left(\frac{1}{T_i} - \frac{1}{T_{ref}}\right)\right]t_i\right) = k_{eff}t_{tot} = k_{ref}\cdot exp\left[-\frac{E_A}{R}\left(\frac{1}{T_{eff}} - \frac{1}{T_{ref}}\right)\right]t_{tot} \quad (11)$$

from which the value of T_{eff} can be calculated.

The rates of quality deterioration calculated by the developed kinetic models were compared to the respective experimental values by calculating the relative error (RE) values (Equation (12),

$$\%RE = \frac{k_{exp} - k_{fitted}}{k_{exp}} \cdot 100 \quad (12)$$

where 20% is the limit of acceptance [43].

2.5. Statistical Analyses

Analysis of variance (ANOVA) was conducted by STATISTICA 12.0 software (Stat. Soft. Inc. Tulsa, OK, USA), while significant differences in mean values of WL, SG, and a_w values were estimated by Tukey's HSD test ($p < 0.05$). Regarding RSM implementation and desirability approach, the data processing and the statistical analysis were performed using Minitab 17 software (trial version, Minitab LCC, State College, PA, USA).

3. Results and Discussion

3.1. OD Process

3.1.1. Mass Exchange during OD Process

The results for the OD of cherry tomato samples (50–70% w/w glycerol concentration and temperatures 25–45 °C) in terms of water loss (WL) and solid gain (SG) are presented in Figure 1a,b. According to the particular Figures, the dehydration (water loss) as well as solids uptake was accelerated with temperature increase. At OD time t = 0, i.e., the immersion time, there is an almost instantaneous water loss, as the mass transfer starts as soon as the sample meets the osmotic solution (maximum concentration gradient, which is the driving force for OD). The water loss rate increases during the first 30–60 min of the process; then it decreases and tends to stabilize, as reported by Ma et al. (2021) [45]. This effect had been reported for kiwi fruit which was subjected to osmotic dehydration with glycerol and sucrose. The high mass transfer in the early phase of osmotic treatment was due to the large osmotic pressure difference between the apple cube and the surrounding osmotic solution [46]. Glycerol concentration appears to have a significant effect on water loss, which assumes its maximum value for 60% glycerol concentration. Higher glycerol concentration (70%) probably prevents the OD solution components (NaCl, $CaCl_2$) from dissolving completely, as reported by [13,47]. Figure 1b shows the uptake of solids (SG), owing to the transport of glycerol and salts inside the cells during OD, but also to the retention of components on the surface of the plant tissue during immersion. The solids uptake rate is observed to be increased at the beginning of the process (30–60 min). It seems that the temperature increase from 25 °C to 35 °C does not significantly affect the uptake of solids, while the increase from 35 °C to 45 °C significantly increases the solids uptake (almost double) [4].

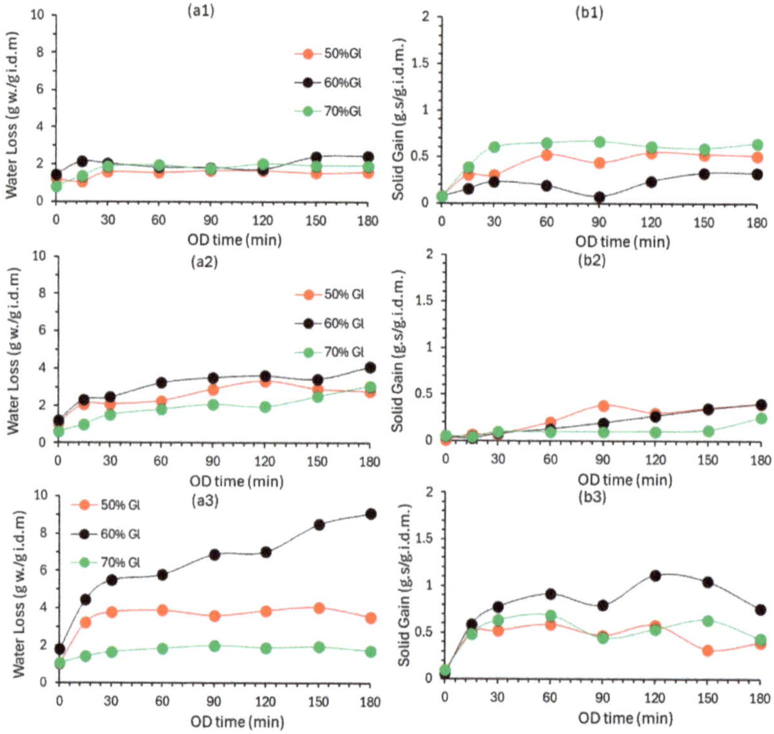

Figure 1. Evolution of (**a**) water loss (WL), and (**b**) solid gain (SG) of OD cherry tomatoes at (**1**) 25, (**2**) 35, and (**3**) 45 °C using 50, 60, 70% w/w glycerol-based solution. (Experimental data points: average ± standard error).

3.1.2. Water Activity during OD Process

Water activity values obtained for OD tomato samples are plotted vs. OD time for all glycerol concentrations and temperatures studied (Figure 2a–c). As expected, water activity immediately decreases during the first minutes of the process and throughout its duration. According to Figure 2a–c, the temperature significantly accelerates water activity decrease ($p < 0.05$), in agreement with published work [37]. A significant a_w reduction seems to be achieved at 45 °C and 60% glycerol solution (t = 180 min => a_w = 0.85). When glycerol concentration increases, the rate of a_w decrease seems to increase, with 60% glycerol solution leading to the highest a_w decrease. In published studies [13,48], it was reported that the high carbohydrate concentration (in the osmotic solution) may present obstacles in the dissolution of the remaining components in the solution; for this reason, a lower water activity reduction rate is observed for the solution with 70% glycerol concentration (compared to 50 and 60%).

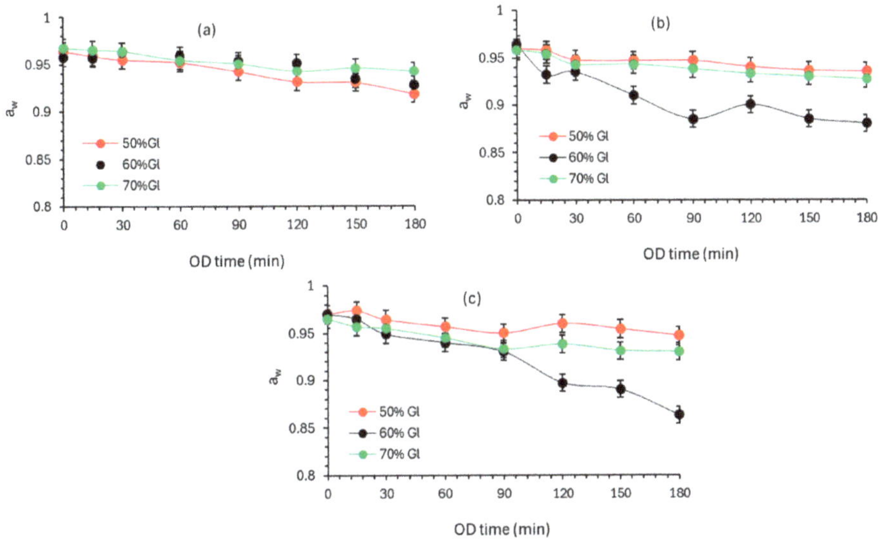

Figure 2. Evolution of water activity of cherry tomatoes during OD at (**a**) 25 °C, (**b**) 35 °C, and (**c**) 45 °C using 50%, 60%, 70% *w*/*w* glycerol-based solution. Markers indicate experimental data points (average ± standard deviation).

3.1.3. Color and Texture Changes during OD

In Figure 3, the total color change (ΔE) and texture (firmness, F_{max}) of the osmodehydrated (OD) cherry tomato samples during OD at 35 °C is representatively depicted. It was observed that OD temperature (from 25 to 45 °C) does not affect significantly color change. For the OD temperatures studied, the color change rates were increased in the early stages (approximately during the 15–60 min) where the mass transfer was higher. Changes in tissue structure (semipermeable membrane in plant cells) caused by OD also affect the color change. During the osmotic dehydration process, carotenoids that are responsible for the color may diffuse into the solution and cause a partial loss of color [49]. The highest ΔE values were calculated for glycerol concentrations of 60 and 70%. The glycerol concentration seemed to have a significant effect on the color change, after 120 min of OD, for the 50–70% glycerol concentration. A L value decrease (lightness), along with a b value increase (yellowness) and a value increase (redness), was observed as OD time increased. Overall color ΔE values ranged from 6 to 12, expressing a significant perceivable color change of tomato samples. According to the interpretation by Choi et al. (2002), the ΔE value > 2 indicates the visible difference [50]. The perceived color change of tomatoes observed during OD could not be only connected to their pigment color but also to the spatial distribution of the pigments within the plant cells, which can change as the cells become more permeable and pigments are released from the chromoplasts into the cytoplasm. Even though there is some loss of dry matter towards the osmotic solution during OD (observed by slight red pigmentation of the solution post-treatment), the removal of water significantly counteracts it, leading to a concentration of the pigments in tomato samples [8].

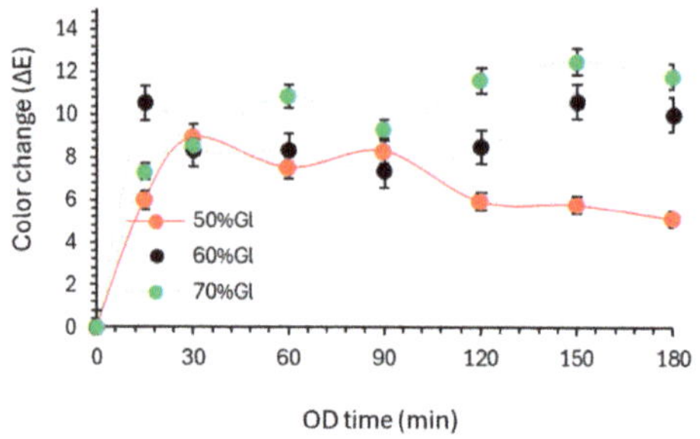

Figure 3. Evolution of color (ΔE index) of cherry tomato samples during OD (with temperature = 35 °C), using 50%, 60%, 70% w/w glycerol-based solution. Markers indicate experimental data points and error bars indicate the standard deviation from replicates.

Regarding tomato texture during OD, the firmness was significantly affected by the OD temperature (Figure 4). The highest firmness values were calculated for 70% glycerol concentration. The 50 and 60% glycerol concentrations did not affect significantly the texture of the osmodehydrated tomato. Ninety-one hundred twenty (90–120)% higher firmness values were reported for osmotically dehydrated tomatoes compared to the non-osmodehydrated ones by Dermesonlouoglou et al. [4]. Nowacka et al. (2014) observed that the OD kiwifruits presented a decrease in firmness during the 60 min of OD, followed by an increase after 120 min [51].

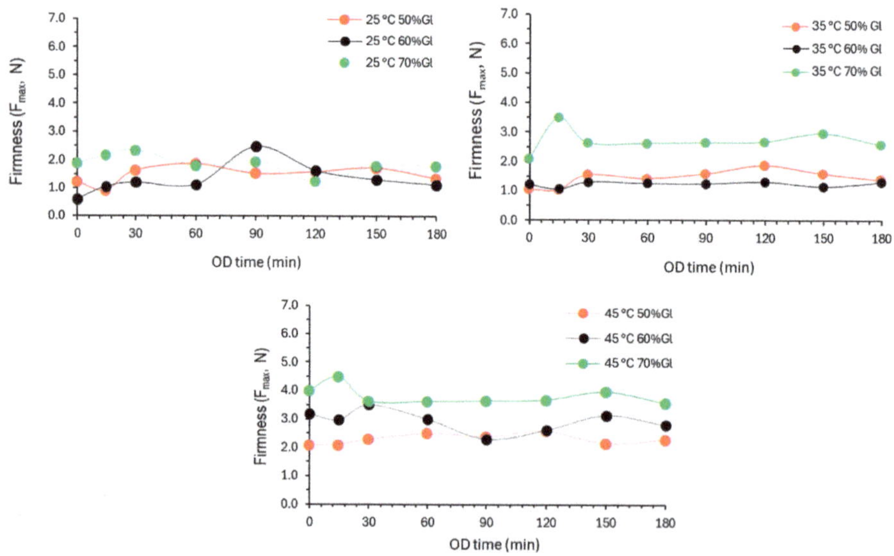

Figure 4. Evolution of texture (hardness, F_{max}, N) of cherry tomato samples at the three different temperatures studied, namely 25, 35 and 45 °C, using, in each case, the three alternative osmotic solutions, containing 50%, 60%, and 70% w/w glycerol. Markers indicate experimental data points.

3.2. OD Process Optimization

According to the results presented in Table 2, it could be concluded that the fitted models adequately represented the data and could be used to describe the design space. The asterisk sign in the table accounts for p value < 0.05 and shows which coefficients (contribution of each factor: linear, quadratic, interaction) are statistically significant, at a confidence level of 95%.

Table 2. Water loss (*WL*), Solid Gain (*SG*), water activity (a_w), hardness (F_{max}) and color (ΔE) factor effects obtained from the second-order polynomial model for the osmo-dehydrated tomatoes (in coded coefficients).

Model Coefficient	*WL* (g w./g i.d.m.)	*SG* (g s./g i.d.m.)	a_w	Firmness (F_{max}, N)	Color (ΔE)
Constant a0	4.889 *	0.130	0.91000 *	1.253 *	7.498 *
Linear					
a1	1.340 *	0.1727	−0.00525	0.398 *	0.355
a2	0.317	0.0228	−0.00437	−0.147	−0.273
a3	−0.378	−0.0033	−0.00288	0.017	0.327
Quadratic					
a11	−0.325	0.394 *	0.02325 *	0.374	−0.9418 *
a22	−0.504	0.055	0.01550 *	1.490 *	2.906 *
a33	−2.251 *	0.086	0.01850 *	−0.055	−1.624 *
Interaction					
a12	0.400	0.044	−0.00225	0.172	0.327
a13	−0.602	−0.008	−0.00325	−0.076	1.511 *
a23	−0.067	−0.077	−0.00100	0.390	0.331
R^2	0.9324	0.9211	0.9544	0.9503	0.9648
R^2 (adj)	0.8835	0.8293	0.8722	0.8609	0.9013
p-value (model)	<0.05	<0.05	<0.05	<0.05	<0.05
Lack of fit (*p*-value)	>0.05	>0.05	>0.05	>0.05	>0.05

* Statistical analysis of the factors of polynomial equations (RSM) with variance analysis (ANOVA). *p* value < 0.05, values assigned an asterisk are statistically significant coefficients at a level of 0.95.

Water loss (*WL*) and solid gain (*SG*) values are influenced mostly by temperature (a1), as shown by the higher values of the corresponding factors (Table 2), and much less by osmosis time (a2) and glycerol concentration (a3); nonetheless, none of the process conditions of time (a2) and glycerol concentration (a3) has a significant effect on *WL* and *SG*, water activity reduction and color/texture change. Regarding synergistic effects, interactions of temperature with glycerol concentration (for color) were found to have a significant effect on the parameters measured. The texture was positively affected by the OD temperature; an increase in OD temperature led to an increase in tomato firmness. The same approach for the modelling and validation of osmotic treatments of fruits (such as tomatoes) was used in the literature [6,7,9,23,24]. The regression equations (in uncoded, actual units) were as follows (Equations (13)–(17)):

$$\text{Water activity } a_w = 1.871 - 0.014408T - 0.001750t - 0.02115C + 0.000232T^2 + 0.000017t^2 + 0.000185C^2 - 0.000007T * t - 0.000032T * C - 0.000003t * C \quad (13)$$

$$\text{Water Loss } WL = -95.8 + 0.643T + 0.044t + 2.887C - 0.00325T2 - 0.000560t2 - 0.02251C2 + 0.00133T * t - 0.00602T * C - 0.00022t * C \quad (14)$$

$$\text{Solid Gain } SG = 6.40 - 0.262T + 0.0183t - 0.085C + 0.00394T2 - 0.000061t2 + 0.00086C2 + 0.000146T * t - 0.00008T * C - 0.000255T * C \quad (15)$$

$$\text{Color } \Delta E = -15.5 - 0.278T - 0.5008t + 1.386C - 0.00941T2 + 0.003228t2 - 0.01624C2 + 0.00109T * t + 0.01511T * C + 0.00110t * C \quad (16)$$

$$\text{Texture } Firmness = 12.90 - 0.210T - 0.3016t + 0.016C + 0.00374T2 + 0.001656t2 - 0.00055C2 + 0.000572T * t - 0.00076T * C + 0.001300t * C \quad (17)$$

In Figure 5, plots of the predicted and the observed (experimental) values are presented. An acceptable level of agreement ($R^2 > 0.8$) was calculated. Moreover, based on the statistical results (ANOVA for Response Surface Quadratic Model, at a confidence level of

95%), it can be observed that p-value of the model, for all different indices investigated, is significant ($p < 0.05$), whereas p-value of lack of fit is non-significant ($p > 0.05$). From all the above information, it can be concluded that the developed models could adequately represent the data.

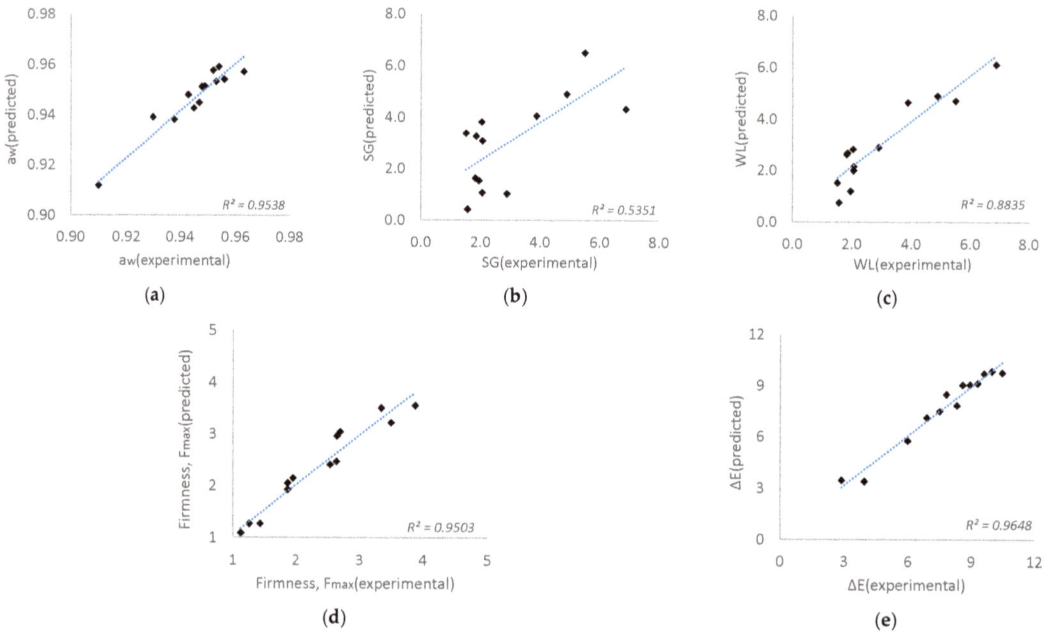

Figure 5. Linear correlation plots between predicted and observed values for (**a**) water activity (a_w), (**b**) water loss (WL), (**c**) solid gain (SG), (**d**) color (ΔE) and (**e**) firmness (F_{max}) responses.

Osmotic Dehydration Process Optimization and Validation

According to mass transfer and quality loss results, of the osmodehydrated tomato (from the previous section), the criteria for the OD process optimization were set as: water loss $WL \leq 5$, color change $\Delta E \leq 8$ and at the same time minimization of the water activity. The upper limit of acceptance for the color change was chosen with the help of the sensory evaluation test results as well as the kinetic study results for the color loss. The upper limit for the water loss was set based on the mass transfer kinetic modelling. To find the OD process parameters that meet the pre-determined criteria, the desirability function method was applied and the corresponding composite desirability profiles were depicted in Figure 5. To apply the methodology, for each of the critical process conditions (% glycerol, process temperature, process time,) the levels were allowed to take values within the experimental range. Optimal OD process conditions were found as follows: 36 °C–72 min, 61.5% glycerol, respectively, based on the simultaneous, multiple criteria set (Figure 6).

Figure 6 shows the desirability diagram of the process, with the criteria set for optimization. How effectively a combined variable satisfies the objectives is shown by the individual level of desirability (represented by d for each response) and the composite level of desirability (represented by D for the integrated response). While the composite factor (D) evaluates how effectively the arrangements optimize an integrated criterion, the individual desirability (d) evaluates the degree to which the arrangements optimize a particular response (y). The geometric mean of the individual desirabilities for the chosen responses is used to determine the composite desirability, in accordance with the algorithm used in this work. The composite desirability D = 0.99 of the configurations suggests that they significantly satisfy the required combination of criteria. Given that the individual desirability levels are d = 0.99, this is reasonable. In our study, the synthetic desirability (D = 0.99) indicates that the arrangements appear to significantly satisfy the combination of criteria required. This is reasonable since the individual desirability values are (d = 0.99).

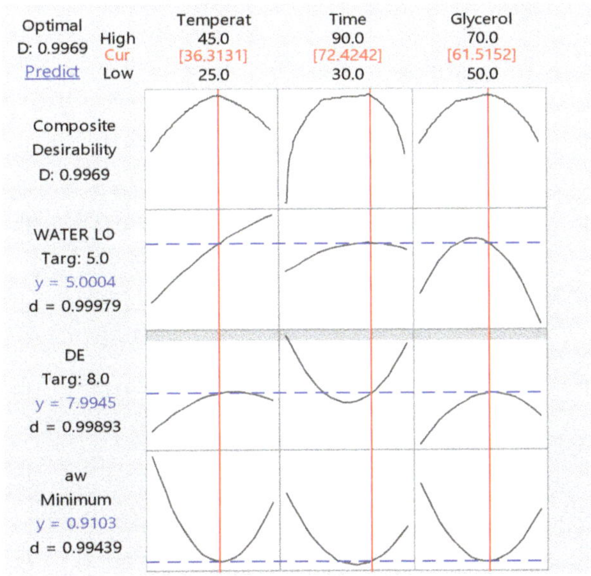

Figure 6. Desirability plot of variables.

In Table 3, predicted values of the dependent variables are estimated at those optimum conditions, calculated by the polynomial equations developed, and compared to values measured by an independent validation experiment (% Absolute Error < 20%). The data were validated by conducting an independent experimental test in the optimal process conditions (validation test). It is observed the % Absolute Error calculated for all measured values (*WL*, *SG*, a_w, ΔE) (apart from the % Absolute Error for the texture) was below <20%. Various studies have shown that the variety of the tomato can affect its texture (firmness) after osmotic dehydration as different varieties have different textures, structures and chemical compositions. Some tomato varieties (depending on the growing conditions) may have thinner skin and softer cells, while others may have a denser structure or more flesh. These differences in structure and chemical composition can affect how the tomato responds to osmotic dehydration. Different quality parameters have also been reported for the same tomato varieties [52].

Table 3. Experimental and predicted values for the responses at optimum OD process parameters (temperature: 36 °C; time: 72 min; Glycerol concentration: 61.5% *w*/*w*) of cherry tomatoes.

	Experimental Value	Predicted Value	% Error
Water Loss (*WL*, g w./g i.d.m.)	4.881	4.890	0.21
Solid Gain (*SG*, g S./g i.d.m.)	0.130	0.152	16.5
Water activity (a_w)	0.9105	0.9117	0.13
Color change (ΔE)	6.24	7.49	−18.0
Firmness (F_{max})	2.08	1.26	39.4

3.3. Shelf Life of Frozen Tomatoes, OD vs. Untreated, during Storage

3.3.1. Quality Evolution during Storage at Isothermal Conditions

Physicochemical Characteristics

The osmodehydrofrozen cherry tomatoes had water activity values around 0.90 compared to 0.95 for the non-osmodehydrated frozen (untreated). The pH values were reduced from 4.10 to 3.90. After the application of osmotic dehydration, the soluble solids (expressed as °Brix) significantly increased, from 8.81 to 20.01. The frozen storage did not affect a_w, pH, dry matter and soluble solids ($p > 0.05$).

Drip Loss and Texture

In Figure 7a,b the drip loss for untreated and OD-treated frozen cherry tomatoes stored at isothermal conditions is presented. Non-treated samples suffered from significantly higher drip loss and tissue softening at zero storage time. Tomato integrity was retained with storage time for all osmotically pre-treated samples (low drip loss showing a decrease with storage time and increased firmness not showing any clear tendency). Similar results were obtained for osmodehydrofrozen fruit and vegetables such as pineapple, mangoes melon and strawberries, respectively [39,53–55].

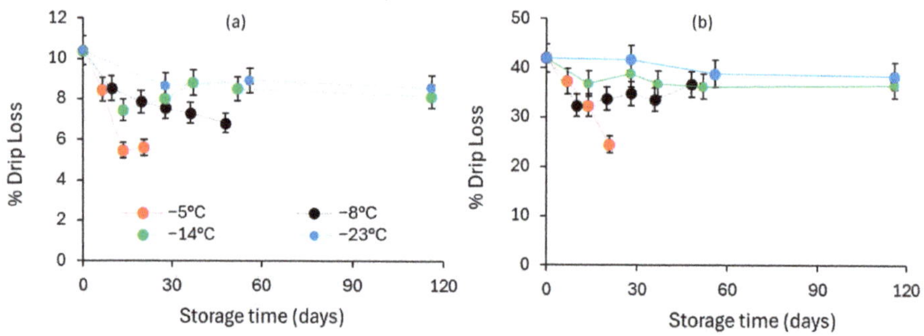

Figure 7. Evolution of drip loss (%) during storage for (**a**) non-pretreated and (**b**) OD-pretreated cherry tomatoes stored at T = −5, −8, −14, −23 °C (error bars indicate the standard deviation from replicates).

The texture properties (mainly firmness) of tomato acceptability are significant. Changes in the composition and arrangement of the cells and the turgor during the maturation process lead to the development of firmness that impacts texture, shelf life, and total acceptance [56]. The firmness values for the osmodehydrated tomatoes were 2.664 ± 0.527 N at all storage temperatures compared to the firmness values of untreated tomatoes which were 0.527 ± 0.106 N. An increasing trend was observed during storage (not statistically significant). In Figure 8, the firmness values of untreated and OD-treated frozen cherry tomatoes stored at isothermal conditions T = −14 °C and −23 °C are representatively presented. The osmodehydrated tomatoes were characterized by the sensory panelists as "not soft", and the same time as "having a more intense flavor, and a sweeter taste (acceptable and desirable)". Osmodehydrated cherry tomatoes presented significantly increased firmness compared to their untreated counterparts ($p < 0.05$). Frozen delicate plant matrices, with high water content (such as tomatoes and fruits) pre-treated by osmotic dehydration have been reported to show improved texture, after thawing [39,55].

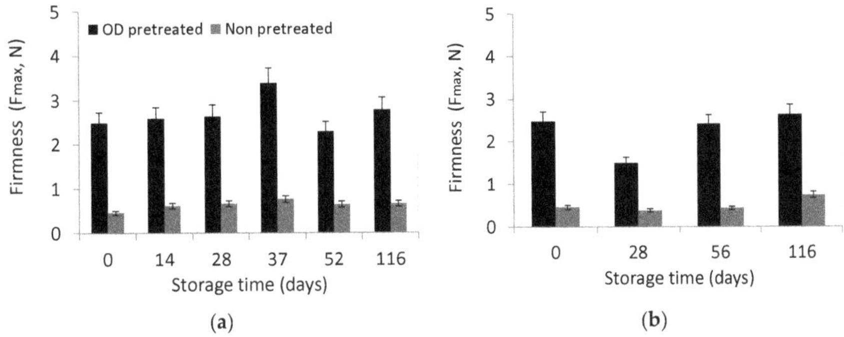

Figure 8. Evolution of firmness (F_{max}, N) during storage for non-pretreated and OD-treated cherry tomatoes stored at (**a**) T = −14 °C and (**b**) T = −23 °C (error bars indicate the standard deviation from replicates).

Color

The color of OD pretreated tomato samples did not present statistically significant differences compared to the color of untreated tomato samples. Color indices L, a, b of non-pretreated and OD pretreated tomato samples were measured as: Non: 28.49 ± 2.97, 18.18 ± 3.10, 13.65 ± 3.20; OD: 28.25 ± 3.41, 19.47 ± 1.30, 13.16 ± 1.82 (at storage time 0 day). According to Gould et al. [30], the color parameter a/b that exceeds the value 0.5 is accepted as a measure of ripeness. United States Department of Agriculture (USDA) recommendation [57], cherry tomatoes are considered attractive for marketing when their a/b ranges from 0.95 to 1.21. In this study, a/b values of untreated and OD pre-treated cherry tomatoes were calculated as 1.33 and 1.47, respectively.

Cherry tomato color change during storage is attributed to lycopene degradation and pigment accumulation including anthocyanins, flavonoids and carotenoids in the tomato skin [47]. In this study, the total color change of all frozen cherry tomatoes was expressed by ΔE (Figure 9a,b). The storage of frozen cherry tomatoes significantly changed color. The color change was more intense for the osmotically dehydrated tomato samples. L and b parameters significantly increased; a parameter decreased. The same trend was observed for the non-pretreated tomato samples. Most published works reported a slight color change from red to orange. Urbanyi et al. (1998) reported a similar color change accompanied by a gradual reduction in carotenoids during 24 weeks of frozen storage [58]. The same was observed by Lisiewska et al. (2000) after 6 months of storage at −20 °C [29].

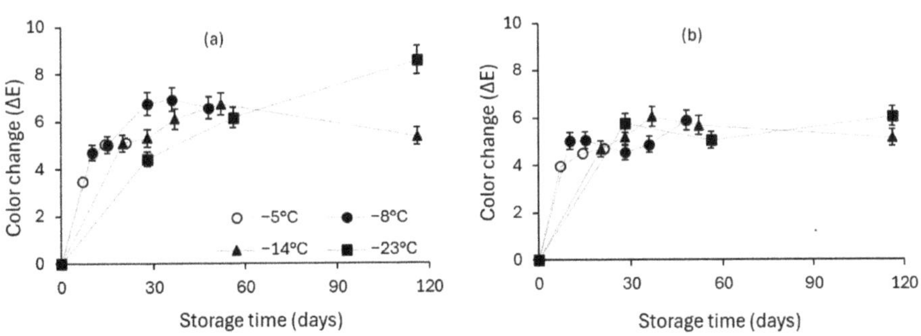

Figure 9. Evolution of color change (ΔE) during storage for (**a**) untreated and (**b**) osmodehyddrofrozen cherry tomatoes stored at T = −5, −8, −14, −23 °C (error bars indicate the standard deviation from replicates).

3.3.2. Vitamin C, Lycopene and Sensory Quality Loss: Mathematical Modelling

The protective effect of OD on vitamin C, lycopene and sensory properties' retention is demonstrated. Vitamin C, total lycopene and sensory quality of the non-pretreated and OD pre-treated cherry tomatoes stored at constant temperature conditions T = −5, −8, −14, −23 °C are presented (Figure 10). Vitamin C and lycopene, owing important health benefits, are considered quality indicators of cherry tomatoes [59]. As shown in Figure 10, throughout frozen storage, Vitamin C and lycopene content decreases. The decrease in Vitamin C and lycopene content can be attributed to oxidation [28]. Compared to the untreated samples, OD pre-treatment delayed the vitamin C and lycopene loss of cherry tomatoes during storage.

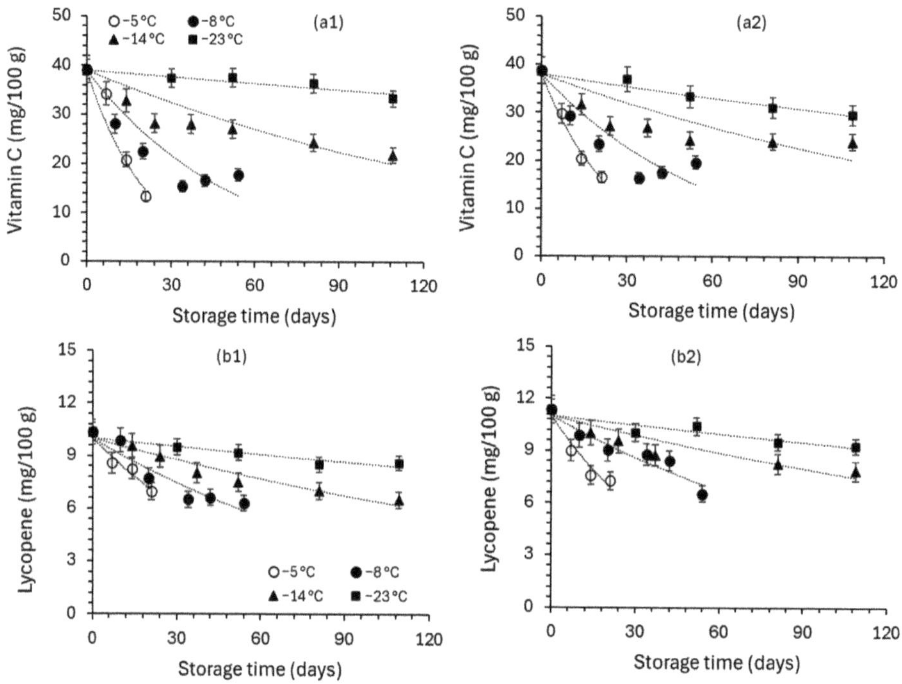

Figure 10. Evolution of (**a**) vitamin C and (**b**) lycopene during storage for (**1**) non-pretreated and (**2**) OD-pretreated cherry tomatoes stored at constant temperature conditions T = −5, −8, −14, −23 °C (error bars indicate the standard deviation from replicates).

Osmodehydrated cherry tomatoes with glycerol-based solutions showed higher lycopene content than fresh cherry tomato, as reported by Heredia et al. [60,61]. According to Heredia et al. (2010), this could be due to the optimal temperature of synthesis of lycopene (40 °C), or to the effect of osmotic stress that encouraged phytochemicals generation and the presence of a precursor like sucrose [60]. Additionally, Heradia et al. [61] showed that the osmotic dehydration limited the isomerization during the later stage of drying, whereas both the total lycopene loss and the trans–cis isomerization, mainly to the 13-cis form, were favored by an increase in osmotic dehydration/drying temperature and the microwave power. In this study, the lycopene content of the osmodehydrated sample was 11.35 mg lycopene/100 g OD tomato compared to the value of the non-osmodehydrated tomato sample 10.35 mg/100 g. During frozen storage, after 3 months of storage at the temperature of −23 °C, the osmodehydrofrozen samples retained better their (total) lycopene content (83.1% of the initial content) compared to the non-osmodehydrofrozen tomato samples (81.5%, respectively of the initial content).

Dermesonlouoglou et al. [4] did not report statistically significant differences between untreated and OD pre-treated tomatoes, values ranged from 24.0 to 30.1 mg L-ascorbic acid/100 g OD tomato compared to the value of the untreated tomato 28.9 mg/100 g. After 12 months of storage at −20 °C, osmodehydrofrozen tomatoes retained better their vitamin content (66.2–88.0% of the initial) compared to the conventionally frozen, untreated and blanched tomato samples (44.1% and 57.8%, respectively). Osmodehydrofrozen tomatoes stored at higher temperatures (e.g., at −12 °C) also exhibited better Vitamin C stability for long storage times. In this study, the vitamin C content of osmodehydrated sample was 39.15 mg L-ascorbic acid/100 g OD tomato compared to the value of the non-osmodehydrated tomato sample 38.79 mg/100 g. During frozen storage, after 3 months of storage at the temperature of −23 °C, the osmodehydrofrozen samples retained better their vitamin content (85.0% of the initial vitamin content) compared to the non-osmodehydrofrozen tomato samples (76.2%, respectively, of the initial vitamin content). An et al., 2013 reported an improved retention of vitamin C in osmotically dehydrated and dried cherry tomatoes compared to dried untreated tomatoes [62]. However, a slight decrease in vitamin C after the osmodehydrofreezing process, possibly owing to a mild leakage of water-soluble compounds out of the cell tissue, has been reported [63]. Zhao et al. (2017) in [53] showed that when using a mixture of sucrose/glucose/fructose as the OD medium, the vitamin C of ODF mangoes was significantly decreased before frozen storage (with OD samples having a water content of approximately 80% w.b.). Nonetheless, as it was pointed out, this loss was counterbalanced by the improved vitamin C retention of OD samples during subsequent storage, possibly owing to rapid leaching via increased drip loss of the untreated samples [58]. Fan et al. (2020) studied osmotic dehydration with ultrasound enhancement before freezing of kiwi and found better retention of ascorbic acid content; however, they underlined that, since Vitamin C is a water-soluble compound, it can be readily lost during thawing, and thus the extent of drip loss in osmodehydrofrozen samples is strongly related to the retention of ascorbic acid [64]. Similar findings can be found in Xin et al. 2013, who observed a milder leak of vitamin C into the osmotic solution (containing 40% trehalose) when an ultrasound pretreatment was applied, mainly attributed to the shorter dehydration time compared with the OD counterparts (with a water content of approximately 65% w.b. [65].

In this study, OD frozen cherry tomatoes presented bright color, good texture and pleasant taste, whereas the untreated tomatoes suffered from a detrimental texture change [high drip loss (Figure 6) and tissue softening] and taste deterioration during storage. The pre-treated samples were preferred in terms of all sensory attributes including taste and they showed increased stability with storage compared to non-treated samples Dermesonlouoglou et al. (2008, 2016, 2018) reported that panel assessments confirmed the analytical results of better-quality retention of somodehydrofrozen strawberries, and kiwi, also judging their flavor as pleasant [55,66]. Other research results showed that osmotic dehydration enhanced the sensory attributes of frozen samples [64,67]. Zhao et al. (2016) found improved color and texture, reduced drip loss, as well as retention of the aromatic compounds in osmodehydrofrozen mangoes [68].

The kinetic models developed for frozen tomatoes, based on Equations (4) and (5), are presented in Table 4 (Equations (2)–(4)).

The rejection of both untreated and osmodehydrofrozen cherry tomatoes was due to sensory quality loss (the shelf life determining factor was the sensory evaluated overall impression). For example, the shelf life at −18 °C for untreated osmodehydrofrozen cherry tomatoes was calculated as: Untreated 42 days (sensory rejection), 166 days (vitamin C loss) and 212 days (lycopene loss) and OD pretreated: 136 days (sensory rejection), 208 days (lycopene loss) and 234 days (vitamin C loss), respectively. OD increased the shelf life of frozen cherry tomato (up to 3.5 times).

Table 4. Estimated kinetic parameters [$k_{asc/lyc/sens}$, k_{ref} (days^{-1}) at T_{ref} = −18 °C, E_a, kJ/mol] for Vitamin C, lycopene and sensory quality of untreated and osmodehydrofrozen cherry tomatoes.

	Kinetic Model Structure and Model Dependent Variables	
	Untreated	OD Treated
Vitamin C C_{asc} = L-ascorbic acid content (mg/100 g)	E_a = 88.8 kJ/mol k_{ref} (−18 °C) = 0.0033 d^{-1} *SL* (50% loss): 20 d (−5 °C); 37 d (−8 °C); 87 d (−14 °C); 402 d (−23 °C)	E_a = 115.6 kJ/mol k_{ref} (−18 °C) = 0.0025 d^{-1} *SL* (50% loss): 15 d (−5 °C); 33 d (−8 °C); 102 d (−14 °C); 737 d (−23 °C)
Lycopene C_{lyc} = Lycopene content (mg/100 g)	E_a = 88.5 kJ/mol k_{ref} (−18 °C) = 0.0025 d^{-1} *SL* (50% Loss): 44 d (−5 °C); 68 d (−8 °C); 130 d (−14 °C); 415 d (−23 °C)	E_a = 86.1 kJ/mol k_{ref} (−18 °C) = 0.0020 d^{-1} *SL* (50% Loss): 46 d (−5 °C); 70 d (−8 °C); 131 d (−14 °C); 395 d (−23 °C)
Sensory evaluation *S* = Score for sensory quality	E_a = 56.58 kJ/mol k_{ref} = 0.017 d^{-1} S_i = 7.1, S_f = 4 *SL*: 14 d (−5 °C); 19 d (−8 °C); 30 d (−14 °C); 67 d (−23 °C)	E_a = 57.10 kJ/mol k_{ref} = 0.009 d^{-1} S_i = 9.0; S_f = 4 *SL*: 50 d (−5 °C); 66 d (−8 °C); 100 d (−14 °C); 207 d (−23 °C)

3.3.3. Validation in Non-Isothermal Storage Conditions

Since storage and transportation temperatures of frozen foods (fruits) can vary, it is important to be able to reliably estimate the effect of this variability and predict product shelf life in the actual frozen food chain. The developed models at isothermal storage experiments were validated at non-isothermal (dynamic) storage conditions (fluctuating temperatures from −5 to −12 °C with effective temperature, T_{eff} = −7.3 °C) [37]. The results showed that these models can be adequately used at variable temperature conditions, within the same range and time of temperature fluctuations, to predict the tomato quality and consequently the remaining shelf life, based on the relative error (*RE*) values, calculated using Equation (5) (Table 5).

Table 5. Quality loss rates (*k*), experimental and predicted, for Vitamin C, lycopene and sensory quality of untreated and OD-treated cherry tomato samples, and the relative error (RE) values.

	Vitamin C		Lycopene		Sensory Quality	
	Untreated	OD	Untreated	OD	Untreated	OD
$k_{experimental}$ (d^{-1})	0.0179	0.0228	0.011	0.011	0.1816	0.0760
$k_{predicted}$ (d^{-1})	0.023 ± 0.023	0.0284 ± 0.019	0.0074 ± 0.0052	0.0123 ± 0.0143	0.1867 ± 0.247	0.0654 ± 0.0305
%RE	20.01	19.71	−47.05	10.56	2.73	−16.2

4. Conclusions

Osmotic dehydration could lead to adequate water removal, solids uptake and water activity reduction, while improving the retention of the main quality characteristics of frozen cherry tomato. Mass exchange (*WL*, *SG*, a_w), firmness, and color change during OD were modeled using a second-order polynomial model describing the effect of the most important OD processing factors. The optimum processing conditions were OD: T_{OD} = 36 °C, t_{OD} = 72 min, and C_{glyc} + 61.5% w/w, based on the triple criterion of obtaining water loss $WL \leq 5$, color change $\Delta E \leq 8$ and at the same time minimizing water activity at the end of the pretreatment. The shelf life determining criteria for both untreated and OD pretreated samples were found to be sensory rejection (expressed by the score given for the overall quality and acceptability). OD pretreated frozen cherry tomatoes presented an a_w decrease from 0.95 to 0.92, acceptable color, increased firmness, low drip loss and high vitamin C/lycopene retention during frozen storage. Based on this study, it is suggested that OD may have the potential as a pre-processing step in the manufacture of sensitive frozen fruit products. However, one should mention the limitations of the RSM implemented in the first part (OD process); being a local analysis, the developed response surface models, as outlined in Equations (13)–(17), are invalid for regions other than the studied ranges of independent factors investigated (temperature, glycerol concentration and osmosis time). Another often reported limitation is that Response Surface Methodology (RSM) is a 'black box' approach [69], and thus, estimating the accuracy of the approximation is rather difficult. Nonetheless, in our approach, an experimental validation of RSM models was performed at the optimal conditions calculated (to alleviate the acknowledged RSM weaknesses) and the applicability of the RSM equations derived was confirmed. Concerning the second part of the integrated analysis (the stability test during frozen storage), one should also bear in mind that the results of the Arrhenius equation should be cautiously used in different conditions, and always within the range of temperature conditions analyzed. Finally, one should also consider possible deviations from the Arrhenius law, if the glass transition of the frozen matrix occurs within the temperature range studied [70]. Especially in the particular study conducted, the tomato matrix was modified due to the OD process (solid impregnation and water activity decrease occurred; therefore, the glass transition temperature is expected to decrease, and thus a thorough investigation of kinetics at the lower temperatures of storage (<-18 °C) would be necessary to challenge the uniform application of the Arrhenius equation.

Author Contributions: Methodology, L.P. and E.D.; validation, L.P.; investigation, L.P. and E.D.; data curation, L.P.; writing—original draft preparation, E.D.; writing—review and editing, M.G., P.T.; visualization, L.P.; supervision, E.D., M.G. and P.T. All authors have read and agreed to the published version of the manuscript.

Funding: This research received no external funding.

Institutional Review Board Statement: Ethical review and approval were waived for this study due to: only adults participated in the recruitment to the sensory team. Participation in the tests and assessments was voluntary. Informed written consent was obtained from the participants in the sensory evaluation study. Each of them could withdraw their consent without providing any justification. Each participant also consented to the processing of their personal data in accordance with Article 6 of Regulation (EU) 2016/679 of the European Parliament and of the Council of 27 April 2016 on the protection of natural persons regarding the processing of personal data and on the free movement of such data, and repealing Directive 95/46/EC (General Data Protection Regulation). All participants obtained a detailed description of the test and were informed about the food samples that would be assessed. Each of the assessors was obliged to report any indispositions and allergies and if such was the case, the subject did not participate in the tests. Recently, we published sensory test results obtained in our Laboratory following the same experimental procedure.

Informed Consent Statement: Informed consent was obtained from all subjects involved in the study.

Data Availability Statement: The original contributions presented in the study are included in the article, further inquiries can be directed to the corresponding author.

Conflicts of Interest: The authors declare no conflicts of interest.

References

1. Ahmed, I.; Qazi, I.M.; Jamal, S. Developments in osmotic dehydration technique for the preservation of fruits and vegetables. *Innov. Food Sci. Emerg. Technol.* **2016**, *34*, 29–43. [CrossRef]
2. Asghari, A.; Assana Zongo, P.; Osse, E.F.; Aghajanzadeh, S.; Raghavan, V.; Khalloufi, S. Review of osmotic dehydration: Promising technologies for enhancing products' attributes, opportunities, and challenges for the food industries. *Compr. Rev. Food Sci. Food Saf.* **2024**, *23*, e13346. [CrossRef] [PubMed]
3. Rastogi, N.K. Developments in osmotic dehydration of foods. In *Drying Technology in Food Processing*; Jafari, S.M., Malekjani, N., Eds.; Woodhead Publishing: Cambridge, UK, 2023; pp. 241–304.
4. Dermesonlouoglou, E.; Giannakourou, M.; Taoukis, P. Stability of dehydrofrozen tomatoes pretreated with alternative osmotic solutes. *J. Food Eng.* **2007**, *78*, 272–280. [CrossRef]
5. Dermesonlouoglou, E.K.; Pantelaiaki, K.; Andreou, V.; Katsaros, G.J.; Taoukis, P.S. Osmotic pretreatment for the production of novel dehydrated tomatoes and cucumbers. *J. Food Process. Preserv.* **2019**, *43*, e13968. [CrossRef]
6. Derossi, A.; Severini, C.; Del Mastro, A.; De Pilli, T. Study and optimization of osmotic dehydration of cherry tomatoes in complex solution by response surface methodology and desirability approach. *LWT* **2015**, *60*, 641–648. [CrossRef]

7. Saleena, P.; Jayashree, E.; Anees, K. A Comprehensive Review on Vacuum Impregnation: Mechanism, Applications and Prospects. *Food Biopr. Technol.* **2024**, *17*, 1434–1447. [CrossRef]
8. Katsimichas, A.; Dimopoulos, G.; Dermesonlouoglou, E.; Taoukis, P. Modelling and Evaluation of the Effect of Pulsed Electric Fields and High-Pressure Processing Conditions on the Quality Parameters of Osmotically Dehydrated Tomatoes. *Appl. Sci.* **2023**, *13*, 11397. [CrossRef]
9. Wu, S.; Wang, J.; Zhang, L.; Liu, S.; Li, C. Effects of Osmotic Dehydration on Mass Transfer of Tender Coconut Kernel. *Foods* **2024**, *13*, 2188. [CrossRef]
10. Assis, F.R.; Morais, R.M.S.C.; Morais, A.M.M.B. Mass Transfer in Osmotic Dehydration of Food Products: Comparison between Mathematical Models. *Food Eng. Rev.* **2016**, *8*, 116–133. [CrossRef]
11. de Farias Aires, J.E.; da Silva, W.P.; Aires, K.L.C.d.A.F.; da Silva Júnior, A.F.; Silva, C.M.D.P.d.S.e. Description of osmotic dehydration of apple using two-dimensional diffusion models considering shrinkage and variations in process parameters. *Dry. Technol.* **2017**, *35*, 815–826. [CrossRef]
12. Pacheco-Angulo, H.; Herman-Lara, E.; García-Alvarado, M.A.; Ruiz-López, I.I. Mass transfer modeling in osmotic dehydration: Equilibrium characteristics and process dynamics under variable solution concentration and convective boundary. *Food Bioprod. Process.* **2016**, *97*, 88–99. [CrossRef]
13. Ramya, V.; Jain, N.K. A Review on osmotic dehydration of fruits and vegetables: An integrated approach. *J. Food Process Eng.* **2017**, *40*, e12440. [CrossRef]
14. Pandiselvam, R.; Tak, Y.; Olum, E.; Sujayasree, O.; Tekgül, Y.; Çalışkan Koç, G.; Kaur, M.; Nayi, P.; Kothakota, A.; Kumar, M. Advanced osmotic dehydration techniques combined with emerging drying methods for sustainable food production: Impacton bioactive components, texture, color, and sensory properties of food. *J. Texture Stud.* **2022**, *53*, 737–762. [CrossRef]
15. Kaur, D.; Singh, M.; Zalpouri, R.; Singh, I. Osmotic dehy-dration of fruits using unconventional natural sweeteners andnon-thermal-assisted technologies: A review. *J. Food Process. Preserv.* **2022**, *46*, e16890. [CrossRef]
16. Kaur, B.; Rana, P.; Sridhar, K. Mass transfer kinetics andprocess optimization of osmotic dehydration of Kinnow man-darin (*Citrus reticulata*) peel. *J. Food Process. Preserv.* **2022**, *46*, e16318.
17. Reyes-Alvarez, C.A.; Lanari, M.C. Effect of freezing, osmodehydro-freezing, freezedrying and osmodehydro-freezedrying on the physicochemical and nutritional properties of arazá (*Eugenia stipitata* McVaugh). *Food Chem. Adv.* **2023**, *3*, 100496. [CrossRef]
18. Manzoor, A.; Jan, B.; Rizvi, Q.U.E.H.; Junaid, P.M.; Pandith, J.A.; Dar, I.H.; Bhat, S.A.; Ahmad, S. Osmotic dehydration technology for preservation of fruits and vegetables. In *Quality Control in Fruit and Vegetable Processing*; Apple Academic Press: Palm Bay, FL, USA, 2023; pp. 167–184.
19. Aachary, A.A.; Prapulla, S.G. Value addition to spent osmotic sugar solution (SOS) by enzymatic conversion to fructooligosaccharides (FOS), a low calorie prebiotic. *Innov. Food Sci. Emerg. Technol.* **2009**, *10*, 284–288. [CrossRef]
20. Fernández, P.R.; Lovera, N.; Ramallo, L.A. Sucrose syrup reuse during one- and multi-stage osmotic dehydration of pineapple. *J. Food Process Eng.* **2020**, *43*, e13399. [CrossRef]
21. Yadav, B.S.; Yadav, R.B.; Jatain, M. Optimization of osmotic dehydration conditions of peach. *J. Food Sci. Technol.* **2012**, *49*, 547–555. [CrossRef]
22. Vieira, G.S.; Pereira, L.M.; Hubinger, D.M. Optimization of osmotic dehydration process of guavas by response surface methodology and desirability function. *Int. J. Food Sci. Technol.* **2012**, *47*, 132–140. [CrossRef]
23. Rahman, S.M.A.; Sharma, P.; Said, Z. Application of response surface methodology based D-optimal design for modeling and optimisation of osmotic dehydration of zucchini. *Digit. Chem. Eng.* **2022**, *4*, 100039. [CrossRef]
24. Romero, I.P.; Rodríguez Gomez, M.J.; Sanchez Iniguez, F.M.; Magro, P.C. Optimization of the osmotic dehydration process of plums (*Prunus Salicina* Lindl.) in solutions enriched with inulin, using response surface methodology. *LWT* **2022**, *157*, 113092. [CrossRef]
25. Tang, N.; An, J.; Deng, W.; Gao, Y.; Chen, Z.; Li, Z. Metabolic and transcriptional regulatory mechanism associated with postharvest fruit ripening and senescence in cherry tomatoes. *Postharvest Biol. Technol.* **2020**, *168*, 111274. [CrossRef]
26. Li, Y.; Nie, J.; Shi, L.; Xie, Y.; Tan, D.; Yang, X.; Zhang, C.; Zheng, J. Transcriptomic and metabolomic profiling reveals the mechanisms of color and taste development in cherry tomato cultivars. *LWT* **2022**, *167*, 113810. [CrossRef]
27. Zhu, M.; Yang, P.; Zhu, L. Preparation of modified atmosphere packaging based on the respiratory characteristics of cherry tomato and its freshness preservation application. *Sci. Hortic.* **2024**, *333*, 113286. [CrossRef]
28. Zoo, F.; Tan, C.; Chang, Z.; Shinali, T.S.; Zhang, B.; Zhang, L.; Han, Z.; Wu, W.; Shang, N. New strategy for improving postharvest quality of cherry tomatoes: Synergy of plasma-activated water and Welsh onion leaf protein extracts. *Food Control* **2024**, *164*, 110592. [CrossRef]
29. Lisiewska, Z.; Kmiecik, W. Effect of storage period and temperature on the chemical composition and organoleptic quality of frozen tomato cubes. *Food Chem.* **2000**, *70*, 167–173. [CrossRef]
30. Gould, W.A. Chapter 16 Quality Evaluation of Processed Tomatoes and Tomato Products. In *Tomato Production, Processing and Technology*, 3rd ed.; Elsevier: Cambridge, UK, 1992; pp. 293–296.
31. Parniakov, O.; Bals, O.; Mykhailyk, V.; Lebovka, N.; Vorobiev, E. Unfreezable water in apple treated by pulsed electric fields: Impact of osmotic impregnation in glycerol solutions. *Food Bioprocess Technol.* **2016**, *9*, 243–251. [CrossRef]
32. Guiamba, I.; Ahrne, L.; Khan, M.A.M.; Svanberg, U. Retention of β-carotene and vitamin C in dried mango osmotically pretreated with osmotic solutions containing calcium or ascorbic acid. *Food Bioprod. Process.* **2016**, *98*, 320–326. [CrossRef]

33. Gras, M.L.; Vidal, D.; Betoret, N.; Chiralt, A.; Fito, P. Calcium fortification of vegetables by vacuum impregnation: Interactions with cellular matrix. *J. Food Eng.* **2003**, *56*, 279–284. [CrossRef]
34. Panagiotou, N.M.; Karathanos, V.T.; Maroulis, Z.B. Mass transfer modelling of the osmotic dehydration of some fruits. *Int. J. Food Sci. Technol.* **1998**, *33*, 267–284. [CrossRef]
35. Terkmane, N.; Krea, M.; Moulai-Mostefa, N. Optimisation of inulin extraction from globe artichoke (*Cynara cardunculus* L. subsp. *scolymus* (L.) Hegi) by electromagnetic induction heating process. *Int. J. Food Sci. Technol.* **2016**, *51*, 1997–2008. [CrossRef]
36. Taoukis, P.S.; Labuza, T.P. *Handbook of Food Engineering Practice*; Valentas, K.J., Rotstein, E., Singh, R.P., Eds.; CRC: New York, NY, USA, 1997; p. 361.
37. Giannakourou, M.C.; Taoukis, P.S. Kinetic modelling of vitamin C loss in frozen green vegetables under variable storage conditions. *Food Chem.* **2003**, *83*, 33–41. [CrossRef]
38. Rodriguez, A.; Soteras, M.; Campañone, L. Effect of the combined application of edible coatings and osmotic dehydration on the performance of the process and the quality of pear cubes. *Int. J. Food Sci. Technol.* **2021**, *56*, 6474–6483. [CrossRef]
39. Alharaty, G.; Ramaswamy, H.S. Microwave-Osmo-Dehydro-Freezing and Storage of Pineapple Titbits—Quality Advantage. *Processes* **2023**, *11*, 494. [CrossRef]
40. Andreou, V.; Dimopoulos, G.; Dermesonlouoglou, E.; Taoukis, P. Application of pulsed electric fields to improve product yield and waste valorization in industrial tomato processing. *J. Food Eng.* **2020**, *270*, 109778. [CrossRef]
41. *ISO 6658:2017*; Sensory Analysis—Methodology—General Guidance. International Organization for Standardization: Geneva, Switzerland, 2017.
42. *ISO 20613:2019*; Sensory Analysis—General Guidance for the Application of Sensory Analysis in Quality Control. International Organization for Standardization: Geneva, Switzerland, 2019.
43. Dermesonlouoglou, E.K.; Giannakourou, M.C.; Taoukis, P.S. Kinetic modelling of the degradation of quality of osmo-dehydrofrozen tomatoes during storage. *Food Chem.* **2007**, *103*, 985–993. [CrossRef]
44. Cánovas, J.A.; Gea-Botella, S.; Borrás, F.; Martí, N.; Valero, M.; Saura, D.; Martínez-Madrid, M.C.; Laencina, J. Vitamin C loss kinetics and shelf life study in fruit-based baby foods during post packaging storage. *Food Pack. Shelf Life* **2020**, *23*, 100453. [CrossRef]
45. Gougouli, M.; Angelidis, A.S.; Koutsoumanis, K. A study on the kinetic behavior of *Listeria monocytogenes* in ice cream stored under static and dynamic chilling and freezing conditions. *J. Dairy Sci.* **2008**, *97*, 523–530. [CrossRef]
46. Ma, Y.; Yi, J.; Bi, J.; Zhao, Y.; Li, X.; Wu, X.; Du, Q. Effect of ultrasound on mass transfer kinetics and phenolic compounds of apple cubes during osmotic dehydration. *LWT* **2021**, *151*, 112186. [CrossRef]
47. Prithani, R.; Dash, K.K. Mass transfer modelling in ultrasound assisted osmotic dehydration of kiwi fruit. *Innov. Food Sci. Emerg. Technol.* **2020**, *64*, 102407. [CrossRef]
48. Madaeni, S.S.; Khodabakhshi, A. Dehydration of alcohols using osmotic concentration—Dehydration of aqueous glycerol solution. *J. Food Eng.* **2008**, *86*, 49–54. [CrossRef]
49. Kroehnke, J.; Szadzińska, J.; Radziejewska-Kubzdela, E.; Biegańska-Marecik, R.; Musielak, G.; Mierzwa, D. Osmotic dehydration and convective drying of kiwifruit (*Actinidia deliciosa*)—The influence of ultrasound on process kinetics and product quality. *Ultrason. Sonochem.* **2021**, *71*, 105377. [CrossRef]
50. Choi, M.H.; Kim, G.H.; Lee, H.S. Effects of ascorbic acid retention on juice color and pigment stability in blood orange (*Citrus sinensis*) juice during refrigerated storage. *Food Res. Int.* **2002**, *35*, 753–775. [CrossRef]
51. Nowacka, M.; Tylewicz, U.; Romani, S.; Dalla Rosa, M.; Witrowa-Rajchert, D. Influence of ultrasound-assisted osmotic dehydration on the main quality parameters of kiwifruit. *Innov. Food Sci. Emerg.* **2017**, *41*, 71–78. [CrossRef]
52. Diane, B.M.; John, B.C.; Rob, S. Color, flavor, texture, and nutritional quality of fresh-cut fruits and vegetables: Desirable levels, instrumental and sensory measurement, and the effects of processing. *Crit. Rev. Food Sci. Nutr.* **2010**, *50*, 369–389.
53. Zhao, J.-H.; Xiao, H.-W.; Ding, Y.; Nie, Y.; Zhang, Y.; Zhu, Z.; Tang, X.-M. Effect of osmotic dehydration pretreatment and glassy state storage on the quality attributes of frozen mangoes under long-term storage. *J. Food Sci. Technol.* **2017**, *54*, 1527–1537. [CrossRef]
54. Ayala-Aponte, A.; Cadena-G, M.I. The influence of osmotic pretreatments on melon (*Cucumis melo* L.) quality during frozen storage. *Dyna* **2014**, *81*, 81–86. [CrossRef]
55. Dermesonlouoglou, E.K.; Giannakourou, M.; Taoukis, P.S. Kinetic study of the effect of the osmotic dehydration pretreatment with alternative osmotic solutes to the shelf life of frozen strawberry. *Food Bioprod. Process.* **2016**, *99*, 212–221. [CrossRef]
56. Huang, Y.; Lu, R.; Chen, K. Prediction of firmness parameters of tomatoes by portable visible and near-infrared spectroscopy. *J. Food Eng.* **2018**, *222*, 185–198. [CrossRef]
57. Yun, J.; Fan, X.; Li, X.; Jin, T.Z.; Jia, X.; Mattheis, J.P. Natural surface coating to inactivate *Salmonella enterica* serovar Typhimurium and maintain quality of cherry tomatoes. *Int. J. Food Microbiol.* **2015**, *193*, 59–67. [CrossRef] [PubMed]
58. Urbanyi, G.; Horti, K. Colour and carotenoid content of quick-frozen tomato cubes during frozen storage. *Acta Aliment.* **1998**, *18*, 247–267.
59. Wu, S.J.; Lu, M.S.; Wang, S.J. Effect of oligosaccharides derived from Laminaria japonica-incorporated pullulan coatings on preservation of cherry tomatoes. *Food Chem.* **2016**, *199*, 296–300. [CrossRef] [PubMed]
60. Heredia, A.; Peinado, I.; Rosa, E.; Andrés, A. Effect of osmotic pre-treatment and microwave heating on lycopene degradation and isomerization in cherry tomato. *Food Chem.* **2010**, *123*, 92–98. [CrossRef]

61. Heredia, A.; Peinado, I.; Barrera, C.; Andrés, A. Influence of process variables on colour changes, carotenoids retention and cellular tissue alteration of cherry tomato during osmotic dehydration. *J. Food Compos. Anal.* **2009**, *22*, 285–294. [CrossRef]
62. An, K.; Zhao, H.D.; Tao, S.D.H.; Wang, Z. Effect of osmotic dehydration with pulsed vacuum on hot-air drying kinetics and quality attributes of cherry tomatoes. *Dry. Technol.* **2013**, *31*, 698–706. [CrossRef]
63. Xu, B.; Zhang, M.; Bhandari, B.; Cheng, X. Influence of ultrasound-assisted osmotic dehydration and freezing on the water state, cell structure, and quality of radish (*Raphanus sativus* L.) cylinders. *Dry. Technol.* **2014**, *32*, 1803–1811. [CrossRef]
64. Fan, K.; Zhang, M.; Wang, W.; Bhandari, B. A novel method of osmotic-dehydrofreezing with ultrasound enhancement to improve water status and physicochemical properties of kiwifruit. *Int. J. Refrig.* **2020**, *113*, 49–57. [CrossRef]
65. Xin, Y.; Zhang, M.; Adhikari, B. Freezing characteristics and storage stability of broccoli (*Brassica oleracea* L. var. *botrytis* L.) under osmodehydrofreezing and ultrasound-assisted osmodehydrofreezing treatments. *Food Bioprocess Technol.* **2013**, *7*, 1736–1744. [CrossRef]
66. Dermesonlouoglou, E.K.; Zachariou, I.; Andreou, V.; Taoukis, P.S. Quality assessment and shelf life modeling of pulsed electric field pretreated osmodehydrofrozen kiwifruit slices. *Int. J. Food Stud.* **2018**, *7*, 34–51. [CrossRef]
67. Ando, Y.; Watanabe, T. Mechanical properties of dehydrofrozen carrots as a function of cell membrane damage, ice crystal development, and rehydration. *J. Food Eng.* **2023**, *357*, 111643. [CrossRef]
68. Zhao, J.H.; Liu, F.; Pang, X.L.; Xiao, H.W.; Wen, X.; Ni, Y.Y. Effects of different osmo-dehydrofreezing treatments on the volatile compounds, phenolic compounds and physicochemical properties in mango (*Mangifera indica* L.). *Int. J. Food Sci. Technol.* **2016**, *51*, 1441–1448. [CrossRef]
69. Cox, D.C.; Baybutt, P. Methods for Uncertainty Analysis: A Comparative Survey. *Risk Anal.* **1981**, *1*, 251–258. [CrossRef]
70. Peleg, M.; Normand, M.D.; Corradini, M.G. The Arrhenius equation revisited. *Crit. Rev. Food Sci. Nutr.* **2012**, *52*, 830–851. [CrossRef]

foods

MDPI

Article

Study on Quality Changes of Kelp Gel Edible Granules during Storage

Tingru Chen [1,†], Ying Li [2,†], Yin Wang [1], Jicheng Chen [2], Lin'ao Fan [2] and Zhiyu Liu [1,*]

[1] Key Laboratory of Cultivation and High-Value Utilization of Marine Organisms in Fujian Province, National and Local Joint Engineering Research Center of Marine Biological Seed Industry Technology, Fisheries Research Institute of Fujian, Xiamen 361013, China; ctr1003533034@163.com (T.C.); wangyin_83@163.com (Y.W.)

[2] College of Food Science, Fujian Agriculture and Forestry University, Fuzhou 350002, China; 15180261868@163.com (Y.L.); newtaicjc@163.com (J.C.); 15359131221@163.com (L.F.)

* Correspondence: negroliu@163.com; Tel.: +86-6010170

† These authors have contributed equally to this work.

Abstract: The kelp gel edible granules developed utilizing the gel properties of alginate are prone to quality deterioration if improperly stored during the storage process. This study comprehensively investigated the quality changes of kelp gel edible granules stored at 4 °C and 25 °C by evaluating indicators such as total bacterial count, coliform bacteria, pH, relaxation time, color difference, appearance, texture characteristics, gel strength, and sensory scoring. The results showed that during the storage at 4 °C, the total bacterial count remained within the national standard range, the hardness and chewiness increased, the gel strength first increased and then decreased, the partial exudation of the bound water in the product occurred, and the sensory score slightly decreased, with an overall minor change in quality. During the storage at 25 °C, significant quality changes were observed, with the total bacterial count exceeding the national standard on the 20th day; additionally, the hardness, chewiness, and gel strength all initially increased and then decreased, both the bound water and the restrained water in the product exuded, the moisture stability decreased, and the sensory score significantly decreased between 16 to 20 days. The spoilage of the product was characterized by a significant water loss, reduction in volume, color change from bright green to dark yellow-brown, and a distinct smell of decaying algae. No coliform bacteria was detected in all products during the storage period. In summary, the shelf life endpoint of the product stored at 25 °C is 16 days, and the shelf life of the product stored at 4 °C is greater than 20 days. Storage at 4 °C can better maintain product quality, extend the shelf life, and effectively maintain the overall color of the product.

Keywords: kelp; gel; edible granules; storage; quality changes

1. Introduction

Kelp (*Laminaria japonica*) belongs to the brown algae and is one of the important economic algae in large-scale mariculture in China. It is rich in polysaccharides, proteins, vitamins, minerals, iodine, iron, zinc, and more than 60 other nutrients [1], and has various physiological activities such as anti-inflammatory, anti-obesity, anti-thrombotic, and immunomodulatory effects [2–5]. Kelp has a delicious taste and has a long history of consumption in China, being well-loved by consumers. However, the current kelp processing products in China are mainly primary processed products such as salted, dried, and seasoned kelp strips, with relatively low economic value added. To enhance the economic benefits of kelp processing products, it is necessary to develop new and innovative processed food products. In recent years, the emerging beverage category of milk tea has developed rapidly in the food and beverage industry and has subsequently driven the development of the milk tea ingredient industry. Common milk tea ingredients include pearls, coconut gels, crispy boo boo, and popping bobas. The crispy boo boo is made of konjac gum, and the outer skin of the popping bobas is made of calcium alginate gel. Therefore, the polysaccharide with gel characteristics has the potential to be developed into beverage ingredients. Kelp contains a higher amount of alginate [6], which has unique gel properties. Utilizing this characteristic, a new type of kelp gel edible granules can be developed. These granules have a uniform light green color, an elastic texture, good taste, and a unique flavor of kelp. They are also rich in natural dietary fiber polysaccharides, such as alginate, which are beneficial to intestinal health and can be further processed into ready-to-eat products or added to beverages as ingredients.

The product needs to be circulated, and during this process, issues related to the deterioration of product quality arise. At the same time, temperature plays a crucial role. Appropriate temperature can extend the shelf life of the product, while unsuitable storage temperature can accelerate the spoilage and deterioration of the product [7]. Therefore, studying the quality changes of kelp gel edible granules under different storage temperatures is of great significance for their future practical production. In addition, relying solely on a single evaluation index to reflect the quality change pattern during the storage process is too one-sided and has a low prediction accuracy. It is necessary to measure multiple quality change indicators for comprehensive evaluation. In summary, clarifying the quality characteristics and change patterns of kelp gel edible granules under different storage conditions is the foundation for predicting their shelf life and improving product safety.

This study comprehensively investigated the quality changes of kelp gel edible granules stored at 4 °C and 25 °C by evaluating indicators such as total bacterial count, coliform bacteria, pH, relaxation time, color difference, appearance, texture characteristics, gel strength, and sensory scoring. The aim is to provide basic data for the subsequent industrial application of the product and to provide a theoretical reference for similar products.

2. Materials and Methods

2.1. Materials

Salted kelp was purchased from Fujian Tianyuan Fisheries Group Co., Ltd. (Fuzhou, Fujian, China); sodium bicarbonate was of food grade and was purchased from Tianjin Bohua Yongli Chemical Co., Ltd. (Tianjin, China); calcium chloride was of food grade and was purchased from Jiangsu Kelunduo Food Ingredients Co., Ltd. (Lianyungang, Jiangsu, China); citric acid was of food grade and was purchased from Huaifang Yingxuan Industrial Co., Ltd. (Huaifang, Shangdong, China); silver nitrate was purchased from Shenzhen Bolinda Technology Co., Ltd. (Shenzhen, Guangdong, China); sodium hydroxide was purchased from Sinopharm Chemical Reagent Co., Ltd. (Shanghai, China); physiological saline was purchased from Fuzhou Haiwang Fuyao Pharmaceutical Co., Ltd. (Fuzhou, Fujian, China); 0.85% sterile physiological saline tubes were purchased from Qingdao High-tech Industrial Park Haibo Biotechnology Co., Ltd. (Qingdao, Shangdong, China); total bacterial count plates and coliform bacteria test plates were purchased from Guangdong Dayuan Oasis Food Safety Technology Co., Ltd. (Guangzhou, Guangdong, China); and pH precision test paper was purchased from Shanghai San Ai Si Reagent Co., Ltd. (Shanghai, China).

2.2. Preparation of Kelp Gel Edible Granules

The preparation process of kelp gel edible granules developed by the laboratory in the early stage is as follows [8,9]:

Cleaning: The salted kelp was cleaned using a sponge brush in tap water, then rinsed with pure water.

Desalting: The salted kelp was desalted by soaking. Cleaned salted kelp (150 g) was soaked in 4 L of pure water at 25 °C for 12 h. The pure water was changed every 3 h (stirred once every hour). $AgNO_3$ solution (0.1 mol/L) was added to an appropriate amount of kelp soaking liquid to observe if there is a white precipitate, and a small piece of desalted kelp was tasted. The desalting process was completed when no significant precipitate formed with the addition of 0.1 mol/L $AgNO_3$ solution, and the desalted kelp was tasted without obvious saltiness.

Deodorizing: The desalted kelp was cut into evenly sized strips and deodorized for 90 tmin at 25 °C with 3% β-cyclodextrin, then rinsed three times with pure water to remove excess deodorizing liquid.

Mincing: The deodorized kelp was minced into small pieces using a meat grinder (TK-12, Jinhua, Zhejiang, China).

Heating and converting: A total of 3% of the minced kelp's weight of sodium bicarbonate and 100 mL of pure water were added into minced kelp (100 g). The mixture was placed in a constant temperature water bath pot (HWS-28, Shanghai, China) at 100 °C and stirred continuously for 11 min.

Fine grinding: Pure water at a ratio of 1: 1.1 (kelp to water, m:v) was added into the heated kelp mixture and grinded for 5 min using a wall breaker (L18-Y68S, Jinan, Shangdong, China).

Solidification and molding: An appropriate amount of kelp paste was placed into a round mold with a pipette and demolished in 2% calcium chloride solution at 20 °C and pH 8 and stood for 4 min to set.

Rinsing: After solidification, the kelp gel edible granules were quickly removed and placed in pure water to wash off excess calcium chloride.

Vacuum packaging: The rinsed kelp gel edible granules and 30 mL of pure water were vacuum sealed in a pure aluminum foil high-temperature steaming bag for subsequent experiments.

Heat treatment: The kelp gel edible granules of equal weight packaged in aluminum foil cooking bags were sterilized at 90 °C for 5 min.

Storage conditions: The kelp gel edible granules were stored at 4 °C and 25 °C, respectively. The total bacterial count, coliform bacteria, pH, relaxation time, color difference, appearance, size, texture characteristics, gel strength, and sensory scoring of the products were measured every four days.

2.3. Determination of Sterilization Parameters

The products of equal weight packaged in aluminum foil cooking bags were sterilized using different sterilization parameters (80, 85, 90, and 95 °C for 5 min and 10 min) in hot water in a constant temperature water bath pot (HWS-28, Shanghai, China). After sterilization, the products were immediately cooled in running water for 30 min. Finally, the products were placed in an incubator at 37 °C for 7 days, after which the bacterial count, coliform bacteria, hardness, and appearance were measured.

2.4. Determination of Total Bacterial Count

The total bacterial count was determined according to GB 4789.2-2022 [10]. Minced kelp gel edible granules (5 g) were placed into a sterile homogenization bag containing 45 mL of physiological saline. The mixture was homogenized and evenly patted for 90 s to create a 1:10 (g:mL) sample homogenate. A total of 1 mL of the homogenate was added to a tube containing 9 mL of 0.85% physiological saline to create a 1:100 sample homogenate. This procedure was repeated to prepare 2–3 additional appropriate dilution samples. A total of 1 mL of the appropriately diluted sample homogenate was dropped vertically and uniformly onto the center of the total bacterial count test plate. The test plates were incubated at 37 °C for 24 h. After incubation, the total bacterial count was calculated.

2.5. Determination of Coliform Bacteria

The coliform bacteria was determined according to the second method outlined in GB 4789.3-2016 [11]. The sample homogenate was prepared using the same method as described in the total bacterial count determination. The pH of the sample homogenate was adjusted to between 6.5 and 7.5 using 1 mol/L NaOH. A total of 1 mL of the appropriately diluted sample homogenate was dropped vertically and uniformly onto the center of the coliform bacteria test plate. The test plates were incubated at 37 °C for 24 h.

2.6. Determination of pH

The pH measurement was conducted with slight modifications based on GB 5009.237-2016 [12]. Minced kelp gel edible granules (1 g) were mixed with 9 mL of distilled water. The mixture was shaken for 30 s and then extracted for 30 min. The mixture was filtered to obtain the filtrate. The pH of filtrate was measured using a pH meter (Five Easy Plus FE28, Mettler Toledoand, Zurich, Switzerland) and the value was recorded.

2.7. Low-Field Nuclear Magnetic Resonance Analysis

LF-NMR relaxation tests were performed according to the method of Li, with some modifications [13]. A kelp gel edible granule (about 0.65 g) was dried to remove surface moisture and placed in a 40 mm NMR tube (NM120-069H-1, Suzhou, Jiangsu, China). The T_2 relaxation time was measured using the Carr–Purcell–Meiboom–Gill (CPMG) sequence. The echo time, wait time, and the number of scans were set to 0.5 ms, 5000 ms, and 4, respectively. A total of 4000 echoes were acquired for analysis. The obtained data were inverted using Multi Exp Inv Analysis software Version 4.0.

2.8. Color Difference Determination

The color of individual kelp gel edible granule was measured using a colorimeter (CR9, Guanzhou, Shenzhen, China) in Specular Component Include (SCI) mode and the color values were represented with L* (lightness), a* (red-green value; positive for red, negative for green), and b* (yellow-blue value; positive for yellow, negative for blue). The total color difference (ΔE) was evaluated using Equation (1).

$$\Delta E = \sqrt{(L - L_0)^2 + (a - a_0)^2 + (b - b_0)^2} \quad (1)$$

where L_0, a_0, and b_0 are the values on the 0th day of storage.

2.9. Texture Profile Analysis (TPA)

Texture profile analysis tests were performed according to Cui with some modifications [14]. The TPA tests were performed using a texture analyzer (TA-XT plus, Stable Micro System, Godalming, UK) with a cylindrical steel P/36R probe at 30% compression. The pre-test speed, test speed, and post-test speed were all set to 1.0 mm/s. The trigger force was set to 5 g. The test was performed twice with a 5 s interval between measurements. The gel strength was evaluated using Equation (2) according to Liu [15].

$$\text{Gel Strength}\left(g/cm^2\right) = \frac{\text{Hardness}}{\text{Contact Area}} \quad (2)$$

2.10. Sensory Quality Determination

The sensory quality of the kelp gel edible granules was evaluated using a quantitative descriptive analysis (QDA) method. An evaluation panel consisting of eight food science graduate students (4 males and 4 females) was organized to taste the kelp gel edible granules. The evaluation was based on three dimensions: appearance, texture, and flavor, with weights set at 30%, 40%, and 30%, respectively. The comprehensive score was calculated by removing the highest and lowest scores and then averaging the remaining scores (Table 1).

Table 1. Sensory standard of kelp gel edible granules during storage.

Attribute	Scoring Criteria	Score Range
Color (30%)	Green	8.1~10
	Slightly faded green or deeper green	6.1~8
	Light brown or dark green	4.1~6
	Brown	2.1~4
	Dark brown	0~2

Table 1. *Cont.*

Attribute	Scoring Criteria	Score Range
Odor (40%)	Strong kelp aroma, no off-flavor	8.1~10
	Mild kelp aroma, no off-flavor	6.1~8
	No kelp aroma, no off-flavor	4.1~6
	Slight off-flavor	2.1~4
	Strong off-flavor	0~2
Texture (30%)	Good elasticity, firm texture	8.1~10
	Fair elasticity, firm texture	6.1~8
	Average elasticity, relatively firm texture	4.1~6
	Poor elasticity, soft or hard texture	2.1~4
	No elasticity, very soft or very hard texture	0~2

2.11. Data Processing

Data were organized using EXCEL software Version 2021. Single-factor ANOVA tests were performed using SPSS 27.0.1 software. Graphs were created using Origin 2023b software. Each measurement was conducted with at least three replicates, and results are presented as mean ± SEM. A correlation analysis was performed using Pearson's test to determine the direct and indirect contributions of these indicators to the quality deterioration of products during storage.

3. Results

3.1. Determination of Sterilization Process Parameters

The products were sterilized under different parameters and stored at 37 °C for 7 days. It was observed that at a sterilization temperature of 80 °C, both 5 min and 10 min treatments resulted in significant quality deterioration (Table 2). The products showed a deepened color, volume shrinkage, and increased hardness. At a sterilization temperature of 85 °C, the products maintained a normal appearance, but hardness still increased. When the sterilization temperature was above 85 °C, the products retained their light green color, volume remained unchanged, hardness did not significantly increase, and no total bacterial count was detected. With the increase in sterilization temperature and sterilization time, the less the total bacterial count, the better the quality of products. Therefore, a sterilization temperature of 90 °C for 5 min was selected as the parameter for subsequent product storage tests.

Table 2. Effect of different sterilization parameters on products quality.

Sterilization Conditions		Microbial Indicators		Quality Changes	
Sterilization Temperature (°C)	Sterilization Time (min)	Total Bacterial Count (CFU/g)	Coliform Bacteria (CFU/g)	Hardness (g)	Appearance
80	5	142.50 ± 41.13^{a}	-	162.68 ± 5.25^{a}	Darkened color, volume shrinkage
	10	115.00 ± 17.32^{ab}	-	155.40 ± 1.16^{b}	
85	5	67.50 ± 35.94^{bc}	-	140.74 ± 8.17^{c}	Darkened color, volume shrinkage
	10	60.00 ± 40.99^{c}	-	133.97 ± 2.61^{c}	Light green, no volume change
90	5	-	-	116.99 ± 1.32^{d}	Light green, no volume change
	10	-	-	110.32 ± 1.28^{d}	Light green, no volume change
95	5	-	-	115.38 ± 4.40^{d}	Light green, no volume change
	10	-	-	115.74 ± 2.50^{d}	Light green, no volume change

"-" indicates not detected. Each value is the mean ± standard deviation (n = 8). Superscripts: different small letters within rows indicate significant difference among treatments ($p \leq 0.05$).

3.2. Changes in Total Bacterial Count and Coliform Bacteria during Storage

The growth rate of the total bacterial count in products stored at 4 °C was slower compared to those stored at 25 °C. By the end of the storage period, the total bacterial count in the 4 °C stored products did not exceed the limit of 4.48 log10 (CFU/g) as per GB 19643-2016 [16]. For products stored at 25 °C, the total bacterial count exhibited two distinct phases. In the first phase, microbial growth was relatively slow, while in the second phase, the growth rate accelerated. From 4 to 8 days, the bacterial count increased slowly, but from 8 to 20 days, the growth rate rapidly accelerated, reaching 4.62 log10 (CFU/g) on the 20th day, surpassing the GB 19643-2016 limit [16]. At this point, the product showed signs of spoilage, including severe moisture loss, reduced volume, color change from bright green to dark yellow-brown, and a foul algae-like odor (Figure 1). Throughout the entire storage period, the coliform bacteria was not detected, indicating that the product was not contaminated by fecal matter during production and maintained good sanitary conditions

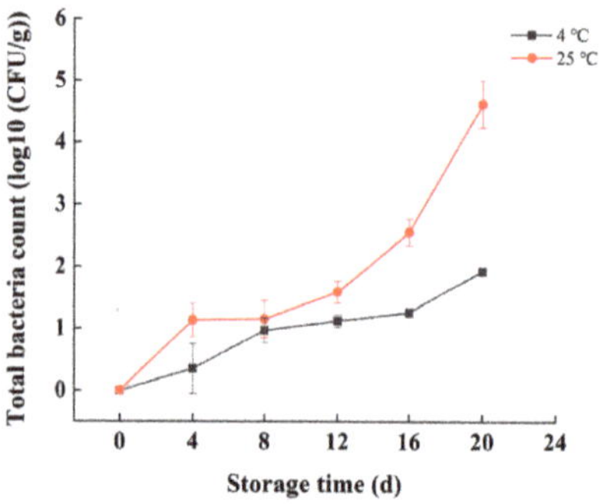

Figure 1. Changes in the total bacteria count of kelp gel edible granules during storage times.

3.3. Changes in pH during Storage

The initial pH of all groups was around 9.35. The pH value of the products stored at 4 °C showed minimal fluctuation throughout the storage period. In contrast, the pH of products stored at 25 °Cexhibited two distinct phases: a slight decrease from 0 to 8 days and a significant drop from 8 to 20 days, ultimately reaching 6.96 on the 20th day (Figure 2).

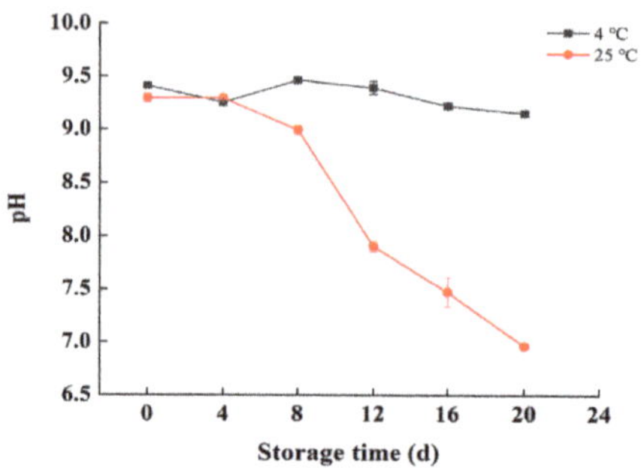

Figure 2. Changes in the pH of kelp gel edible granules during storage times.

3.4. Changes in Moisture Distribution during Storage

T_{21} represents the chemically bound water in the product (bound water), and T_{22} represents the water bound within the gel network (restrained water) of the product (accounting for most of the product), which is more unstable compared to T_{21} and can transform into free water and be lost. P_{21} and P_{22} indicate the proportion of water with relaxation times T_{21} and T_{22} in the total moisture of the product, respectively. It could be seen that T_{21} of the product stored at 4 °C did not change significantly throughout the storage period, consistent with the findings of Huang [17] (Figure 3a). However, T_{22} shortened from day 4 to day 20 compared to day 0, indicating that restrained water within the gel network in the 4 °C product became more stable during storage. P_{21} in the product stored at 4 °C tended to exude, forming $P_{21\text{-}1}$, which was less stable than P_{21} but more stable than P_{22} (Figure 3b). This suggested that the stability of bound water decreased while the stability of restrained water within the gel network increased during storage at 4 °C. It could be seen that T_{21} of the product stored at 25 °C gradually shifted to the right, while T_{22} first shifted to the left and then to the right (Figure 4a). The number of water types in the product stored at 25 °C changed from two to four. During the storage period from day 4 to day 16, T_{21} exuded, forming the $P_{21\text{-}1}$ component. In the later storage period (day 16 to day 20), the stability of P_{22} decreased, and it exuded as $P_{22\text{-}1}$, whose relaxation time significantly increased, indicating a rapid decrease in stability and a tendency to be lost (Figure 4b).

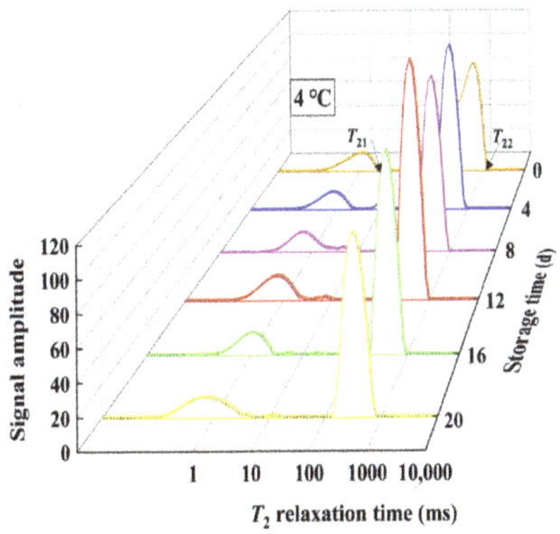

(a)

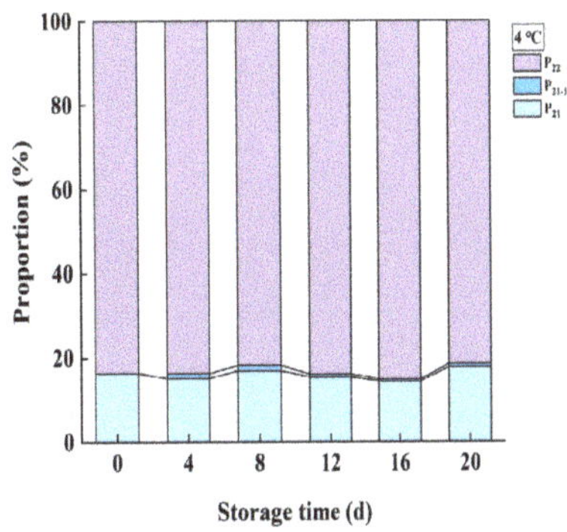

(b)

Figure 3. Changes in the water distribution of kelp gel edible granules under 4 °C during storage times. (**a**) Change in relaxation time of the product during storage at 4 °C; (**b**) change in water distribution ratio during storage at 4 °C.

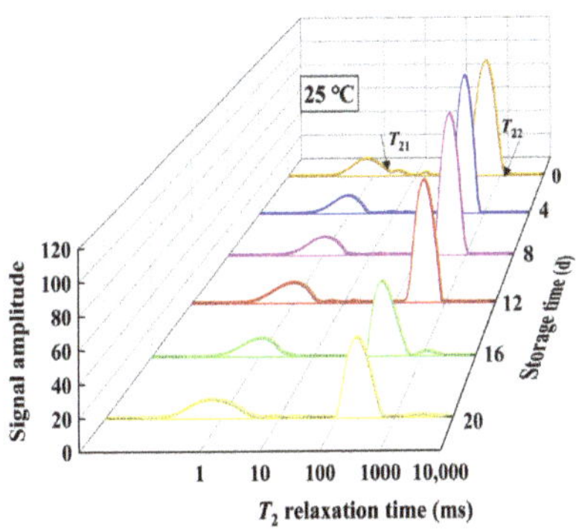

(a)

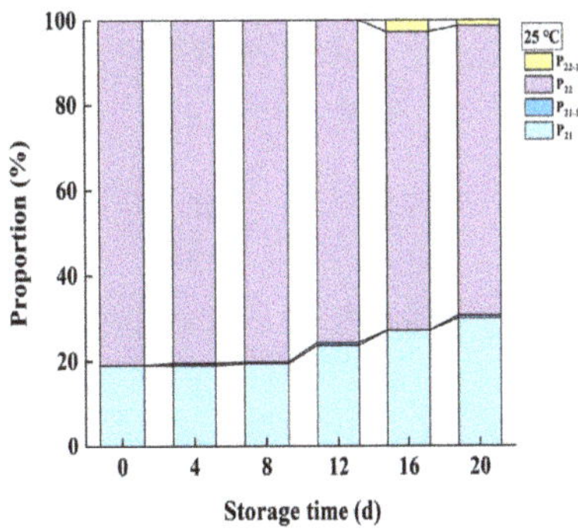

(b)

Figure 4. Changes in the water distribution of kelp gel edible granules under 25 °C during storage. times. (**a**) Change in relaxation time of the product during storage at 25 °C; (**b**) change in water distribution ratio during storage at 25 °C.

3.5. Changes in Color Difference during Storage

ΔE represents the overall color change, with larger ΔE values indicating more severe deterioration. The ΔE values for products stored at both 4 °C and 25 °C increased over time, indicating a greater overall color change during storage. On the 20th day, the total color difference of products stored at 25 °C was nearly twice than products stored at 4 °C. The decrease in L* and a* values in products stored at 4 °C contributed to the increase in ΔE, while b* remained unchanged. For products stored at 25 °C, L*, a*, and b* all changed significantly, reflecting a shift in appearance from bright green to dark yellow-brown (Table 3). This demonstrated that low-temperature storage better preserved the overall color of the product.

Table 3. Color parameters and changes of kelp gel edible granules during storage times.

Storage Temperature	Storage Days(d)	L*	a*	b*	ΔE
4 °C	0	27.09 ± 1.46^{a}	-11.36 ± 1.03^{ab}	12.57 ± 2.39^{a}	
	4	22.67 ± 0.46^{b}	-13.28 ± 1.03^{bc}	14.63 ± 1.41^{a}	5.32 ± 1.30^{c}
	8	22.91 ± 0.35^{b}	-13.07 ± 0.84^{bc}	13.26 ± 2.16^{a}	4.95 ± 0.32^{c}
	12	21.73 ± 0.57^{b}	-14.23 ± 0.11^{c}	14.80 ± 1.26^{a}	6.55 ± 0.08^{bc}
	16	19.74 ± 0.92^{c}	-13.57 ± 0.75^{c}	11.65 ± 1.33^{a}	7.82 ± 0.98^{ab}
	20	23.21 ± 0.12^{b}	-10.94 ± 1.34^{a}	9.65 ± 2.62^{a}	8.44 ± 0.88^{a}
25 °C	0	31.20 ± 1.89^{a}	-10.95 ± 0.67^{b}	13.14 ± 0.68^{ab}	
	4	25.38 ± 1.42^{b}	-13.48 ± 0.31^{c}	16.27 ± 2.55^{a}	7.47 ± 0.34^{d}
	8	22.47 ± 2.40^{bc}	-13.90 ± 0.94^{c}	15.18 ± 4.59^{a}	10.37 ± 0.17^{c}
	12	20.50 ± 1.00^{cd}	-13.51 ± 0.72^{c}	11.69 ± 3.96^{abc}	11.58 ± 0.94^{c}
	16	17.17 ± 0.47^{de}	-11.79 ± 1.72^{b}	8.93 ± 0.52^{bc}	14.73 ± 0.20^{b}
	20	15.60 ± 2.36^{e}	-8.93 ± 0.40^{a}	6.60 ± 0.55^{c}	17.08 ± 1.94^{a}

Each value is the mean ± standard deviation ($n = 6$). Superscripts: different small letters within rows indicate significant difference among treatments ($p \leq 0.05$).

3.6. Changes in Appearance during Storage

The color of the products stored at 4 °C changed from the initial green to dark green, with reduced brightness, consistent with the changes in a* and ΔE values. For products stored at 25 °C, the color initially shifted from green to dark green and eventually turned dark brown by the 16th day, aligning with the changes in a*, b*, and ΔE values. The significant color change in appearance was related to the significant increase in total bacterial count and the decrease in pH discussed earlier (Figure 5).

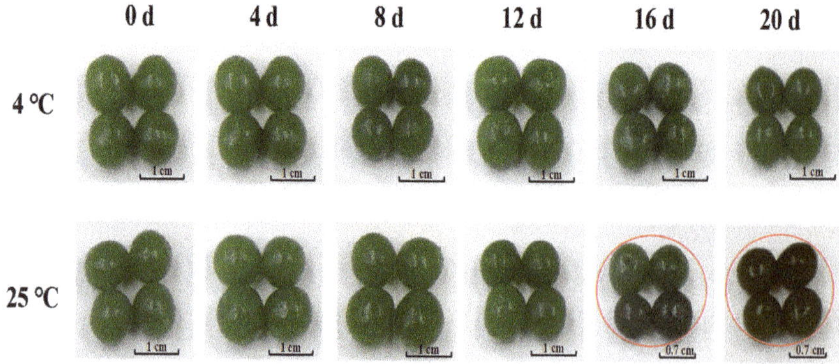

Figure 5. Changes in the appearance of kelp gel edible granules during storage time. The pictures circled in red indicated that the appearance of the products had changed significantly.

By the 20th day of storage, the volume of spoiled products significantly decreased compared to normal products. This observation aligned with the changes in relaxation time and moisture ratio. During spoilage, the T_{22} relaxation time gradually increased, indicating the decreased stability of restrained water within the gel network. As restrained water, which constitutes the majority of the moisture of products, leaked out and transformed into less stable free water, a reduction in restrained water in the gel network and consequently a decrease in product volume occurred (Figure 6).

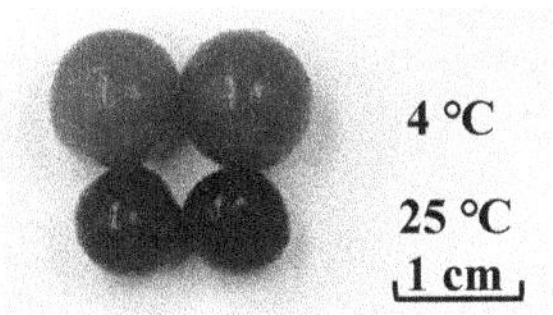

Figure 6. Appearance comparison of kelp gel edible granules on the 20th day of storage.

3.7. Changes in Texture and Gel Strength during Storage

The hardness of products stored at 4 °C significantly increased from 0 to 4 days and remained relatively stable from 4 to 20 days, with an overall increase of nearly 50% compared to day 0. Chewiness, elasticity, resilience, and gel strength generally showed a trend of first increasing and then decreasing, while cohesiveness showed no significant difference. For products stored at 25 °C, the changes in hardness and chewiness exhibited three phases: First, from 0 to 4 days, hardness and chewiness rapidly increased, with both being 1.5 times higher on day 4 compared to day 0. Second, from 4 to 12 days, hardness slightly decreased, and chewiness remained relatively unchanged. Third, from 12 to 20 days, the products underwent spoilage, resulting in a rapid decrease in hardness and chewiness, while elasticity and cohesiveness increased, and resilience showed no significant difference. Gel strength increased gradually from 0 to 8 days and then decreased after 8 days (Table 4).

Table 4. Changes in TPA and gel strength of kelp gel edible granules during storage times.

Storage Temperature	Storage Days (d)	Hardness (g)	Elasticity	Cohesiveness	Chewiness (g)	Recoverability	Gel Strength (g/cm^2)
4 °C	0	107.31 ± 3.62 ^{c}	0.74 ± 0.03 ^{c}	0.72 ± 0.02 ^{a}	57.00 ± 4.09 ^{c}	0.46 ± 0.01 ^{a}	96.8 ± 10.29 ^{d}
	4	145.30 ± 6.54 ^{ab}	0.81 ± 0.01 ^{a}	0.73 ± 0.02 ^{a}	85.89 ± 4.92 ^{ab}	0.47 ± 0.01 ^{a}	135.21 ± 6.08 ^{b}
	8	151.54 ± 8.27 ^{a}	0.81 ± 0.01 ^{a}	0.73 ± 0.01 ^{a}	89.05 ± 5.33 ^{a}	0.47 ± 0.01 ^{a}	128.17 ± 12.65 ^{bc}
	12	141.47 ± 2.00 ^{b}	0.81 ± 0.01 ^{a}	0.72 ± 0.01 ^{a}	82.34 ± 2.38 ^{b}	0.46 ± 0.01 ^{a}	157.30 ± 10.15 ^{a}
	16	146.33 ± 5.05 ^{ab}	0.79 ± 0.01 ^{ab}	0.71 ± 0.01 ^{a}	81.48 ± 5.28 ^{b}	0.45 ± 0.01 ^{ab}	128.81 ± 9.67 b ^{c}
	20	151.42 ± 3.03 ^{a}	0.78 ± 0.02 ^{b}	0.70 ± 0.02 ^{a}	83.19 ± 3.38 ^{b}	0.44 ± 0.01 ^{b}	117.90 ± 5.52 ^{c}
25 °C	0	96.38 ± 3.06 ^{c}	0.75 ± 0.02 ^{d}	0.71 ± 0.01 ^{c}	50.69 ± 1.39 ^{d}	0.46 ± 0.00 ^{a}	89.51 ± 4.03 ^{c}
	4	144.33 ± 4.31 ^{a}	0.80 ± 0.02 ^{c}	0.73 ± 0.02 ^{c}	84.47 ± 5.54 ^{b}	0.46 ± 0.02 ^{a}	134.31 ± 4.01 ^{a}
	8	150.59 ± 4.78 ^{a}	0.81 ± 0.01 ^{c}	0.72 ± 0.01 ^{c}	87.30 ± 4.28 ^{ab}	0.45 ± 0.01 ^{a}	135.13 ± 11.32 ^{a}
	12	135.86 ± 8.20 ^{b}	0.84 ± 0.02 ^{ab}	0.80 ± 0.03 ^{b}	90.73 ± 4.83 ^{a}	0.47 ± 0.02 ^{a}	112.92 ± 7.28 ^{b}
	16	86.07 ± 3.42 ^{d}	0.82 ± 0.01 ^{bc}	0.82 ± 0.02 ^{b}	58.04 ± 4.59 ^{c}	0.45 ± 0.02 ^{a}	115.17 ± 11.05 ^{b}
	20	86.67 ± 7.41 ^{d}	0.86 ± 0.02 ^{a}	0.84 ± 0.01 ^{a}	62.96 ± 6.15 ^{c}	0.46 ± 0.01 ^{a}	120.71 ± 10.39 ^{b}

Each value is the mean ± standard deviation (n = 6). Superscripts: different small letters within rows indicate significant difference among treatments ($p \leq 0.05$).

3.8. Changes in Sensory Quality during Storage

The color, odor, texture, and comprehensive sensory scores of the product stored at 4 °C changed little, and the comprehensive sensory score remained between 8.2 and 7.5; throughout the entire storage process, the kelp flavor of the 4 °C product was lighter, but the texture was tight, and the color remained green. Although ΔE indicated that the overall color of the product had changed, the change was not significant enough for the average observer to discern (Figure 7a).

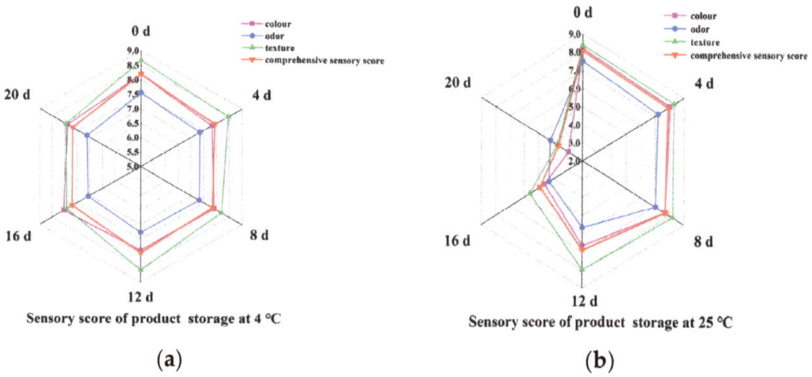

Figure 7. Changes in sensory quality of kelp gel edible granules during storage times. (**a**) Change in quality of the product during storage at 4 °C; (**b**) change in sensory quality of the product during storage at 25 °C.

The color score of the 25 °C product decreased from 8.17 to 2.97; the odor score decreased from 7.48 to 4.23; the texture score decreased from 8.37 to 3.75; and the comprehensive sensory score decreased from 8.04 to 3.66 (Figure 7b).

3.9. Correlation Analysis of Quality Indicators of Kelp Gel Edible Granules

The total bacterial count and ΔE showed a highly significant positive correlation with storage time and the comprehensive sensory score showed a significant negative correlation with the storage time of the 4 °C stored product. For the 25 °C stored product, the total bacterial count and ΔE had a highly significant positive correlation with storage time; pH and overall sensory score had a highly significant negative correlation with storage time.

The above results indicated that the total bacterial count, ΔE, and comprehensive sensory score of the 4 °C stored product all changed significantly with the change in storage time. In addition to the above indicators, the pH of the 25 °C stored product could also reflect the change in product quality with storage time. Moreover, the key quality indicator of the product stored at different temperatures was different except for the comprehensive sensory evaluation. The key indicator for the 4 °C stored product was the total bacterial count (correlation coefficient was 0.977), and the key indicator for the 25 °C stored product was pH (correlation coefficient was −0.965) (Table 5).

Table 5. Correlation analysis results of various indexes of kelp gel edible granules.

Storage Temperature	Measurement Indicator	Storage Time	Total Bacterial Count	ΔE	pH	Hardness	Chewiness	Gel Strength	Comprehensive Sensory Score
4 °C	Storage time	1	0.977 **	0.907 **	−0.632	0.682	0.515	0.310	−0.877 *
	Total bacterial count		1	0.879 *	−0.537	0.725	0.569	0.302	−0.858 *
	ΔE			1	−0.653	0.879 *	0.780 *	0.574	−0.906 **
	pH				1	−0.402	−0.246	0.030	0.690
	Hardness					1	0.972 **	0.597	−0.843 *
	Chewiness						1	0.696	−0.701
	Gel strength							1	−0.236
	Comprehensive Sensory score								1

Table 5. *Cont.*

Storage Temperature	Measurement Indicator	Storage Time	Total Bacterial Count	ΔE	pH	Hardness	Chewiness	Gel Strength	Comprehensive Sensory Score
25 °C	Storage time	1	0.934 **	0.962 **	−0.965 **	−0.423	−0.045	0.243	−0.930 **
	Total bacterial count		1	0.889 **	−0.900 **	−0.484	−0.147	0.267	−0.954 **
	ΔE			1	−0.864 *	−0.200	0.167	0.490	−0.830 *
	pH				1	0.584	0.208	0.006	0.949 **
	Hardness					1	0.912 **	0.600	0.694
	Chewiness						1	0.689	0.375
	Gel strength							1	−0.037
	Comprehensive Sensory score								1

* indicates significant difference at the $p < 0.05$ level; ** indicates extremely significant difference at the $p < 0.01$ level.

4. Discussion

4.1. Microbiological Analysis

The total bacterial count is an important indicator for assessing the hygienic condition of processed products, while coliform bacteria is commonly used as an indicator of fecal contamination. Both indicators reflect the sanitary status of food and are crucial for determining its edibility. Kelp can be contaminated with human pathogens during cultivation, such as *Vibrio* spp., *Salmonella* spp., *Staphylococcus aureus*, *Listeria*, etc. Additionally, the introduction of pathogens uncommon to the marine environment, such as *L. monocytogenes*, during handling and post-harvesting processing into finished products can be another route for kelp contamination [1,18]. The observed changes in the total bacterial count at 25 °C might be due to the initial low bacterial count after pasteurization, which killed most microorganisms mentioned above except for some heat-resistant spores. These spores might be some of the thermostable Bacillus species, such as *Bacillus pumilus* and *B. licheniformis* [19]. As storage time progressed, these spores began to germinate but adapted slowly to the environment, leading to a slower initial growth rate. In the later stages, microorganisms adapted to the environment and rapidly proliferated using the abundant materials such as alginate polysaccharides present in the product. In summary, low temperatures effectively inhibit the metabolic activity and proliferation of microorganisms. Similarly, several studies have reported that microorganisms grew more slowly at lower temperatures [20,21].

4.2. pH Analysis

The changes in products stored at 25 °C could be explained by the initial heat sterilization stage at 25 °C, which killed most microorganisms, resulting in a low acid production and minimal pH change in the early stage. After 8 days, the microbial growth rate increased, the total bacterial count rose, and the microorganisms utilized organic substances such as alginate polysaccharides, producing organic acids that led to a decrease in pH. It was noteworthy that the pH changes did not exhibit the typical "V" shape (initial decrease followed by an increase) often seen in food products [22]. This was primarily due to the low protein content in the kelp gel particles. During spoilage, the decomposition of proteins into alkaline substances like ammonia and amines was insufficient to neutralize the acidic substances produced from the decomposition of polysaccharides and other organic materials. Consequently, the pH value of the product continued to decline throughout the storage period.

4.3. Moisture Distribution Analysis

The observed phenomena of moisture distribution could be attributed to several factors. First, for products stored at 4 °C, the significant increase in hardness during storage might cause the three-dimensional gel network to become finer and more uniform, increasing the capillary binding force of the gel network and thus increasing the stability of restrained water, resulting in the leftward shift of T_{22}. Second, for products stored at 25 °C, the initial leftward shift of T_{22} could be attributed to the same reasons mentioned above. After 16 days of storage, T_{22} began to shift to the right. At this point, the internal microbial proliferation rate increased, and bacteria decomposed alginate, leading to the destruction and disintegration of some parts of the gel network. The damaged gel network reduced its water-binding capacity [23], causing some restrained water to exude from the disintegrated network as free water. Additionally, kelp gel edible granules are hydrogels with a three-dimensional network of an "egg box" model formed by cross-linking COO- and Ca^{2+} in sodium alginate [24]. A large amount of H^+ produced by the bacterial decomposition of alginate and other organic matter might compete with Ca^{2+} (H^+ and COO- will protonate to form carboxylic acid) [25], thereby replacing Ca^{2+} from calcium alginate to produce alginate acid, which will also lead to the disintegration of the gel network. This ultimately resulted in the leftward shift of T_{22} and the formation of $P_{22\text{-}1}$ water in the later storage period.

4.4. Physicochemical Properties

Color can reflect the sensory and physicochemical changes in a product and is an important indicator of product acceptability [26]. This observed phenomenon of color could be attributed to the structure of chlorophyll, which consists of a planar ring formed by one porphyrin ring and four pyrrole subunits. The electrons can migrate freely within this structure, making chlorophyll unstable under conditions such as acid, heat, light, and oxygen [27]. Research has shown that pheophytinisation was associated with the number of acids liberated during processing. Once an acidic environment (pH < 7) surrounded the chlorophyll compounds, the structural change in the chlorophyll molecule affected the overall structure of the macroalgae, which led to a change in the color of the macroalgae [28]. During storage at 25 °C, the products underwent spoilage, and microorganisms decomposed organic matter to produce acids, leading to the replacement of Mg^{2+} in chlorophyll with H^+, forming yellow-brown pheophytin [29]. Conversely, the products stored at 4 °C had slower microbial growth, maintaining a pH above 9, and thus the green retention was good. In summary, low-temperature storage can inhibit microbial growth and reduce the damage caused by acid and heat to chlorophyll in the product, thereby better preserving the overall color.

The reasons for the changes in texture characteristics might be as follows: First, the kelp paste contained dispersed sodium alginate, which, during the demolding and curing process, was attracted by the Ca^{2+} in calcium chloride to the surface of the product for a replacement reaction, resulting in an uneven distribution of sodium alginate inside the product (more on the surface) [30]. As the storage time extended, Ca^{2+} gradually diffused from the surface to the inside of the product, and sodium alginate also gradually distributed evenly in this process. The three-dimensional network gel structure inside the product became more uniform and dense, thereby enhancing the overall deformation resistance, which ultimately led to a significant increase in all product hardness and chewiness in the first 0–4 days [25,31]. Since a stable gel structure had been formed inside the product during this period, the hardness and chewiness of the product stored at 4 °C did not change much from 4 to 20 days. Second, the quality of the 25 °C product deteriorated significantly in the later stage of storage, and the pH decreased significantly during the massive proliferation of microorganisms; the produced H^+ might replace some Ca^{2+} and cause the alginate calcium gel to transform into alginate gel. Since ionic bonds have higher bond energy than hydrogen bonds [32], the mechanical properties of the alginate calcium gel were stronger than alginate gel, and the gel network structure collapsed during the spoilage process of the product, which comprehensively led to a significant decrease in hardness and chewiness. As the hardness decreased, elasticity and cohesiveness increased. Finally, since the gel strength is a composite index based on hardness and product area, its changes during the storage process were consistent with the reasons for the changes in hardness. A previous study has also reported that table grapes could better maintain their hardness under low temperature conditions [33].

4.5. Sensory Quality Analysis

Generally, a mean liking score of ≥7 on a 9-point hedonic scale is associated with a highly acceptable sensory quality [34]. When the products were stored at 4 °C, the color, odor, texture, and comprehensive sensory scores were kept between 8.2 and 7.5, indicating that the products were highly acceptable. While the products were stored at 25 °C, the color, odor, texture, and comprehensive sensory scores were all lower than 7 after 8 days of storage. The main reason for the significant decrease in the sensory score of the 25 °C product in the later stage of storage was the massive proliferation of microorganisms. This was consistent with the above reasons for changes in color, texture, and total bacterial count. A previous study has also reported that under low-temperature conditions, the pH change of beef meatballs was slower, and the starting point of sensory quality change was longer [35].

4.6. Correlation Analysis of Quality Indicators Analysis

The purpose of correlation analysis is to determine the direct and indirect contribution of these quality indicators to the quality deterioration of products in the storage process through the correlation coefficient, and to provide an indicator basis for monitoring product quality changes in the future. The key indicator of products stored at 4 °C was the total bacterial count, which was significantly positively correlated with storage time, indicating that low temperature only slowed down the growth rate of microorganisms. The key indicators of the products stored at 25 °C were pH and total bacterial count, which were significantly positively correlated with storage time. The change in pH was caused by the change in total bacterial count.

5. Conclusions

This study comprehensively investigated the quality changes of kelp gel edible granules stored at 4 °C and 25 °C by evaluating indicators such as total bacterial count, coliform bacteria, pH, relaxation time, color difference, appearance, texture characteristics, gel strength, and sensory scoring. Data gathered from this research show that the shelf life endpoint for products stored at 25 °C is 16 days, while for products stored at 4 °C, it is more than 20 days. Low-temperature storage (4 °C) better maintains the quality of kelp gel edible particles, extends their shelf life, and effectively preserves the overall color of the product. Future studies are warranted to flavor kelp gel edible particles to enrich their flavor substances, develop beverages adding kelp gel edible particles, enrich their application forms, and expand the application range in the market.

Author Contributions: T.C. and Y.L. contributed to the methodology, experiments, data collection, analysis, and prepared the initial draft of the manuscript. Y.W., J.C., L.F. and Z.L. were responsible for funding acquisition, conceptualization, experimental design, data analysis, and manuscript editing. All authors have read and agreed to the published version of the manuscript.

Funding: This study was supported by the Youth Science and Technology Innovation Program of Xiamen Ocean and Fisheries Development Special Funds, grant number 23YYST077QCA05, the Fujian Province Marine Services and High Quality Development of Fisheries Special Project, grant number FJHY-YYKJ-2022-1-3, the Basic Research Special Projects of Fujian Provincial Public Welfare Research Institutes, grant number 2023R1012002, the Fujian Province Marine Services and High Quality Development of Fisheries Special Project, grant number FJHY-YYKJ-2024-1-17, and the Fujian Province Marine Services and High Quality Development of Fisheries Special Project, grant number FJHY-YYKJ-2023-1-2.

Institutional Review Board Statement: Not applicable.

Informed Consent Statement: Not applicable.

Data Availability Statement: The original contributions presented in the study are included in the article, further inquiries can be directed to the corresponding author.

Conflicts of Interest: The authors declare no conflict of interest.

References

1. Akomea-Frempong, S.; Perry, J.J.; Skonberg, D.I. Effects of pre-freezing blanching procedures on the physicochemical properties and microbial quality of frozen sugar kelp. *J. Appl. Phycol.* **2022**, *34*, 609–624. [CrossRef]
2. Deng, Z.Z.; Wu, N.; Wang, J.; Zhang, Q.B. Dietary fibers extracted from *Saccharina japonica* can improve metabolic syndrome and ameliorate gut microbiota dysbiosis induced by high fat diet. *J. Funct. Foods* **2021**, *85*, 104642. [CrossRef]
3. Peng, F.H.; Zha, X.Q.; Cui, S.H.; Asghar, M.N.; Pan, L.H.; Wang, J.H.; Luo, J.P. Purification, structure features and anti-atherosclerosis activity of a *Laminaria japonica* polysaccharide. *Int. J. Biol. Macromol.* **2015**, *81*, 926–935. [CrossRef] [PubMed]
4. Zeng, M.L.; Wu, X.Y.; Li, F.; She, W.S.; Zhou, L.; Pi, B.Y.; Xu, Z.W.; Huang, X.L. *Laminaria Japonica* Polysaccharides effectively inhibited the growth of nasopharyngeal carcinoma cells in vivo and in vitro study. *Exp. Toxicol. Pathol.* **2017**, *69*, 527–532. [CrossRef] [PubMed]
5. Zhao, X.; Guo, F.; Hu, J.; Zhang, L.; Xue, C.; Zhang, Z.; Li, B. Antithrombotic activity of oral administered low molecular weight fucoidan from *Laminaria japonica*. *Thromb. Res.* **2016**, *144*, 46–52. [CrossRef]
6. Jiang, Z.D.; Zhang, X.W.; Wu, L.Y.; Li, H.B.; Chen, Y.H.; Li, L.J.; Ni, H.; Li, Q.B.; Zhu, Y.B. Exolytic products of alginate by the immobilized alginate lyase confer antioxidant and antiapoptotic bioactivities in human umbilical vein endothelial cells. *Carbohydr. Polym.* **2021**, *251*, 116976. [CrossRef]
7. Zhao, Y.; Han, G.; Li, Y.F.; Lv, H.X. Changes in quality characteristics and metabolites composition of wheat under different storage temperatures. *J. Stored Prod. Res.* **2024**, *105*, 102229. [CrossRef]
8. Li, Y.; Wang, Y.; Chen, T.R.; Fan, L.A.; Chen, J.C.; Liu, Z.Y. Study on preparation technology and quality characteristics of kelp alginate edible granules. *Food Ferment. Ind.* **2023**, 1–11. [CrossRef]
9. Li, Y.; Chen, T.R.; Chen, J.C.; Fan, L.A.; Wang, Y.; Liu, Z.Y. Effect of $CaCl_2$ concentration on quality characteristics of kelp alginate edible granules. *Food Ferment. Ind.* **2024**, 1–8. [CrossRef]
10. *GB4789.2-2022;* National Food Safety Standard—Microbiological Examination of Food: Aerobic Plate Count. China Standards Press: Beijing, China, 2022.
11. *GB4789.3-2016;* National food safety standard—Food Microbiological Examination: Enumeration of Coliforms. China Standards Press: Beijing, China, 2016.
12. *GB5009.237-2016;* National Food Safety Standard—Determination of pH Value of Food. China Standards Press: Beijing, China, 2016.
13. Li, J.H.; Li, X.; Wang, C.Y.; Zhang, M.Q.; Xu, Y.L.; Zhou, B.; Su, Y.J.; Yang, Y.J. Characteristics of gelling and water holding properties of hen egg white/yolk gel with NaCl addition. *Food Hydrocoll.* **2018**, *77*, 887–893. [CrossRef]
14. Cui, C.L.; Jiang, H.; HangGuan, M.; Ji, N.; Xiong, L.; Sun, Q.J. Characterization and in vitro digestibility of potato starch encapsulated in calcium alginate beads. *Food Hydrocoll.* **2022**, *126*, 107458. [CrossRef]
15. Liu, S.L.; Zhu, F.; Lin, S.N.; Huang, J.C.; Li, T.J.; Wang, H.L.; Lin, X.Y. Research of strength and relaxation properties of agar gel. *Sci. Technol. Food Ind.* **2017**, *38*, 85–89, 100.
16. *GB19643-2016;* Hygienic Standard for Marine Algae and Algae Products. China Standards Press: Beijing, China, 2016.
17. Huang, Z.R.; PingZhou, H.; Jiang, Q.H.; He, W.Y.; Zhou, X.H.; Chen, H.; Zhou, X.J.; Li, M.; SongZhou, J.; ZhongZhao, L. Study on the quality change and deterioration mechanism of leisure dried tofu under different storage temperature conditions. *LWT* **2022**, *172*, 114257. [CrossRef]
18. Akomea-Frempong, S.; Skonberg, D.I.; Camire, M.E.P.; Jennifer, J. Impact of blanching, freezing, and fermentation on physico-chemical, microbial, and sensory quality of sugar kelp (*Saccharina latissima*). *Foods* **2021**, *10*, 2258. [CrossRef] [PubMed]
19. Blikra, M.J.; Lvdal, T.; Vaka, M.R.; Roiha, I.S.; Lunestad, B.T.; Lindseth, C.; Skipnes, D. Assessment of food quality and microbial safety of brown macroalgae (*Alaria esculenta* and *Saccharina latissima*). *J. Sci. Food Agric.* **2019**, *99*, 1198–1206. [CrossRef] [PubMed]
20. Zhang, L.; Han, L.; Yang, J.Y.; Sun, Q.X.; Li, K.; Prakash, S.; Dong, X.P. Preservation strategies for processed grass carp products: Analyzing quality and microbial dynamics during chilled and ice temperature storage. *Food Chem. X* **2024**, *23*, 101428. [CrossRef] [PubMed]
21. Jie, X.; Xiu, S.Q.; Yang, Z.O.; Fu, W.Z.; Shuai, W.; Yuan, H.Z.; Wu, J.H.; Cheng, L.S. Novel insights into the microbial succession and quality variation of golden pompano (*Trachinotus ovatus*) driven by storage temperatures based on high-throughput sequencing. *LWT* **2024**, *199*, 116087.
22. Gao, S.; Liu, Y.Y.; Fu, Z.X.; Zhang, H.J.; Zhang, L.T.; Li, B.; Tan, Y.Q.; Hong, H.; Luo, Y.K. Uncovering quality changes of salted bighead carp fillets during frozen storage: The potential role of time-dependent protein denaturation and oxidation. *Food Chem.* **2023**, *414*, 135714. [CrossRef] [PubMed]
23. Xiao, Y.; Zhao, J.; Zhang, X.R.; Jiao, Y.; Liu, Y.F. Analysis of quality changes of Hengshan goat hindquarter meat at four storage temperatures. *J. Food Compos. Anal.* **2023**, *117*, 105129. [CrossRef]
24. Fernández Farrés, I.; Norton, I.T. Formation kinetics and rheology of alginate fluid gels produced by in-situ calcium release. *Food Hydrocoll.* **2014**, *40*, 76–84. [CrossRef]
25. Puguan, J.M.C.; Yu, X.H.; Kim, H. Characterization of structure, physico-chemical properties and diffusion behavior of Ca-Alginate gel beads prepared by different gelation methods. *J. Colloid Interface Sci.* **2014**, *432*, 109–116. [CrossRef]
26. Zhu, X.L.; Healy, L.E.; Sevindik, O.; Sun, D.W.; Selli, S.; Kelebek, H.; Tiwari, B.K. Impacts of novel blanching treatments combined with commercial drying methods on the physicochemical properties of Irish brown seaweed Alaria esculenta. *Food Chem.* **2022**, *369*, 130949. [CrossRef]

27. Yang, C.F.; Lai, S.S.; Chen, Y.H.; Liu, D.; Liu, B.; Ai, C.; Wan, X.Z.; Gao, L.Y.; Chen, X.H.; Zhao, C. Anti-diabetic effect of oligosaccharides from seaweed *Sargassum confusum* via JNK-IRS1/PI3K signalling pathways and regulation of gut microbiota. *Food Chem. Toxicol.* **2019**, *131*, 110562. [CrossRef] [PubMed]
28. Wei, C.K.; Roca, M. Cooking effects on chlorophyll profile of the main edible seaweeds. *Food Chem.* **2018**, *266*, 368–374.
29. Chen, K.W.; Roca, M. In vitro bioavailability of chlorophyll pigments from edible seaweeds. *J. Funct. Foods* **2018**, *41*, 25–33. [CrossRef]
30. Liang, X.P.; Ma, C.C.; Yan, X.J.; Zeng, H.H.; McClements, D.J.; Liu, X.B.; Liu, F.G. Structure, rheology and functionality of whey protein emulsion gels: Effects of double cross-linking with transglutaminase and calcium ions. *Food Hydrocoll.* **2020**, *102*, 105569. [CrossRef]
31. Shu, J.X.; Mcclements, D.J.; Luo, S.J.; Ye, J.P.; Liu, C.M. Effect of internal and external gelation on the physical properties, water distribution, and lycopene encapsulation properties of alginate-based emulsion gels. *Food Hydrocoll.* **2023**, *139*, 108499. [CrossRef]
32. Lin, D.Q.; Kelly, A.L.; Miao, S. The impact of pH on mechanical properties, storage stability and digestion of alginate-based and soy protein isolate-stabilized emulsion gel beads with encapsulated lycopene. *Food Chem.* **2022**, *372*, 131262. [CrossRef] [PubMed]
33. Owoyemi, A.; Balaklav, M.; Kochanek, B.; Porat, R.; Koenigstein, N.; Salzer, Y.; Lichter, A. Deviations from optimal storage temperature and its impact on postharvest quality of table grape cv. Scarlotta Seedless. *Postharvest Biol. Technol.* **2024**, *215*, 113013. [CrossRef]
34. Everitt, M. Consumer-Targeted Sensory Quality. In *Global Issues in Food Science and Technology*, 1st ed.; Barbosa-Canovas, G.V., Mortimer, A., Lineback, D., Spiess, W., Buckle, K., Colonna, P., Eds.; Academic Press: Cambridge, MA, USA, 2009; pp. 117–128.
35. Laksanawati, T.A.; Khirzin, M.H.; Meidayanti, K.; Kusherawati, P.A.; Kusuma, H.S.; Darmokoesoemo, H.; Iqbal, M. Prediction of shelf life and sensory qualities of beef meatball with biodegradable taro starch-duck bone gelatin packaging at different storage temperatures. *Appl. Food Res.* **2024**, *4*, 100402. [CrossRef]

foods

MDPI

Review

Development and Application of Mucilage and Bioactive Compounds from Cactaceae to Formulate Novel and Sustainable Edible Films and Coatings to Preserve Fruits and Vegetables—A Review

Viviane Priscila Barros de Medeiros, Kataryne Árabe Rimá de Oliveira, Talita Silveira Queiroga and Evandro Leite de Souza *

Department of Nutrition, Health Sciences Center, Federal University of Paraíba, João Pessoa 58051-900, PB, Brazil; barrosviviane89@gmail.com (V.P.B.d.M.); kataryne.arabe@academico.ufpb.br (K.Á.R.d.O.); talitaasq@gmail.com (T.S.Q.)
* Correspondence: els@academico.ufpb.br

Abstract: The accelerated ripening and senescence of fruits and vegetables is characterized by various biochemical changes that hinder the maintenance of their postharvest quality. In this context, developing edible films and coatings formulated with natural and biodegradable materials emerges as a sustainable strategy for preserving the quality parameters of these products in replacement of conventional petroleum-based packaging. Recently, plant-based polymers, including mucilage from different cactus species and/or their bioactive compounds, have been investigated to develop edible films and coatings. As the available literature indicates, the *Opuntia* genus stands out as the most used for mucilage extraction, with the cladode being the most exploited part of the plant. Conventional extraction methods are widely employed to obtain mucilages, which are applied to fruits and vegetables after being combined with plasticizing and cross-linking agents. In general, these films and coatings have proven effective in prolonging the shelf life and maintaining the nutritional, physical, and sensory quality of fruits and vegetables. Given their preservation potential, combining cactus mucilages with bioactive compounds, probiotics, and prebiotics represents an emerging trend in developing functional films and coatings. However, some limitations have been identified, such as the underutilization of different species and parts of the plant, the lack of standardization in extraction methods, and the absence of studies on the effects of the physicochemical properties of mucilages in the formulation and characteristics of films and coatings. Therefore, overcoming these limitations is essential for developing edible films and coatings with enhanced techno-functional properties and greater commercial viability.

Keywords: active packaging; polymers; quality parameters; vegetable matrices

1. Introduction

Harvested products, especially fruits and vegetables, are metabolically active, undergoing ripening and senescence processes that must be controlled to prolong their postharvest quality [1]. These products have intrinsic properties and are exposed to extrinsic factors that accelerate postharvest losses. Intrinsic factors include physiology, size, and ripeness. In contrast, extrinsic factors include mechanical damage, microbial diseases (especially fungi), and the lack of or inadequate postharvest management [2]. The inefficient management of these processes can result in significant losses of nutritional and quality attributes and financial setbacks [1,3]. Postharvest losses in fruits and vegetables can be as high as 45% due to poor postharvest handling [4].

In this sense, maintaining the quality of fruits and vegetables becomes crucial for the agri-food sector. To improve the postharvest use of these products, the production of minimally processed fresh foods has increased [5]. Minimal fruit and vegetable processing involves applying operations such as washing, sanitizing, peeling, chopping, and partial preservation treatments. This allows for obtaining products ready for consumption with sensory and nutritional characteristics like fresh fruits [6]. However, processing can cause physical damage (e.g., a soft texture and flavor loss), enzymatic browning, increased ethylene production, and respiration for vegetables and fruits. In addition, vegetable and fruit cell rupture caused by cutting releases cellular contents and promotes microbial growth, reducing their potential economic value [7].

These changes in fresh and minimally processed products require conservation techniques. Cold storage, alone or in combination with other treatments, represents the most effective technique for maintaining the quality (physical, nutritional, and sensory properties) of fruits and vegetables, as it reduces the rate of many metabolic processes and controls microbial deterioration over time [8,9]. Although cold storage remains a fundamental postharvest strategy, its efficacy in preserving the optimum quality of fruits and vegetables during transportation and in markets has yet to diminish. Furthermore, some of these materials may suffer chilling injury at temperatures below 7 °C, which favors the development of fungal diseases [4]. These situations have directed integrating postharvest technologies, such as modified and controlled atmosphere packaging, to preserve horticultural products [10]. However, conventional food packaging materials like plastic are not environmentally friendly. For this reason, researchers have focused on developing innovative biodegradable packaging for application in the food industry, aiming to replace packaging made from petroleum derivatives [11]. Active packaging in the form of edible films and coatings could be a sustainable alternative to preserving the quality of fruits and vegetables that are fresh and minimally processed [6,12].

Edible coatings and films are semipermeable barriers that may positively influence vegetable tissue metabolism by protecting against mechanical, physical, and chemical damage (facilitating weight loss control, reducing breathing and perspiration processes, and preserving texture and nutritional properties), in addition to being able to carry antioxidants, antimicrobials, and other stabilizers, controlling microbial growth, and preserving fruit/vegetable quality for a more extended period [9,13]. Furthermore, edible coatings and films can be consumed with food, providing additional nutrients, improving sensory characteristics, and enhancing product quality [13,14].

The main components used for producing edible/biodegradable films and coating are biopolymers, such as carbohydrates–polysaccharides (chitosan, sodium alginate, gums, starch, pectin, and cellulose derivatives); proteins; a solubilizing medium (e.g., water and ethanol); and plasticizers. The advantages of using such components are their capability to provide moisture resistance, a water-soluble nature, gelling properties, and good thermal and mechanical properties [15]. Researchers have recently focused on using natural biopolymers from different plant-based materials, highlighting the mucilage derived from cacti, characterized as a subgroup of hydrocolloids that can form gels in the presence of water [16]. Cactus mucilages have already been investigated for creating films and coatings, demonstrating promising results for improving the shelf life of fruits and vegetables [1,6,9,16].

Cactus mucilage is a complex high-molecular-weight polymeric heteropolysaccharide that contains variable amounts of neutral sugars (D-galactose, D-xylose, l-rhamnose, and l-arabinose) and D-galacturonic acid [12]. These substances swell when mixed with water or form colloidal suspensions, producing a shiny coating on the surface of the fruit (or vegetable). In addition, these substances have hydrophilic properties that create a barrier against water evaporation from the plant tissue, slowing weight loss and increasing firmness [9]. Its application also reduces microbial growth by developing a modified atmosphere around the product and can facilitate the incorporation of additives, such as antimicrobials and antioxidants [12]. Cactus species are also rich in bioactive compounds, such as betalains, phytosterols, polyphenols, vitamins, and minerals, which can improve the nutritional composition of foods when incorporated into coatings applied to foods [17].

Mucilage and its bioactive compounds could be preferred to synthetic materials in edible packages due to their intrinsic properties, such as nontoxicity, biocompatibility, biodegradability, and adaptability. Mucilage also has other advantages, including producing low-cost films compared to most biopolymers and being obtained through an easy extraction process with a high yield [16,18]. However, it has disadvantages, such as low mechanical strength and a high affinity for water. The properties of cactus mucilage films and coatings are closely related to the composition and structure of the polysaccharide, which is highly branched and rich in hydrophilic groups [19].

This review updates the main production methods, chemical composition, and technological and functional properties of edible films and coatings formulated with mucilage and bioactive compounds from different cactus species. Furthermore, the effects of these edible films and coatings on the quality parameters of fruits and vegetables, as well as the market challenges and future perspectives, are presented and discussed.

2. Cactaceas: Genera, Species, and Plant Parts Used for Extraction of Mucilages and Formulation of Films and Coatings

Cacti are xerotrophic plants cultivated in Central and South America, Asia, and Southern Europe [20]. The family species have a high concentration of polysaccharides, especially mucilage [21]. Mucilage is a subgroup of hydrocolloids that can create gels in the presence of water [22]. *Opuntia* species are the most potential sources of mucilage among plants due to their abundance, biodegradability, and renewability. *Opuntia* mucilage is a complex material formed from carbohydrates, including L-arabinose, D-galactose, D-xylose, L-rhamnose, and D-galacturonic acid, in variable proportions [2]. In the *Opuntia* genotype, the cladode age, cultivation area, and pruning season affect its composition [13,19]. Furthermore, seasonal variations contribute to the physicochemical changes of mucilage, such as the relative amount of carbohydrates related to water availability to the plant or the effects of changes in pH and electrical conductivity [23].

According to the data shown in Table 1, the *Opuntia* genus is widely used to extract and formulate edible films and coatings applied to fruits and vegetables. The most employed species are *O. ficus-indica* [1–3,6,8–10,12,14,15,19,24–29], *Nopalea cochenillifera* [17,30], *O. dillenii* [31,32], *O. stricta L.* [4], *O. robusta* [21], *O. stenopetala* [13], and *O. oligacantha* [33]. The *Hylocereus undatus* species was also used to extract mucilage in a study [34]. *O. ficus-indica* has gained popularity as a potential source of additives for the food industry due to its nutritional properties, such as high calcium and potassium contents, a high water-retention capacity, and bioactive phytochemicals [12]. The mucilage in *O. ficus-indica* represents about 10% of the dry weight of cladodes and can retain more than 30% of the total water in the storage parenchyma. This anatomical characteristic contributes to its high drought resistance and remarkable water-retention capacity in the tissues, which is explained by the polymerization of sugars, the elasticity of the storage parenchyma cells, and the high water-retention capacity of the mucilage [26].

Table 1. Species and plant parts and methods and conditions for extracting mucilages and bioactive compounds from cacti.

Species of Cacti	Plant Part	Extraction Method	Extraction Procedure	References
O. ficus-indica	Cladodes	Hot extraction	Scraping of cladodes; pressing; sieved to separate pure mucilage from slurry; extraction in water at 25%, 50%, and 75%, (*w*/*v*); and pasteurization at 70 °C for 45 min.	[1]
O. stricta L.	Stems	Aqueous extract	Hygienization; peeling and cutting; addition of distilled water (1:1 *w*/*v*); blended for 3 min; filtration through cloth; precipitation of the filtrate using 20% isopropyl alcohol in a 1:1 (*v*/*v*) ratio; centrifugation of the precipitate at 2.683× *g* for 10 min; and drying of the precipitate at 70 °C for 4 h.	[4]

Table 1. *Cont.*

Species of Cacti	Plant Part	Extraction Method	Extraction Procedure	References
O. ficus-indica	Cladodes	Hot extraction	Court of cladodes; crushing; homogenization in water in a 1:1.5 (*w*/*v*) ratio and at 20 °C; heating at 40 °C for 90 min; centrifugation at 1450× *g* for 20 min; heating of the supernatant until half of the initial volume; addition of ethanol (1:2 *v*/*v*); storage of the solution at 4 °C for 48 h; and elimination of the supernatant and use of pure mucilage.	[9,24,26]
O. robusta	Cladodes	Hot extraction	Removal of chlorenchymatous tissue and parenchymatous tissue; homogenization with ethanol or water in a 1:1 (*w*/*v*) ratio; agitation at 50 °C for 2 h; crushing; resting; filtering; drying at 50 °C; and milling and storage at 25 °C.	[21]
O. ficus-indica	Cladodes	Hot extraction	Hygienization; removal of the parenchyma; immersion in hot water for 30 s; homogenization with water for 5 min; and filtration.	[6]
O. stenopetala	Cladodes	Hot extraction	Homogenization with distilled water in a 1:2 (*w*/*v*) ratio; magnetic stirring at 60 °C for 90 min; centrifugation for 10 min at 246× *g*; addition of 96% ethanol in a 1:2 (*v*/*v*) ratio with the supernatant; centrifugation at 61× *g* for 10 min; drying of precipitate at 40 °C for 24 h; and grinding and storage.	[13]
O. ficus-indica	Cladodes	Hot extraction	Mucilage used as a gel: peeling and slicing of cladodes; homogenization in a blender; heating of the paste at 40 °C for 90 min; centrifugation at 10,000× *g* for 10 min; recovery of the supernatant; pasteurization at 77 °C for 1 min; and storage at 4 °C. Mucilage used as a polysaccharide: peeling and slicing of cladodes; homogenization in a blender; heating of the paste at 40 °C for 90 min; centrifugation at 10,000× *g* for 10 min; boiling of the supernatant at 100 °C to 50–60% of the initial concentration; centrifugation; precipitation with 96% ethanol (1:1); incubation at 1 °C for 2 h; centrifugation at 3600× *g* for 5 min; drying the pellet at 50 °C for 12 h; and grinding and sieving.	[20]
N. cochenellifera	Cladodes	Alcoholic extraction	Removal of the epidermis; slicing of the parenchyma; homogenization with 99.5% ethanol in a ratio of 2:3 m/v for 60 s; addition of ethanol to precipitate the mucilage; filtration of the mucilage in polypropylene fabric; washing (2×) with 99.5% ethanol for pigment removal; drying at 55 °C for 24 h; and grinding to obtain dry powder.	[17]
O. ficus-indica	Cladodes	Aqueous extraction	Peeling and cutting the cladodes; homogenization; dilution of the extract in water (2:1 *w*/*v*); and obtaining mucilage.	[3]
O. ficus-indica	Cladodes	Microwave-assisted extraction	Hygienization; cutting into cubes; microwave heating (900 W for 2 min); homogenization; centrifugation at 12,500 rpm for 15 min at 4 °C; discarding of insoluble fiber; and decanting and separation of soluble fiber.	[10]
O. ficus-indica	Cladodes	Microwave-assisted extraction	Hygienization; removal of thorns; removal of chlorenchyma; cutting into cubes; heating in microwave oven (900 W) for 3–5 min; homogenization; centrifugation of pulp at 8117× *g* for 15 min at 4 °C; and decantation of mucilage.	[14,24]
O. ficus-indica	Pear fruits	Aqueous extraction	Hygienization; cutting; immersion in water (5:1 *w*/*v*); overnight incubation at room temperature; filtration of the mixture; and concentration of mucilage by evaporation.	[28]

Table 1. *Cont.*

Species of Cacti	Plant Part	Extraction Method	Extraction Procedure	References
N. cochenillifera	Cladodes	Acid extraction	Hygienization; removal of glochids; cutting of cladodes into cubes; immersed in a solution of citric acid (5 mg/L) for 10 min; and mesh filtration.	[30]
O. ficus-indica	Cladodes	Aqueous extraction	Peeling and cutting of cladodes; immersion in distilled water (5:1 *w*/*v*); incubation overnight; filtration; and concentration of the solution by evaporation.	[2]
O. oligacantha	Fruit (xoconostle)	Ultrasound assisted extraction	Addition of orange essential oil to soy lecithin and aqueous extracts of 20% and 10% xoconostle; sonication of the mixture for 30 min with a 6 mm probe at an amplitude of 80% and a frequency of 20 kHz; distribution of the droplets via a dynamic laser light; scattering technique with an angle of 90 °C; and storing the nanoemulsion at 6 °C.	[33]
H. undatus	Dragon fruits	Alcoholic extraction	Cutting and peeling; removal of the seeds; removal of the pulp; precipitation of the mucilage with ethanol in a ratio of 2:3 (*v*/*v*) for 24 h at 4 °C; filtration; and drying in an oven at 40 °C for 24 h.	[34]
Opuntia spp.	Stems	Aqueous extraction	Homogenization of mucilage (20% *w*/*v*) in distilled water; centrifugation at 4500 rpm for 10 min; and pasteurization of the supernatant at 70 °C for 45 min to obtain pure mucilage.	[35]
Opuntia spp.	Cladodes	Hot extraction	Hygienization; removal of thorns; cutting; heating in three buffer solutions (pH 7, 4, and 9); repose for 5 h in pH 10; separation of the precipitate of the supernatant; resting of supernatant in a refrigerator overnight; separation of the precipitate; and drying in an oven at 105 °C for 12 h.	[36]
O. ficus-indica	Cladodes	Microwave-assisted extraction	Peeling; slicing; microwave heating at 800 W for 4 min; separation of insoluble fibers in gauze; centrifugation of the supernatant at 8000× *g* for 15 min at 4 °C; lyophilization for 72 h; and storage at 25 °C.	[12,29]
O. ficus-indica	Cladodes	Hot extraction	Hygienization; removal of parenchyma; trituration in distilled water in a ratio of 1:1.5 (*w*/*v*) at 20 °C; heating of the solution at 40 °C for 90 min; centrifugation at 1450× *g* for 20 min; heating of the supernatant to half of the initial volume; precipitation with ethanol in a 1:2 (*v*/*v*) ratio; solution stored at 4 °C for 48 h; discarding of the supernatant; and obtaining hydrated mucilage.	[8]
Nopal cactos	Cladodes	Hot extraction	Homogenization for 30 min with distilled water in a 1:1 (*w*/*v*) ratio; heating of the suspension at 90 °C for 30 min; centrifugation at 1500× *g* for 20 min; addition of 96% ethanol to the supernatant in a ratio of 2:1 (*v*/*v*); centrifugation at 1500× *g* for 20 min; drying of the mucilage for 24 h at 70 °C; and pulverization to obtain the mucilage.	[5]
O. ficus-indica	Pads	Aqueous extraction	Hygienization; removal of the peel and thorns; cutting the pulp; extraction in distilled water in a 1:2 *v*/*v* (pulp:distilled water) ratio; resting the solution for 24 h at room temperature; filtration through nylon cloth; centrifugation at 12,880× *g* at 4 °C for 30 min; precipitation with 95% *v*/*v* ethanol; and evaporation of ethanol in a water bath at 50 °C.	[15]

Table 1. *Cont.*

Species of Cacti	Plant Part	Extraction Method	Extraction Procedure	References
O. dillenii	Cladodes	Hot extraction	Peeling and slicing of cladodes; drying at 80 °C for 6 h; sifting to obtain a fine powder; extraction of fat (Soxhlet); dispersion in distilled water under agitation; obtaining a suspension at a concentration of 5% (*w*/*v*); extraction of the suspension in a water bath at 80 °C for 4 h; filtration; removal of proteins; concentration of filtrates; precipitation with three volumes of ethanol; filtration; and lyophilization.	[31]
O. dillenii	Cladodes	Hot extraction	Peeling and slicing of cladodes; homogenization with distilled water in a 1:1 *w*/*v* ratio; heating in a water bath at 75 °C for 6 h; filtration; removal of proteins from the filtrate; addition of three volumes of absolute ethanol to the precipitate; filtration; and lyophilization.	[32]

The mucilage of *O. ficus-indica* found in cladodes is a complex anionic polyelectrolyte polysaccharide formed by alternating galacturonic acid linked to rhamnose residues. Branches are galactose residues that carry sugars, such as xylose and arabinose, as substituents [5]. This compound can be an attractive natural edible coating with high nutritional value, besides helping to preserve fruits and vegetables [29]. In addition to the complex mixture of carbohydrates, it presents quantities of phenolic compounds (flavonoids and phenolic acids); vitamins (mainly ascorbic acid, vitamin B, and α-tocopherol); polyunsaturated fatty acids; and amino acids [15,27,29]. The hydrophilic characteristic of *O. ficus-indica* coatings could act as a barrier to water transfer, retarding transpiration and browning; maintaining fresh weight, firmness, and nutritional attributes; and controlling microbial growth, resulting in a more extended storage period for coated fruits and vegetables [27,28].

Although *O. ficus-indica* is the most studied species, other cactus species have also been characterized as sources of mucilage and other bioactive components. Among the other cactus species producing mucilage is the cactus pear (*Nopalea cochenillifera* Salm-Dyck), a plant with an acid crassulacean metabolism. This mechanism confers an efficient use of water, enabling the plant to thrive in arid and semiarid regions, becoming a vital forage resource for ruminant animals, especially during the dry seasons of the year [17]. Using edible coatings based on forage palm is a way to naturally incorporate antioxidants, such as soluble phenolics, into the treated product [17]. Belonging to the cactus family, *O. dillenii* is mainly distributed in tropical and subtropical regions, growing mainly in the arid or semi-desert areas of America and Mexico. *O. dillenii* polysaccharide is one of the main active compounds of this plant and is extracted mainly from the fruit and plant stem. The main constituents are arabinose, xylose, fructose, glucose, galacturonic acid, and rhamnose [31,32]. A previous study reported high aggregation in the mucilage obtained from *O. robusta*, which indicated that this species has a higher fiber content than other *Opuntia* species [21]. The species *O. stenopetala*, traditionally used for human consumption and livestock fodder, showed potential for use in the formulation of multilayer coatings [13]. Xoconostle (*O. oligacantha* CF Först), a traditional fruit from Central Mexico, contains phenolic compounds, flavonoids, and betalains, which confer antioxidant and antibacterial activities and can thus be ideally incorporated into edible coatings [33]. Pitahaya (*H. undatus*), commonly known as "dragon fruit", is a native fruit from tropical regions of America that belongs to the climbing cacti family. The fruit has white or red flesh with numerous tiny black seeds. Edible films produced with mucilage extracted from dragon fruit peel, combined with glycerol and pectin, demonstrated improved mechanical properties [16]. The addition of pectin in cactus mucilage-based edible coatings is particularly associated with an increase in solid content in the film/coating-forming solution, enhancing the mechanical and barrier properties of the produced film/coating [2,13]. Glycerol, in turn, contributes to a more efficient interaction among the components in the film-forming solution, resulting in films with smooth, homogeneous, and compact microstructures [16]. Pitahaya films have strong potential to aid sustainable food production by reducing food waste (pitahaya peel), which can be used as coating and packaging to extend the shelf life of fruits and vegetables [16].

As shown in Table 1, cactus mucilage can be obtained from different plants and plant parts, including cladodes or pads [1,3,6,9,13,15], stems [3,4,35], and fruits [28,33]. The cladodes represent the most abundant plant portion in cactus cultivation areas and are widely used for the extraction of mucilage and its subsequent application as an edible coating [16,18]. Consisting of branches that exhibit variable morphology, cladodes can be globular, cylindrical, or flattened, and their dimensions differ according to the species, plant age, and the cultivation environment conditions [37]. These structures are covered by a waxy layer that reduces transpiration and are arranged to form a tree without a defined trunk and branches. The internal structure of the cladodes consists of a dense network of cells that contain substances capable of absorbing and retaining water, allowing the plant to survive in arid conditions. Additionally, the fine spines on the outer surface are essential to defend against animals [38].

Mucilage is stored in the cladodes in mucilaginous cells located in the chlorenchymatous and parenchymatous tissues, with the parenchymatous being the tissue with the highest mucilage content [20]. The mucilage extracted from the chlorenchyma, the outer layer of the cladodes, is characterized by a fibrous nature with an irregular structure and dimensions. For this reason, it is removed during the mucilage extraction process [17,20]. In contrast, the mucilage obtained from the parenchyma, which constitutes the inner portion of the cladodes, has a higher concentration of pectic substances. This characteristic enhances its effectiveness in formulating edible coatings for fruits and vegetables [6,8,17,20].

Regardless of the part used in extraction, in general, the life cycle of cactus mucilage encompasses the stages of biosynthesis by the plant, extraction and purification, storage, industrial applications, and its biodegradation in the ecosystem [1,4,9,16,39]. After extraction, the preservation of ideal physicochemical characteristics of the mucilage was observed for up to eight months of storage at low temperatures [39]. As for the biodegradability rate, mucilage-based films disintegrated after 35 days of soiling in organic compost [16]. However, it is noteworthy that, despite its importance, few studies address these aspects during their analyses, which is an essential fact for the valorization of the polymer as a sustainable strategy for food storage.

3. Main Extraction Methods and Physicochemical and Technological Properties of Cactus Mucilages

Cactus mucilages are complex polymeric molecules composed of branched polysaccharides that act as intricate molecular networks capable of absorbing water [21]. These natural colloidal structures possess technological, nutritional, and functional properties that can be industrially exploited [40,41]. In recent years, different methods for extracting cactus mucilages have been developed or optimized to increase mucilage yields and promote its use at the industrial level. As shown in Table 1, conventional methodologies, such as hot extractions, water immersion, or organic solvent use, remain predominant over more advanced methods, such as ultrasound-assisted extractions and microwave-assisted extractions. However, it was observed that conventional methods require more extended periods and high temperatures, which can favor the degradation of thermosensitive bioactive compounds, such as phenolic compounds and pigments present in cactus mucilage [42,43]. Although their low cost is an attractive feature of these methods, the excessive use of organic solvents can increase extraction costs and negatively impact the environment due to improper disposal [44]. In general, mucilages extracted by conventional and ultrasound-assisted methods show similar rheological behaviors; however, the ultrasound-assisted method results in a higher yield and shorter extraction time [45]. Although several extraction methods are used to obtain cactus mucilage applicable in coatings and edible films, no previous study has addressed the influence of extraction methods on the techno-functional attributes of these films and coatings [4,9,24,26].

The first step preceding mucilage extraction from cacti is washing the cladodes to remove dirt from the field, then sanitization with sodium hypochlorite [10,46]. After this process, the cladodes are peeled and manually cut into smaller fractions [4,10]. The next step involves immersing and homogenizing the cactus pulp in different organic solvent concentrations, such as ethanol [9], citric acid [30], or distilled water [15]. This process can be aided by equipment, such as magnetic stirrers, blenders, or food processors, and may occur under heating or at room temperature (Table 1). Thermal treatment is commonly applied during mucilage extraction to weaken molecular interactions between polysaccharide chains, facilitate water entry, and promote non-covalent intermolecular interactions with mucilage aggregates [47,48]. Next, the obtained mucilage is precipitated and washed with ethanol or isopropyl alcohol in varying concentrations (Table 1) to remove pigments and then directed to the separation stage [13]. Ethanol is the most used organic solvent for precipitating mucilages, while isopropanol is applied for extracting and purifying various biomolecules [47]. The separation stage of the mucilage can be carried out by multiple methods, such as filtration [6], decantation [10], evaporation [15], or centrifugation [10], adjusting the parameters of rotation, time, and temperature (Table 1).

After the separation stage, the wet mucilage is concentrated using convective drying [13] or freeze drying [31,32] at different temperatures and drying times. Conventional oven or tray drying is the most used method in the food industry; however, its low efficiency can cause physicochemical changes in the cacti using different times and temperatures [47]. As shown in Table 1, the temperatures used during the mucilage drying process range from 40 to 105 °C. The influence of drying using two different systems, i.e., convective drying and refractance window drying, on mucilages from *O. ficus-indica* and *Austrocylindropuntia cylindrical* dried at various temperature ranges (45–85 °C) was evaluated. The results show that increasing the temperature reduced the drying time in all experiments. Fourier Transform Infrared Spectroscopy (FTIR) analyses indicated that the mucilage and cellulose did not undergo significant structural changes due to the drying techniques employed [47]. However, the refractance window method reduced the mucilage drying time by half, emerging as a more efficient alternative. The final step for obtaining mucilage is grinding and sieving to achieve a fine powder. The final product of the mucilage extraction is a white, amorphous, and water-soluble material that can formulate films and coatings in liquid form [4] or powder form [49].

As characterized previously, due to their ease of execution, conventional methods using organic solvents or distilled water are the most employed for extracting mucilages from cacti to formulate edible films and coatings (Table 1). However, using these methods requires longer extraction times and high temperatures, which can result in the degradation of thermosensitive bioactive compounds. Although the cost of executing these methods is lower than that of non-conventional methods, the large-scale use of organic solvents can increase the process costs and present significant environmental challenges [42]. To mitigate the use of organic solvents during the mucilage extraction process, some studies performed mucilage extraction using only distilled water under different temperature conditions (25 and 90 °C) and volume proportions (in a general ratio of 1:1 w/v), with or without agitation (using a blender) [3,4]. Despite not being characterized, the mucilages obtained, when applied to fruits and vegetables, proved to be functional as a barrier against oxygen and ethylene, slowing down the rate of respiration and ripening [3,4].

Non-conventional extraction methods, such as ultrasound-assisted extractions [31] and microwave-assisted extractions [10,12,34], have also been employed for obtaining mucilages from cacti (Table 1). A microwave-assisted extraction uses microwave energy to heat the solvent and the plant simultaneously, promoting cell rupture and facilitating the release of bioactive compounds, such as mucilage. The main advantages of this technique include its speed, ease of execution, low cost, and reduced use of organic solvents. These factors make this method a sustainable and efficient alternative [18]. In turn, an ultrasound-assisted extraction is based on acoustic cavitation, generated by high-frequency ultrasonic waves applied to samples immersed in extracting liquids. This process causes the formation of microbubbles that, upon collapsing, release the energy responsible for breaking down cell walls and releasing bioactive compounds. In addition to increasing the efficiency of the extraction process, this technique also allows for a significant reduction in the volume of solvents used, making it a sustainable and economically viable alternative without compromising yields or process efficiency [42].

The yield of mucilage from cacti after extraction can vary depending on the part of the plant used, the extraction method, the solvents employed, and the pre-processing and purification steps [49]. In addition, other factors, such as the season, the plant genotype, the maturation stage, and the harvest time, can influence the yield and the physicochemical, technological, and functional properties of the cactus mucilage [23,38]. In particular, the stage of maturity and time of harvest can modify the yield, which is reduced with advancing maturity and the mucilage composition in terms of the concentration of carbohydrates, pigments (such as chlorophyll), and minerals (such as calcium) [22]. Young cladodes have a high content of water, phenolic compounds, and, mainly, soluble fibers, contributing to a higher mucilage yield. In contrast, in older cladodes (which face more significant climatic variations), an increase in the content of insoluble fibers and minerals is commonly observed [43]. However, galacturonic acid was identified as the main component in mature cladodes, being essential for film formation due to their viscosity properties, water retention, and ability to chelate Ca^{2+} ions [50].

Therefore, considering the biotic and abiotic factors involved in cultivating cactus species and selecting the appropriate extraction method is necessary to obtain mucilages from cacti with desirable physicochemical and technological characteristics. Mucilages of seven native cactus species from Brazil (*O. ficus-indica, O. cochenillifera, C. jamacaru, C. hildmannianus, P. gounellei, P. pachycladus,* and *T. inamoena*), obtained through aqueous extraction and precipitated with ethanol, showed high yields, ranging from 8.9% to 21.54%. The extraction process involved aqueous extraction followed by precipitation with ethanol. More specifically, the species *P. gounellei* (21.54%), *P. pachycladus* (20.80%), *O. ficus-indica* (19.05%), and *O. cochenillifera* (18%) exhibited the highest yields, suggesting the potential of these cacti as sources of mucilages applicable to the food industry [22].

The physicochemical properties of cactus mucilages determine the functionality and behavior of coatings during food storage and handling [51]. An early study analyzed the physicochemical, structural, and technological properties of mucilages from seven cactus species, including species native to the semiarid region of Brazil (*O. ficus-indica, Opuntia cochenillifera, Cereus jamacaru, Cereus hildmannianus, Pilosocereus gounellei, Pilosocereus pachycladus,* and *Tacinga inamoena*). The mucilages exhibited a light coloration, with luminosity (L) values ranging from 69.29 to 81.99 and low a* values, close to negative, indicating a tendency toward green [22]. The authors emphasize that the color of the mucilages can vary depending on the plant maturation stage, the amount of chlorophyll, the extraction and purification methods, and the solvents used in precipitation. Technologically, the mucilages demonstrated a high water-retention capacity, ranging from 2.40 to 23.01 g/g. In contrast, the oil-retention capacity was relatively low, between 2.14 and 3.02 g/g, which was attributed to the hydrophilic properties already known of mucilages. The mucilages exhibited satisfactory emulsifying, stabilizing, and foaming properties, highlighting their potential applications in food [22].

The particle size of mucilage can also influence the distribution of nutrients and their functional groups. Grinding and sieving techniques help obtain particles with a better water- and oil-retention capacity and optimize the release of bioactive compounds during processing [48]. Scanning electron microscopy (SEM) tests conducted on mucilages extracted from the parenchyma of *O. robusta* revealed aggregates of small particles with irregular shapes and fibrous dimensions more significant than the mucilage extracted from the chlorenchyma [48]. A study showed that reducing the particle size of the flour from the cladode of *O. ficus-indica*, obtained through grinding and sieving, altered the content of carbohydrates and lipids present in the mucilage [52]. Another study identified changes in the physicochemical and technological properties of mucilage powders obtained from two cultivars of *O. ficus-indica* and one cultivar of *O. robusta* (Robusta), collected during two distinct periods. The powders from mucilages harvested in February exhibited higher porosity, oil absorption, and retention capacity while showing a lower water-retention and swelling capacity. In contrast, the powders of mucilages collected in August exhibited smaller impermeable particles and superior hydrophobic properties and a better emulsifying capacity, stability, and viscosity [53].

The viscosity of these mucilages is influenced by the concentration of monovalent and divalent ions in their composition, and they may exhibit pseudoplastic behavior [39,54]. Due to their hydrophilic nature and the presence of calcium, cactus mucilages have a higher water-retention capacity compared to oil [23]. This characteristic allows for a more efficient interaction with water, promoting water retention and the homogeneous distribution of the mucilage during its application in films and coatings [55]. Edible films and coatings with homogeneous characteristics indicate a structural integrity associated with improved mechanical properties [56]. Several factors, such as the mucilage concentration, temperature, pH, porosity, and particle size, can influence the viscosity of cactus mucilages, potentially hindering their diffusion during the application of edible films and coatings on fruits and vegetables [23,52,54]. Thus, the viscosity of the mucilage should be adjusted to the application method, as less viscous solutions are preferable for spray applications. In contrast, more viscous solutions are suitable for other methods, such as immersion and spreading [57,58].

The barrier capacity of films against water molecules is associated with their efficacy in minimizing moisture loss or gain in foods or food products. Water vapor permeability (WVP) can be measured to evaluate moisture barrier properties [59]. This parameter reflects the film's water barrier property between the environment and the food matrix, contributing to increased shelf life [23]. Films with low WVP are ideal, as they prevent moisture transfer and water loss from the food during storage, thus helping to maintain food quality [11,59]. However, due to their hydrophilic nature and chemical composition, edible coatings and films formulated with cactus mucilage generally exhibit high WVP, which may be related to the mucilage extraction method, type and concentration of the additive used, and the coating processing method [23,49,55]. To improve the techno-functional properties of coatings and films formulated with cactus mucilage, using additives to reduce the spaces within the polymer structure, thus hindering water vapor diffusion, has been a common approach [51,54,60].

The mucilage from the genus *Opuntia*, commonly used for formulating edible coatings and films, is a complex polymer with a branched structure composed of various sugar units. These sugars interact with the hydroxyl groups of uronic acids, imparting the mucilage with its technological and functional characteristics [31]. The spectra obtained through FTIR analyses are commonly used to evaluate the main functional groups present in cactus mucilages, relating the absorption frequencies obtained to known absorption frequencies of recognized chemical bonds [23]. The bands commonly identified are OH groups from polysaccharides, carboxylic acids, uronic acids, and α- and β-D-glucose configurations [23]. In addition to carbohydrates, cactus mucilage is an excellent source of proteins; minerals (calcium, copper, iron, zinc, and manganese); and various phenolic compounds that can act in preserving fruits and vegetables when applied in edible films and coatings [22,23].

4. Processing Methods and Characterization of Films and Coatings Formulated with Cactus Mucilages and Their Bioactive Compounds

Postharvest losses of fruits and vegetables represent a significant challenge for the agricultural sector. In recent years, edible packaging in films and coatings formulated with different biomaterials (carbohydrates, proteins, and lipids) has been widely used as an innovative strategy to minimize postharvest losses [46]. In general, the application of edible coatings helps maintain the firmness of the food, control the respiration rate, and regulate the ripening process, playing a crucial role in determining the shelf life, marketability, nutritional value, and functionality of horticultural products [17,29].

From this perspective, various polymers obtained from natural sources, such as starches, cellulose derivatives, pectin, carrageenan, chitosan, sodium alginate, proteins (of animal or plant origin), and plant-based mucilages have been explored for the formulation of edible films and coatings [10,13,60]. The biodegradability, biocompatibility, bioavailability, low processing cost, and high extraction efficiency, combined with their nutritional, physicochemical, and technological properties, make cactus mucilage a widely used raw material in the formulation of effective edible coatings and films aimed at extending the shelf life of various fruits and vegetables [3,4,17].

Although edible films and coatings based on cactus mucilage offer several advantages, they present some limitations compared to synthetic polymers, including a low yield in the extraction process [22], the use of organic solvents in conventional mucilage extractions [17,34], and variations in mucilage chemical compositions, which make standardizing coating formulations challenging [23]. Additionally, the high hydrophilicity of mucilage reduces the mechanical strength and barrier properties, resulting in films that are fragile and have a granular texture, necessitating the combination of mucilages with other compatible polymers and plasticizing agents to optimize their physical and functional properties [15,16,61]. These challenges highlight the need for research developments that advance current knowledge and provide insights to overcome these limitations, paving the way for more effective coatings suited to the food industry [39].

4.1. Techniques and Components Used to Formulate Films and Coatings with Cactus Mucilages and Their Bioactive Compounds

Table 2 summarizes the main methods and components used to develop edible coatings and films based on cactus mucilage and bioactive compounds to preserve fruits and vegetables. The main components used for formulating edible films and coatings include mucilage in powder or liquid form [4,6]; plasticizing agents (glycerol, Tween, or sorbitol) [15,17,24]; and cross-linking agents, such as calcium chloride [3]. Generally, the formulation process involves diluting and homogenizing the mucilage powder in distilled or deionized water in concentrations ranging from 0.004% (w/v) to 75% (w/v), followed by the addition of plasticizing agents, such as glycerol (0.1% up to 50% v/v), sorbitol (3.4% v/v), or Tween 20 (10% v/v), depending on the methodology used (Table 2). These solutions may or may not be subjected to heating and stirring to ensure a complete incorporation of the mucilage with the other components in the solution [34]. Thus, the method for developing coatings and films is simple and easy to execute.

Table 2. Development and application of coatings and edible films formulated with cactus mucilage and its bioactive compounds for preserving fruits and vegetables.

Cacti	Coating/Film Material		Formulation	Application Method	Fruits/Vegetables	Reference
O. ficus indica	Coating	-Mucilage powder (6% w/v) -Glycerol (10% v/v) -Tween 20 (10% v/v)	Homogenization of pure mucilage extract in distilled water and glycerol or Tween 20.	Immersion	Minimally processed kiwifruit	[24]
O. ficus indica	Coating	-Liquid mucilage (6% v/v) -Glycerol (10% w/v)	Homogenization of pure mucilage extract with distilled water and glycerol.	Immersion	'Dottato' figs	[26]
O. ficus indica	Coating	-Mucilage powder (0%,25%, 50%, and 75% w/v) -Aloe gel (0%, 25%, 50%, and 75% w/v)	Hydration of the cactus mucilage with distilled water and aloe gel in different combinations.	Immersion	Mangoes	[1]

Table 2. *Cont.*

Cacti	Coating/Film Material		Formulation	Application Method	Fruits/Vegetables	Reference
O. stricta L.	Coating	-Mucilage powder (1%, 2%, and 3% *w*/*v*) -Glycerol (2% *v*/*v*)	Homogenization of the mucilage in distilled water and addition of glycerol. Then, the solution was dissolved for 3 min and centrifuged at 2683× *g* for 10 min to use the supernatant.	Immersion	Peppers	[4]
O. robusta	Coating	-Mucilage powder (10% *w*/*v*) -Ethanol (50% *v*/*v*)	Addition of water or ethanol in parenchyma or chlorenchyma mucilage powder (to obtain a solution with 12% soluble solids).	Brushing	Tomatoes	[21]
O. ficus-indica	Coating	-Liquid mucilage (* Np)	Application of liquid mucilage.	Immersion	Minimally processed rambutans	[6]
* Np	Coating	-Mucilage powder (1% *w*/*v*)	Homogenization of mucilage in distilled water, followed or not by ultrasound treatment (40 kHz and 480 W) for 10 min.	Immersion	Minimally processed potatoes	[7]
O. ficus-indica	Coating	-Mucilage gel (25% and 50% *v*/*v*) -Polysaccharide of *O. ficus-indica* (1% *w*/*v*) -Glycerol (5% *v*/*v*)	Homogenization of the mucilage gel or polysaccharide in deionized water, followed by the addition of glycerol.	Immersion	Cherries	[20]
N. cochenellifera	Coating	-Mucilage powder (0.004% *w*/*v*) -Glycerol (30% *v*/*v*)	Homogenization of the mucilage in distilled water, followed by the addition of glycerol.	Immersion	Sweet potatoes	[17]
O. stenopetala	Coating	-Polysaccharides from *O. stenopetala* (0.1% *w*/*v*) -Sodium alginate (0.5% *v*/*v*) -Lactic acid (1% *v*/*v*) -Chitosan (0.5% *w*/*v*) -Extract of *F. microphylla* leaves (0.5 % *v*/*v*)	Preparation of different solutions: -Chitosan or alginate solution with added polysaccharides from *O. stenopetala* isolated or in combination with the extract of *F. microphylla* leaves; -Homogenization of the mucilage gel or polysaccharide in deionized water, followed by the addition of glycerol.	Multilayer system	Tomatoes	[13]
N. cochenellifera	Coating	-Mucilage powder (0.004% *w*/*v*) -Glycerol (30% *v*/*v*)	Homogenization of cactus mucilage powder in distilled water and glycerol.	Immersion	Sweet potatoes	[17]
O. ficus-indica	Coating	-Liquid mucilage (50% *v*/*v*) -Calcium chloride (6% *w*/*v*)	Dissolution of pure mucilage in distilled water; previous application with calcium chloride solution.	Immersion	Tomatoes	[3]

Table 2. *Cont.*

Cacti	Coating/Film	Material	Formulation	Application Method	Fruits/Vegetables	Reference
O. ficus-indica	Coating	-Liquid mucilage (60% and 67% v/v) -Glycerol (30% and 40% v/v) -Glutamine (3% w/v)	Homogenization of mucilage to glycerol and glutamine.	Immersion	Minimally processed white-flesh loquats	[10]
O. ficus-indica	Coating	-Liquid mucilage (* Np) -Ascorbic acid (5% w/v)	Homogenization of mucilage with ascorbic acid.	Atomizing spray	Strawberries	[28]
O. ficus-indica	Coating	-Liquid mucilage (* Np)	Application of liquid mucilage.	Atomizing spray	Minimally processed loquats	[14]
O. ficus-indica	Coating	-Liquid mucilage (* Np)	Application of liquid mucilage.	Atomizing spray	Minimally processed cactus pear fruits	[27]
O. ficus-indica	Coating	-Mucilage powder (3% w/v) -Ascorbic acid (0.03%, 0.009%, 0.015% v/v) -Glycerol (50% v/v)	Homogenization of mucilage with distilled water under magnetic stirring for 4 h at 30 °C. Mixture of varying concentrations of ascorbic acid and glycerol.	Immersion	Pecan nuts	[12]
N. cochenilifera	Coating	-Liquid mucilage -Cassava starch (3% w/v) -Glycerol (1% v/v)	Homogenization of mucilage in a previously heated cassava starch solution (70 °C), followed by the addition of glycerol.	Immersion	Minimally processed yams	[30]
O. ficus-indica	Film	-Mucilage powder (20% w/v) -Chitosan (2% w/v) -Glycerol (3% v/v)	Homogenization of mucilage in distilled water by vortexing, followed by the addition of chitosan and glycerol. Addition of the solution to a Petri dish for drying at room temperature.	Direct application	Tomatoes	[2]
Opuntia spp.	Coating	-Mucilage powder (20% w/v) -Glycerol (25% v/v)	Homogenization of mucilage in distilled water and addition of glycerol.	Immersion	Carica papaya	[35]
O. oligacantha	Coating	-Bioactive extract of *O. oligacantha* (10 and 20% v/v) -Mineral oil (70% v/v) -Soy lecithin	Homogenization of orange essential oil (continuous phase) with soy lecithin (surfactant), followed by the addition of the aqueous extract of *O. oligacantha* (dispersed phase). The mixture was sonicated at 30 °C.	Sprinkling	Tomatoes	[33]
H. undatus	Coating	-Mucilage powder white dragon fruits (3% w/v)	Homogenization of mucilage in distilled water for 30 s.	Immersion	Cherry tomatoes	[34]

Table 2. *Cont.*

Cacti	Coating/Film	Material	Formulation	Application Method	Fruits/Vegetables	Reference
Opuntia spp.	Coating	-Cactus polysaccharide (0.5%, 1%, and 2% *w*/*v*) -Acetic acid (1% *v*/*v*) -Ascorbic acid (2% *w*/*v*) -Glycerol (1.5% *v*/*v*) -Sunflower oil (0.025% *v*/*v*)	Homogenization of cactus polysaccharide with acetic acid, ascorbic acid, citric acid, glycerol, sunflower oil, and distilled water.	Immersion	Citrus fruits	[36]
O. ficus-indica	Coating	-Mucilage powder (6% *w*/*v*) -Glycerol (10% *v*/*v*)	Homogenization of mucilage in distilled water and addition of glycerol	Immersion	Sweet cherries	[8]
O. ficus-indica	Coating	-Mucilage powder (1%, 2%, and 3% *w*/*v*) -Glycerol (0.1% *v*/*v*)	Homogenization of mucilage in distilled water utilizing a magnetic stirrer at 5000 rpm at 30 °C for 4 h, followed by adding glycerol.	Immersion	Bananas	[29]
O. ficus-indica	Coating	-Mucilage powder (15% *w*/*v*) -Glycerol (10% *v*/*v*) -Aloe arborescens gel (2% *v*/*v*)	Homogenization of mucilage in distilled water, followed by the addition of glycerol and aloe arborescens gel.	Immersion	Figs	[9]
Nopal cactus	Coating	-Nopal mucilage (4% *w*/*v*) -Glycerol (0.5% *v*/*v*) -Chitosan solution (1.5% *w*/*v*)	Homogenization of mucilage in distilled water and addition of glycerol. Immersion of fruit in mucilage solution; placed in drying for 2 min; and immersion in chitosan solution and drying.	Layer by layer	Minimally processed pineapples	[5]
O. ficus-indica	Film	-Liquid mucilage (50% *w*/*v*) -Pig skin gelatin (2.9% *w*/*v*) -Glycerol (3.4% *w*/*v*) -Sorbitol (3.4% *w*/*v*) -Probiotic (10^8 CFU/mL)	Homogenization of mucilage, gelatin, glycerol, or sorbitol on a magnetic stirrer at 70 °C for 10 min. The film-forming mixtures were dispersed in Petri dishes and dried at 45 °C for 24 h.	Direct application	Apples	[15]
O. dillenii	Coating	Mucilage powder (0.5% 1% and 1.5% *w*/*v*)	Homogenization of mucilage in distilled water.	Immersion	Potatoes	[31]
O. dillenii	Coating	-Mucilage powder (1% *w*/*v*) -Glutathione (0.4% *w*/*v*)	Homogenization of mucilage in distilled water, followed by the addition of glutathione.	Immersion	Chestnuts	[32]

* Np: data not provided.

The choice of plasticizer and its respective concentration is fundamental for obtaining coatings and films with satisfactory mechanical properties. As shown in Table 2, glycerol is the most used plasticizer for formulating edible films and coatings [4,17], followed by sorbitol [15] and Tween 20 [26]. Glycerol is widely used as a plasticizing agent in formulating edible films and coatings due to its structural characteristics, such as a low molecular weight, hydroxyl (OH) functional groups, and high compatibility with biological polymers [13,15,62]. These properties facilitate the diffusion of glycerol within the polymer matrix, promoting the breaking of intermolecular interactions to increase mobility and interactions among the compounds, contributing to a homogeneous distribution and maintenance of the film's mechanical properties [62]. Furthermore, glycerol has hydrophilic characteristics, acting as a water-retention agent in the film-forming matrix during drying and storage processes [60]. A high solubility of the films is desirable, as this allows for rapid dissolution in the oral phase of digestion or easy removal during consumption [62].

Films formulated with mucilage extracted from *O. ficus-indica* cladodes show higher values of moisture, solubility, water vapor permeability (WVP), and elongation at break (EB). In contrast, films produced with *O. ficus-indica* mucilage and sorbitol show higher tensile strength [62]. The higher tensile strength of films with sorbitol may be associated with the higher molecular weight of this plasticizer, which increases intermolecular spacing, reduces mobility, and promotes cohesion among molecules, thus enhancing film strength [15]. Considering these aspects, glycerol-based plasticizing agents are widely applicable in edible-film formulations for products requiring greater flexibility, such as fruits and vegetables [5,8,15]. However, the glycerol concentration added to the coating solution should be considered, as an anti-plasticizing effect may occur when higher concentrations are used [16]. The increase in parameters, such as the thickness, elongation at break, moisture content, water solubility, and water vapor transmission rate (WVTR), is associated with a higher glycerol content [62]. Therefore, the selection of the plasticizing agent and the applied concentration should consider the desired properties of the film, as plasticizers like sorbitol, despite providing less elasticity, increase mechanical strength, a favorable attribute in formulations where mechanical resistance is essential [15].

Plasticizers in concentrations ranging from 10% to 30% (*w*/*w*) resulted in films firmly adhered to the plate, making removal difficult. On the other hand, films formulated with 50% plasticizers exhibited an excessively sticky texture. Films containing 40% (*w*/*w*) glycerol demonstrated good handling properties and an adequate appearance, making this concentration ideal for film development [61]. However, the concentrations of plasticizers added to coating solutions and edible films formulated with cactus mucilage vary substantially (Table 2).

Edible coatings are defined as a thin layer of semi-liquid material applied to the outer surface of foods, using techniques such as spraying, immersion, or brushing [36]. As presented in Table 2, the most used methods for applying mucilage-based coatings to fruits and vegetables include immersion, followed by an atomizing spray, a direct application multilayer system, brushing, sprinkling, and layer by layer. In contrast, the preparation of films involves the homogenization of mucilage with the other components of the film-forming solution and the dispersion of the solution in Petri dishes, which are then subjected to drying and directly applied to the fruit [15]. Due to its simplicity, immersion is the most widely used method for coating fruits and vegetables. This process is divided into three distinct steps: first, immersion, where the fruit or vegetable remains in contact with the coating solution for a specific time; deposition, which involves the adhesion of the solution to the food surface; and drying, corresponding to the evaporation of the solution, resulting in the formation of the coating adhered to the surface of the fruit or vegetable [3,8,9,30].

To achieve efficiency during the process, it is essential to consider the viscosity, density, and surface tension of the coating solution [57,58]. Fruits and vegetables with irregular shapes can hinder the uniform application of coatings, resulting in areas with inadequate coverage that promote unfavorable biochemical changes, compromising their shelf life. On the other hand, prolonged exposure to the solution can lead to excessive moisture absorption, while shorter contact times may result in an uneven coating [63]. During immersion in the coating solution, cross-linking agents can be used to enhance the adhesion of the coating to the food surface [49]. Cross-linking agents are widely used to improve the formulation of edible coatings intended to preserve fruits and vegetables [22,64]. Among the agents commonly employed for the formulation of coatings made with cactus mucilage are plant extracts [1,9], organic acids [12,28,49,64], and metal ions [9,65].

Calcium ion (Ca^{2+}) is widely incorporated in $CaCl_2$ solutions as a cross-linking agent, given its compatibility with anionic polymers [65]. This interaction promotes improvements in the techno-functional attributes of edible coatings, enhancing their stability and effectiveness in preserving fruits and vegetables [3,65]. When in contact with the citric acid present in the cell wall of fruits, Ca^{2+} prevents the alteration of the intracellular gel layer, helping to preserve the texture and firmness of the fruit. In addition, Ca^{2+} plays an essential role in regulating fruit ripening and senescence, contributing to its resistance to biological stress through the activation of secondary metabolism and stimulating the synthesis of bioactive compounds necessary for the nutritional quality of fruits [65,66]. The combined application of calcium chloride as a cross-linking agent in the cactus mucilage-based coating favored the creation of a modified atmosphere around the food's surface (namely, tomatoes), which resulted in the preservation of the overall quality attributes and extended its shelf life for three weeks under room-storage conditions [3].

Other methods not applied in the reviewed articles have also been studied to prepare coatings, emphasizing electroplating. This is a technique in which metal ions are reduced to metal atoms using an electrolytic solution with the passage of an electric current and then deposited on the surface of another. Its ability to deposit various metals, alloys, and non-metallic materials (such as plastics and ceramics) without hazardous chemicals and with a lower environmental impact makes it extremely valuable. This technique has been used in several industries, such as food packaging [67].

The development of film and coating formulations involving the combination of cactus mucilage with other biopolymers, such as aloe vera gel [1,9], gelatin [15], and cassava starch [30], has been explored to preserve fruits and vegetables (Table 2). The mixture of mucilage, cassava starch, and glycerol was reported to result in a negative interaction with the plant tissue of minimally processed yam, as this combination was not effective in maintaining sensory quality during refrigeration storage, unlike yam coated only with mucilage and glycerol [30]. In contrast, the edible coating made with *O. ficus-indica* mucilage combined with gelatin, probiotic strains (*Enterococcus faecium*), and glycerol demonstrated efficacy in preserving freshly cut apple slices, extending their shelf life and maintaining quality attributes during storage [15].

In addition to these biopolymers, bioactive compounds have also been used in coating formulations based on cactus mucilage to enhance the functionality and antioxidant action of the coatings [12]. An early study reported that adding glycerol and L-glutamine to the coating improved the nutritional profile of loquat fruit during cold storage [10]. Similarly, adding glutathione to the polysaccharide-based coating of *O. dillenii* improved quality parameters and extended the shelf life of freshly cut Chinese cabbage by inhibiting physiological processes [32]. In another study, the coating enriched with ascorbic acid effectively preserved phytochemicals and antioxidant activity in pecan nuts [12]. Edible coatings and films formulated with cactus mucilage are strongly affected by the cactus species, cultivation conditions, extraction method, and the concentration of mucilage and other components used in the formulation [27,28,49]. For this reason, understanding these aspects and knowledge of cactus mucilage's physicochemical and morphological properties is essential for formulating edible coatings and films with satisfactory technological attributes [49].

4.2. *Characterization of Microstructural, Functional, and Thermal Properties of Films and Coatings Formulated with Cactus Mucilages and Their Bioactive Compounds*

The microstructural, functional, and thermal properties of films and coatings formulated with cactus mucilages and their bioactive compounds are strongly associated with the mucilage's chemical composition and structural characteristics [49]. Although their importance is recognized, studies investigating these parameters during the formulation of coatings applied to fruits and vegetables are still scarce. This gap hinders the integrated interpretation of the coating's morphological, functional, and thermal characterization variables [12,16,32]. In contrast, most studies emphasize the effects of these coatings on the overall quality attributes of fruits and vegetables. Therefore, this section will also present information on developing cactus mucilage-based films and coatings not applied in food matrices to deepen knowledge.

The mucilage's particle size significantly influences the film's mechanical properties. When used alone, the *O. ficus-indica* mucilage, in its smaller fraction (NC-120), did not result in adequate film formation, leading to a fragile and brittle matrix [52]. To overcome this limitation, cassava starch was incorporated into the formulation, which resulted in a homogeneous, easy-to-handle film. In another study, *O. ficus-indica* mucilage used for film production exhibited an amorphous shape with a random distribution of irregular particles in the form of aggregates due to calcium and other polyvalent ions. Scanning electron microscopy (SEM) images of the films made with mucilage revealed a porous, network-like structure capable of storing particles, gases, ions, and moisture, which are essential characteristics for the coating functionality [49]. On the other hand, films formulated with *O. ficus-indica* mucilage, with the addition of chitosan (2% v/v) and glycerol (3% v/v), exhibited a homogeneous, non-porous surface, although with a slightly darker color [2]. Color is a crucial attribute to consider when making edible films, as it directly influences consumer acceptance and must be compatible with the food to which the film will be applied [16].

The thickness of the films is considered an important physical parameter, as it affects water vapor permeability and mechanical strength. This parameter is directly related to the volume of the film-forming solution applied and the concentration of the dry matter in the solution [11]. Additionally, increasing thickness can reduce film transparency and decrease light transmission, which is advantageous for protecting photosensitive constituents in fruits and vegetables [23]. Increasing the concentration of mucilage typically results in increased water vapor permeability, which can directly impact the film's functionality. In general, carbohydrate-based films exhibit a weak water vapor barrier due to their structure and composition, which provide excellent affinity for water vapor molecules [49]. Barrier properties are essential in preventing oxidation and microbial spoilage, directly contributing to extending the food's shelf life [19]. High water solubility, low tensile strength, and low water vapor permeability can limit the use of mucilage alone, making the incorporation of additives and other polymers necessary [23].

The gas permeability of edible films and coatings plays a crucial role as a barrier in gas exchanges, particularly for carbon dioxide (CO_2) and oxygen (O_2), which are critical factors in the deterioration of fresh fruits and vegetables [60,63]. Low O_2 levels and high CO_2 levels help control microbial and enzymatic activities, extending the shelf life of fresh fruits and vegetables. This modified atmosphere also reduces the transmission of ultraviolet (UV) light, which limits oxidative processes, preserves sensory attributes, and slows nutrient degradation, promoting product quality conservation [28,56,64]. Due to the polymeric structure of mucilage and its compact network of hydrogen bonds, cactus mucilage-based films and coatings exhibit good barrier properties against gas exchanges [20]. These coatings reduce oxygen permeability and limit carbon dioxide accumulation around fruits and vegetables, helping to preserve quality and minimize economic losses along the supply chain [19,20,68].

Glycerol is the most used plasticizing agent for formulating mucilage-based films, as it improves the mechanical and thermal properties of the film. However, its addition can also negatively affect the barrier properties. The effects of applying plasticizers, such as glycerol, sorbitol, and polyethylene glycol, on the properties of films formulated with mucilage extracted from *O. ficus-indica* cladodes were investigated. All plasticized films exhibited lightness; yellow-green coloration; and a homogeneous, smooth, tasteless, and odorless appearance. The films with plasticizers showed good extensibility but poor resistance regarding mechanical properties. Furthermore, the use of polyols did not affect the thickness and solubility of the films in water. However, it increased the moisture content and reduced thermal stability, with the highest thermal stability obtained in the film formulated with polyethylene glycol due to its lower hydroxyl-group content [61].

Mechanical properties, such as tensile strength and elongation at break, are important indicators of the performance of edible films. Tensile strength refers to the maximum force a film can withstand per unit area before breaking, while elongation at break measures the change in the film's length in the longitudinal direction at the point of rupture [11]. The glycerol concentration added to the formulation must be considered, as high concentrations can decrease tensile strength and increase elongation at the break of the formulated films [16]. The harvest of *O. stricta* clones (Orelha de Elefante Mexicana) and *Nopalea cochenillifera* (L.) Salm-Dyck (Miúda), at 6 a.m., positively impacted the films' mechanical properties, permeability, and thermal stability. Specifically, the films formulated with the Orelha de Elefante Mexicana clone, harvested during the rainy season and the transition between the rainy and dry seasons, showed a darker color, improved the mechanical properties and water permeability resistance, and had a more compact microstructure and higher thermal stability [23].

Another study identified mucilage and films developed from the cactus mucilage of *N. cochenillifera* (L.) Salm-Dyck (Miúda) and *Opuntia stricta* (Haw.) (Orelha de Elefante Mexicana), which exhibited physicochemical stability over eight months of storage. After this period, changes in the appearance of the films were identified [39]. The results indicate that both species were water-soluble; however, *N. cochenillifera* demonstrated superior water- and oil-retention capacities and a higher concentration of phenolic compounds, resulting in films with enhanced mechanical properties compared to those formulated with *O. stricta*. These findings highlight the potential of these species for industrial applications, especially in the formulation of edible films and coatings with extended preservation properties.

Film biodegradation is also frequently assessed under anaerobic conditions through weight loss analyses. This method evaluates the degradation of polymeric materials when buried in organic matter-rich soil, determining the time required for complete decomposition [49]. Biodegradation occurs through the fragmentation and mineralization of the material caused by external factors, such as heat, humidity, and the enzymatic activity of microorganisms present during composting [49]. Films formulated from pitaya began to degrade after 28 days, with complete disintegration occurring after 35 days [16]. Similar results were reported for films made from *O. ficus-indica* mucilage, which started biodegrading after 24 days, with complete degradation achieved after 40 days [49].

As highlighted in the studies above, cactus mucilage-based films and coatings stand out for their flexibility, gas barrier properties, thermal stability, and biodegradability. However, these composite materials have some limitations, such as low mechanical strength and a high affinity for water, which may compromise their performance. Therefore, developing studies to improve these aspects is essential to expand their applicability and efficacy.

5. Effects of Films and Coatings Formulated with Cactus Mucilages and Their Bioactive Compounds on the Postharvest Quality and Storability of Fruits and Vegetables

Quality parameters, such as weight and firmness, color, soluble solid content, acidity, pH, ascorbic acid (vitamin C), and total phenolic content, change in fruits and vegetables (fresh and minimally processed) during storage, mainly as a result of physiological disorders (alterations in ethylene production, respiration, transpiration etc.) and infections caused by several pathogens [14,29]. Therefore, the impacts of applying coatings and edible films based on cactus mucilages and/or their bioactive compounds on the quality attributes of different fruits and vegetables, both fresh and minimally processed, have also been investigated (Figure 1), and the main results are summarized in Table 3.

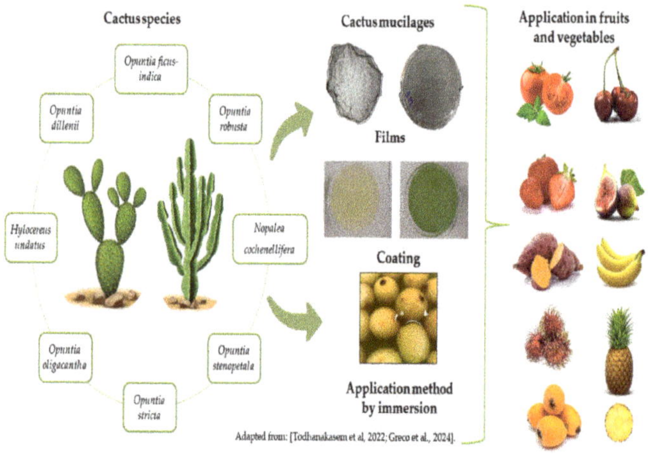

Figure 1. Applying cactus mucilage films and coatings to fresh and minimally processed fruits and vegetables [10,15].

Table 3. The main effects of applying coatings and edible films formulated with cactus mucilage or its bioactive compounds on extending the shelf life and preserving the nutritional, physicochemical, and sensory qualities of fruits and vegetables.

Composition of Coatings and Edible Films	Applied Fruits and Vegetables	Storage Conditions	Quality Parameters Evaluated	Main Effects	References
Aloe (*Aloe debrana*) gel and *O. ficus-indica* mucilage	Mango in natura	25 °C, Np RH, for 16 days.	Sensory evaluation, pH, total soluble solids (TSSs), titratable acidity (TA), and sugar-to-acid ratio.	The application of the aloe gel influenced the sensory attributes, TSS, TA, and the sugar-to-acid ratio, while the cactus mucilage primarily affected the color, appearance, and overall acceptance. Application of the Aloe gel and cactus mucilage (50 and 75%), in combination and alone, had a tendency to retard quality deterioration and maintained a good appearance of the mangoes.	[1]
O. ficus-indica mucilage extract mixed with glycerol	Papaya in natura	27 °C, 55–60% RH, for 6 weeks.	Weight loss, ascorbic acid content, pH, firmness, TSS and microbial qualities (total mesophilic microorganisms (TMM), total psychrotrophic microorganisms (TPMs), yeast, and mold.	Mucilage presented a protective effect to the firmness in coated samples. The *O. ficus-indica* mucilage with glycerol was more effective than only mucilage extract in extending the shelf life of fruits, reducing their ascorbic acid content, pH, and TSS when compared to the control in the storage period.	[35]

Table 3. *Cont.*

Composition of Coatings and Edible Films	Applied Fruits and Vegetables	Storage Conditions	Quality Parameters Evaluated	Main Effects	References
O. ficus-indica mucilage	Minimally processed kiwifruit	5 °C, 90% RH, for 12 days.	Firmness, weight loss, TSS, TA, sensory evaluation, package O_2 and CO_2 analysis, ascorbic acid, microbiological analysis (TMM and spoilage and/or pathogenic groups), and pectin analysis.	Mucilage edible coating presented positive effects on the physical maintenance, ascorbic acid, pectin content, visual quality, and flavor score of the storaged fruits. Samples coated with mucilage and Tween 20 presented an increase in microbial growth, mostly of yeasts. The coating without Tween 20 was more effective for quality parameters in kiwifruits.	[24]
O. ficus-indica mucilage	'Breba' fig	4 °C, for 14 days.	Firmness, weight loss, respiration rate, ethylene production, color, visual appearance score, sensory analysis, TSS, pH, TA, total phenolic compounds (TPCs), total carotenoids, and microbiological analysis (TMM and spoilage and/or pathogenic groups).	The application of the mucilage edible coating was effective in maintaining the fresh fruit's brightness, weight, visual appearance score values, firmness, and total carotenoid content. The coated fruit revealed an expressively lower growth of *Enterobacteriaceae* than a control. The association of temperature and mucilage was effective in reducing water transpiration and retaining the mechanical and chemical properties of the fig through 10 days of storage.	[26]
O. stricta L. mucilage	Pepper in natura	25 °C, 65% RH, for 6 days.	Weight loss, TSS, and iron and ascorbic acid content.	The cactus mucilage coating reduced the weight loss; maintained the TSS, iron, and ascorbic acid; and extended the shelf life of the peppers.	[4]
O. robusta mucilage	Tomato in natura	20 °C, Np RH for 21 days.	Weight loss, texture measurement (firmness), and determination of lycopene.	The mucilage obtained from the parenchymatous tissue was more effective as an edible coating than the mucilage of the chlorenchymatous tissue. Tomatoes coated with mucilage revealed significantly higher firmness and reduced weight loss and lycopene content than the control.	[21]
O. ficus-indica mucilage	Minimally processed rambutam	5 °C for 10 days.	Weight loss, color, firmness, TSS, and sensorial evaluation.	The mucilage coating effectively reduced weight loss and preserved the fruit's firmness during storage. Additionally, the coating helped retain the fruit's color and overall appearance. Sensory evaluations indicated higher acceptability and consumption intentions for coated fruits. Overall, the cactus mucilage coating extended the shelf life and maintained the fruit's quality	[6]

Table 3. *Cont.*

Composition of Coatings and Edible Films	Applied Fruits and Vegetables	Storage Conditions	Quality Parameters Evaluated	Main Effects	References
Chitosan and *O. stenopetala* with alginate and *F. microphylla* extract	Cherry tomato in natura	20 °C, Np RH, for 15 days.	Weight loss, pH, TA, TSS, firmness, color, and microbiological assays (aerobic mesophilic microorganisms, molds, and, yeasts).	The multilayer coating applied controlled weight loss and maintained the pH, TA, and TSS. Further, the coatings did not impact the visual appearance of the fruit. In addition, a visual evaluation of the peduncle scar showed that the coatings incorporating *F. microphylla* prevented *F. oxysporum* growth, prolonging the shelf life of the cherry tomatoes to 6 days.	[13]
Cactus polysaccharides (CPs) and ultrasound (US)	Minimally processed potato	4 °C, Np RH, for 8 days.	Microbiological assay, color, texture analysis, TSS, pH, cooking loss, water status and distribution, enzyme extractions and assays, TPC, malondialdehyde (MDA) content, membrane permeability, antioxidant capacity, catalase activity, and volatile organic compounds.	The CP-and-US combined treatment improved the bacteriostatic effect, browning inhibition effect, and antioxidant capacity in the vegetable. Furthermore, it reduced the mobility of water and the degree of membrane lipid peroxidation, maintaining the appearance quality, texture properties, and cell membrane integrity. The combination of CP and US improved the shelf life of potatoes.	[7]
O. ficus-indica mucilage and chitosan	Cherry in natura	1 °C, 90% RH, for 28 days.	Weight loss, TSS, TA, pH, peel color, moisture/dry matter, respiration, firmness, microbial decay, resistance to pedicel removal, extraction of phytochemicals, TPC, total flavonoid content (TFC), anthocyanins, and antioxidant capacity.	The application of the coating minimised weight loss and respiration rates, while enhancing the firmness of the fruits. Additionally, the coatings increased the antioxidant content, including phenolics, flavonoids, and anthocyanins, compared to the control group. The application of the edible coatings preserved the quality and prolonged the shelf life of cherries during storage.	[20]
Cactus pear (*Napolea cochenellifera* Salm Dick.) mucilage	Sweet potato	8 and 23 °C, Np RH, for 14 and 26 days, respectively, (simulated refrigerated transport and shelf conditions, respectively).	Visual appearance, weight loss, firmness, color, TSS, starch, antioxidant capacity (DPPH and FRAP), total phenolic compounds (TPCs), ascorbic acid, and total carotenoids.	The packed vegetables, with or without coating, maintained greater visual scores after transference to room conditions, exhibiting a lower weight loss for up to 26 days. Additionally, by visual analysis, the firmness, antioxidant activity, and phenolic and starch contents were stable in raw and cooked sweet potatoes packaged (coated or not) for up to 26 days.	[17]

Table 3. *Cont.*

Composition of Coatings and Edible Films	Applied Fruits and Vegetables	Storage Conditions	Quality Parameters Evaluated	Main Effects	References
Cactus mucilage and calcium chloride ($CaCl_2$)	Tomato in natura	21 °C, 45% RH, for 20 days.	Weight loss, fruit decay, firmness, TA, TSS, color, and ripening index.	The combined coating significantly reduced weight loss and decay and retained firmness, TA, and TSS compared to untreated fruits during storage. Additionally, the coatings delayed color changes, indicating a slower ripening process. The combination of mucilage and $CaCl_2$ effectively preserved the quality and extended the shelf life of tomatoes under ambient conditions.	[3]
O. ficus-indica mucilage with glycerol and l-glutamine	Minimally processed loquat	5 °C, Np RH, for 11 days.	TSS, TA, antioxidant activity, extractable juice, weight loss, color, TPC, antioxidant activity, sensory analysis, and visual scores.	The formulation of a mucilage edible coating enriched with 30% glycerol and 10% L-glutamine provided better postharvest resistance than only a mucilage coating, maintaining the TSS, TA, and extractable juice; reducing weight loss; and increasing antioxidant activity and the nutritional profile in treated fruits throughout the entire cold-storage period. The coating did not compromise the sensory and visual quality of loquat fruits after the storage period.	[10]
O. ficus-indica mucilage with ascorbic acid	Strawbery in natura	4 °C, 85% RH, for 12 days.	Weight loss, color; overall quality; firmness; TSS; TA; ascorbic acid content; sensory analysis; and microbiological analysis (TMM, TPM, *Pseudomonas*, *Enterobacteriaceae* family, yeasts, and molds).	The coated fruits showed a linear increase in weight loss during storage. The ascorbic acid content and TSS increased in coated strawberries, and the overall visual quality decreased, being affected by storage. However, visual quality and sensorial analyses verified higher scores in the coated samples at the end of the storage period. Furthermore, the mucilage coating did not negatively interfere with the natural taste of the strawberries. The coating applications were not able to prevent microbial growth, but their development decreased in coated fruits.	[27]

Table 3. *Cont.*

Composition of Coatings and Edible Films	Applied Fruits and Vegetables	Storage Conditions	Quality Parameters Evaluated	Main Effects	References
O. ficus-indica mucilage	Minimally processed cactus pear fruits	5 °C, Np RH, for 9 days.	Firmness; TSS; TA; color; weight loss; bioactive compounds and radical scavenging activity; and microbiological parameters (TMM, TPM, *Pseudomonas*, *Enterobacteriaceae*, and yeasts); sensorial analysis; and visual score.	Mucilage had a barrier effect on minimally processed fruit during cold storage, which was revealed by the lower weight loss, higher firmness, maintenance of TSS, ascorbic acid and betalain contents, sensorial qualities, visual scores, lower respiration rates, and significantly less microbiological growth of the coated samples compared to the control during the storage period.	[28]
O. ficus-indica mucilage	Minimally processed loquat fruit	5 °C, Np RH, for 13 days.	Firmness; SST, TA, color, extractable juice, ascorbic acid content, weight loss, and sensorial analysis.	Mucilage coating preserved the quality, nutraceutical value, and sensorial parameters and improved the postharvest life of minimally processed fruits. The treatment did not impede microbial growth, but considerably reduced its development in coated fruits.	[14]
O. ficus-indica mucilage and Calcium ascorbate	Cactus pear fruits	5 °C, Np RH, for 9 days.	TSS; TA; carbohydrate; color; weight loss; headspace gas composition; nutraceutical attributes; sensory analysis and visual score; and microbiological analyses (TMM, TPM, *Pseudomonas*, members of the *Enterobacteriaceae* family, *Listeria monocytogenes*, and yeasts).	The coating treatment maintained the quality parameters, nutritional value, and sensorial profiles and enhanced the postharvest life of fruits. The coating application restricted the development of bacteria and yeasts and did not negatively affect the natural taste of fruits during refrigerated storage.	[46]
N. cochenillifera mucilage	Minimally processed yam	5 °C, Np RH, for 10 days.	Fresh mass loss, visual assessment, and sensory analysis.	The coating reduced dehydration, maintained sensory quality, and increased the amount of phenolic compounds in the yam during the storage period.	[30]
O. ficus-indica mucilage enriched with ascorbic acid	Pecan nut	60 °C, Np RH, for 25 days.	Color, hardness, microbiological analyses (total mold and yeast), TPC, TFC, antioxidant activity, and enzyme activity assays.	The enriched coatings effectively preserved the color, hardness, and overall appearance of the pecan nuts throughout storage. Additionally, the coatings maintained greater levels of TPC and TFC, besides antioxidant activities. The coatings reduced the activities of polyphenol oxidase and peroxidase enzymes, which was associated with quality deterioration. Overall, the coating protected the quality and extended the shelf life of pecan nuts under storage conditions.	[12]

Table 3. *Cont.*

Composition of Coatings and Edible Films	Applied Fruits and Vegetables	Storage Conditions	Quality Parameters Evaluated	Main Effects	References
O. ficus-indica mucilage (OM), chitosan, and glycerol	Tomato in natura	25 °C, for room RH, for 30 days.	Color, texture, and firmness.	The coated tomatoes displayed a uniform color and firmness comparable to the fruits under initial storage conditions and did not show symptoms of a microbial disease. The film revealed a strong antifungal effect against *Rhizopus stolonifer* in vitro and in situ and improved the shelf life of tomatoes.	[2]
O. oligacantha fruit	Tomato in natura	6 °C, Np RH, for 21 days.	Weight loss, firmness, color, pH, TA, TSS, texture, bioactive compounds, and antioxidant activity.	The nanoemulsion coatings, which included orange essential oils and xoconostle, effectively reduced weight loss and maintained the firmness, color, pH, TA, and TSS of the tomatoes during storage. Additionally, the coatings enhanced the antioxidant activity and preserved the histological structure of the fruit pericarp, indicating slower maturation. The coating application was capable for extending the shelf life and maintaining fruit quality.	[33]
Dragon fruit mucilage coating and UV-C radiation	Cherry tomato	4 °C, 95% RH, for 21 days.	Color; weight loss; TPC; TFC; ascorbic acid; total soluble solids; TA; and microbial analyses (aerobic bacteria, coliform, yeast, and mold).	The application of a combination of UV-C and an edible coating provided a higher ascorbic acid content, antioxidant capacity, and antimicrobial effect. The hurdle treatment reduced weight loss, and despite slightly modifying the color, it maintained the fruit's visual appearance and extended its shelf life during the storage period.	[34]
Polysaccharide extract of *Opuntia* spp.	Citrus fruit in natura	5 °C, 90% RH, for 35 days.	pH, TA, and sensory evaluation.	The maximum moisture content and pH value were observed in fruits coated with *Opuntia* spp. polysaccharides. The coating application increased the shelf life and retained the sensorial quality of the citrus fruits.	[36]
O. ficus indica mucilage	Banana in natura	25 °C, Np RH, for 12 days.	Weight loss, decay incidence, ethylene production, respiration rate, TSS, TA, ion leakage, MDA content, color, total chlorophyll, degrading cell wall enzyme activities, total carotenoid content, protopectin, and firmness.	Mucilage coating positively delayed the fruits' ripening process. The coated fruits presented higher firmness, chlorophyll content, and TA values and a lower TSS content, ethylene production, respiration rate, MDA concentration, ion leakage, and protopectin content than uncoated fruits.	[29]

Table 3. *Cont.*

Composition of Coatings and Edible Films	Applied Fruits and Vegetables	Storage Conditions	Quality Parameters Evaluated	Main Effects	References
O. ficus indica mucilage	Sweet cherry in natura	2 °C, 92% RH, for 14 days.	Fresh weight, TSS, TA, firmness, color parameters nutraceutical, and sensory analysis.	The treated fruits showed better acceptability, and quality parameters, such color, compared to the control group. The mucilage coating considerably reduced weight loss during storage and pedicel browning, enhancing the visual appeal of the cherries. The mucilage effectively increased the storage quality and shelf life of the coated fruits.	[8]
O. ficus-indica mucilage and aloe gel	Figs, 'San Giovanni' and 'Melanzana' cultivars	4 °C, 85% RH, for 12 days.	Weight loss; TSS; TA; Maturation Index; color; firmness; microbiological analyses (TMMs, yeasts, and *Pseudomonas* counts); TPC; and visual appearance score.	The combined application of *O. ficus-indica* mucilage and aloe gel edible coatings improved the fruits' visual appearance, maintained firmness, and reduced weight loss during storage. The coatings extended the shelf life of the figs by maintaining their quality and reducing microbial contamination.	[9]
O. ficus-indica mucilage and probiotic strain *Enterococcus faecium* FM11-2	Minimally processed apple	4 °C, Np RH, for 7 days.	Weight loss	The application of the edible film resulted in improved preservation and shelf life of minimally processed apples, minimizing weight loss and maintaining freshness during storage.	[15]
Nopal cactus mucilage, pullulan, and chitosan	Minimally processed pineapple	4 °C, Np RH, for 18 days.	Weight loss; firmness; color; TA; pH; ascorbic acid content; and microbiological (molds, yeasts, and total aerobic and psychotropic microorganisms) and sensorial analyses.	The coating application reduced weight loss and preserved the firmness of the pineapple cubes throughout storage. Additionally, the coatings helped in retaining the color and overall appearance. Microbiological analyses showed a reduction in microbial load, indicating enhanced safety and an extended shelf life.	[5]
O. dillenii polysaccharides	Minimally processed potato	5 °C, Np RH, for 5 days.	Weight loss, firmness, color, respiratory rate, microbiological analyses (total viable counts), and total sugar.	The application of different concentrations of *O. dillenii* polysaccharides considerably reduced weight loss and preserved firmness and retained the color and overall appearance of the potatoes. The polysaccharide-based coatings presented antibacterial and antioxidant activities, contributing to the quality and extended shelf life of the fresh-cut potatoes.	[31]

Table 3. *Cont.*

Composition of Coatings and Edible Films	Applied Fruits and Vegetables	Storage Conditions	Quality Parameters Evaluated	Main Effects	References
O. dillenii polysaccharides with glutathione	Chestnut	3 °C, Np RH, for 10 days.	Weight loss, firmness, color, respiration rate, TSS, and sensory evaluation.	The application of coatings significantly dropped the respiration rate and weight loss, maintained the firmness of and reduced the browning of the chestnuts during storage. The coatings' application was capable of preserving quality and extending the product shelf life of chestnuts by at least 4 days.	[32]

Malondialdehyde (MDA); data not provided (Np); relative humidity (RH); titratable acidity (TA); total phenolic content (TPC); total soluble solids (TSSs); total mesophilic microorganisms (TMMs); total psychrotrophic microorganisms (TPMs); 2,2-diphenyl-1-picrylhydrazyl (DPPH); ferric-reducing antioxidant power (FRAP).

5.1. Effects on Shelf Life, Color, Firmness, Weight Loss, Total Soluble Solids, pH, Acidity, Ascorbic Acid (Vitamin C), Total Phenolic Content, Antioxidant Activity, and Sensory Parameters

Weight loss is essential in determining the postharvest quality and shelf life of fruits and vegetables. Weight loss is usually caused by the continuous loss of dry matter through respiration and water-loss transpiration processes [15]. Notably, the internal water fluidity of fruits and vegetables increases during storage, which increases water loss and decreases quality [7]. Applying edible coatings and films minimizes water loss by creating a thin semi-permeable layer over the external surface as a barrier and maintaining high relative humidity [2,29,30]. The slowed weight loss and extended shelf life in fruits and vegetables coated with mucilage from cacti might be attributed to the improved water retention facilitated by the hydrophilic nature of this material [10].

Studies also suggest that adding active ingredients, such as mucilage, does not affect the coating barrier's functionality [4,6,13]. On the contrary, the combination of a mucilage coating and film with different treatments, such as UV-C irradiation, chitosan, glycerol, glutamine, probiotics, and aloe gel, were effective in decreasing transpiration and the consequent decline in weight loss, maintaining freshness and prolonging the shelf life of products [9,10,13,15,20].

Texture or firmness is a critical quality attribute in fruits and vegetables. It may be modified during the postharvest period due to high metabolic activity and physical damage, such as the breakage of cell wall components and shocks during postharvest handling [14]. Studies indicate that physical barriers generated by cactus mucilage coatings effectively reduce pectin degradation, with reduced weight loss and firmness [32]. This effect could be attributed to the calcium content in cactus mucilage, which interacts with pectic acid in cell walls to form calcium pectate, thus maintaining the integrity of cell walls in fruit tissues [46].

Greater firmness of tissues preserved with mucilage was also attributed to greater concentrations of pectin and protopectin during storage, suggesting lower cell wall hydrolyzing enzyme activity [20]. Enzymes, such as pectin methylesterase (PME) and polygalacturonase, responsible for the solubilization of pectic substances in plant cells, require high levels of O_2 and ethylene for their action, parameters that are reduced/modified with the application of coatings [20,33]. Studies have reported the potential of cactus mucilage to act as a gas barrier by decreasing O_2 permeability and stimulating CO_2 accumulation around coated fruits and vegetables during storage, significantly reducing the respiration rate [27,28]. Additionally, the incorporation of *Fuchsia microphylla* (flowering shrub) into mucilage coatings also had a positive effect on controlling water loss, thereby increasing fruits' firmness and, consequently, physicochemical quality [13].

Color is an essential indicator of the shelf life of fruits and vegetables, being a factor that directly impacts consumer choice and acceptability and the marketing of these products [7,10,28]. During ripening, changes in the colors of fruits and vegetables are observed due to the degradation of the pigment chlorophyll and the biosynthesis/cumulation of carotenoids [3]. In fresh-cut fruits and vegetables, color changes are also caused by the enzymatic oxidation of carotenoids and polyphenols, forming brown pigments (quinones) [27]. The peroxidase (POD) and polyphenol oxidase (PPO) activities causing the oxidation of phenolic compounds are related to the browning and loss of quality of many foods [12]. The stability of pigments provided by applying edible coatings formulated with cactus mucilage occurs due to reduced gas transfers [12]. This reduces respiration and ethylene production rates, delaying ripening and its associated modifications in fruits' color development [3]. In addition, the application of mucilage-based coatings could retard the oxidation reactions and enhance the brightness and overall visual quality of fruits [12,17,46]. However, decreased brightness was observed in cold storage, which is probably correlated with a decrease in specific pigments, such as the content of betalains [27].

Changes in the content of total soluble solids (TSSs) in fruits and vegetables are associated with the biosynthesis of polysaccharides and accumulation of sugars during the ripening, volatilization of soluble compounds, evaporation of water, and variations in storage conditions, such as temperature and humidity [13]. At advanced stages of ripening, the disassociation of some molecules and structural enzymes in soluble compounds also increases TSS levels [4]. The effect of cactus mucilage edible coatings limiting the increase in TSS could be attributed to the role of coatings acting as a semi-permeable barrier against O_2, CO_2, and ethylene, reducing respiration, transpiration, metabolic activity, and ripening, leading to a decrease in accumulated sugars and polysaccharides [27].

Due to the ripening process, organic acid content declines in fruits and vegetables as storage time progresses, reflected in a reduction in titratable acidity [1]. The decreasing trend of titratable acidity could be due to metabolic activities that lead to the biosynthesis of sugars from organic acids in the postharvest period [29,53]. However, coating applications, especially when combined with cold storage, can interfere with metabolic activities by reducing the use of organic acid during respiration, delaying the postharvest ripening process in fruits [3,29].

pH is also associated with acidity, and its modification may result from respiration and the increasing concentration of CO_2 during storage [7]. The reduction in the concentration of organic acids, caused by the ripeness and water loss of fruits, also causes variations in this parameter [20,53]. However, a gradual increase in the pH of coated fruits has been reported, indicating less accelerated metabolic processes [33]. It was suggested that including a *F. microphylla* extract in the coatings could control water loss, thus delaying the maturity of fruits and changes in ripening parameters [13].

Ascorbic acid is an essential quality indicator for fruits and vegetables due to its sensitivity to oxidation during food processing and storage [4,27]. Usually, ascorbic acid content declines during storage, with the rate affected by the genotype, maturity stage, and storage conditions [5,34]. In minimally processed fruits, physical injuries can accelerate ascorbic acid degradation [46]. Low oxygen levels and low storage temperatures are other factors that can mitigate vitamin C degradation [28]. Applying cactus mucilage coatings effectively reduces losses in ascorbic acid content during storage, mainly by forming an oxygen barrier [46]. However, despite these effects, its preservation is compromised by its susceptibility to rapid oxidation, and a decrease in the content of this compound is evident in senescence [33]. Another fact related to this compound was that adding ascorbic acid to cactus mucilage also helps mitigate the loss of hardness, possibly due to cross-links forming a firmer structure that restricts moisture transfer. This effect was attributed to reduced cell wall degradation and cell collapse in the coated samples [12].

Phenolic compounds, such as anthocyanins and flavonols, are crucial in fruit quality, influencing its appearance, taste, and nutritional value [33]. Studies have shown that phenolic compounds act as non-enzymatic antioxidants, which scavenge free radicals and protect cells against oxidative injury in fruits and vegetables [19,46]. As an elicitor, the cactus mucilage could enhance antioxidant activity, scavenging reactive oxygen species, protecting cell membranes, and peroxidation, which could further extend the shelf life of fruits and vegetables and preserve their nutritional quality [10,29]. Moreover, Cactaceae species are abundant in bioactive compounds, such as betalains, phytosterols, polyphenols, vitamins, and minerals, which may also increase the nutritional composition of foods [17].

Malondialdehyde (MDA) quantification is another parameter for evaluating the degree of lipid peroxidation in fresh-cut product tissues and cell membranes. The antioxidant effects of the cactus mucilage coating were reflected in the elimination of free radicals generated by cells, preventing the oxidation of membrane lipids, reducing the formation of MDA, and effectively delaying enzymatic browning [7].

Sensory quality directly influences the consumption of fruits and vegetables and is a significant indicator for measuring their commercial value [32]. Due to the possible sensory impact that films and coatings can cause, this parameter was evaluated in some of the retrieved studies in this review. The mucilage coating and combined components, such as glutathione and glycerol, enhanced the sensory profile and preserved the overall appearance but did not interfere with or alter the natural taste of products, an essential characteristic concerning edible coatings [6,46]. Cactus mucilage coatings also exalt some critical parameters, such as firmness, aroma, sweetness, and taste, that are particularly appreciated by consumers [10,27,36]. These aspects can also be related to the changes the coatings cause in the fruit ripening process, emphasizing the content of phenolics, flavonoids, anthocyanins, and volatile compounds [7,17,20,30].

5.2. Effects on Microbiological Parameters

Fruits and vegetables are very exposed to microbial decay [19]. Although there is a high presence of carbohydrates and low acidity in cactus mucilage, studies have found that applying coatings based on this polysaccharide can reduce the microbial count [9,24,46]. A reduced oxygen supply and oxidative stress induced by the coating may be responsible for inhibiting the growth of fungi and yeasts, the primary fruit contaminants [12]. The activity of bioactive compounds present in Cactaceae can also increase microbial efficacy and scavenge free radicals, mainly by altering membrane permeability, damaging the cell wall, and thus blocking the growth and infection processes, mainly of fungi [29].

This antimicrobial activity can also be attributed to combined applications with other compounds, such as chitosan, glycerol, and aloe gel, which are known to prevent the microbial spoilage of fresh produce [9,46]. Thus, combining treatments could further enhance the microbiostatic effect of mucilage [32]. Furthermore, the addition of probiotics in cactus mucilage has also been shown to be an alternative method for controlling microorganisms in fruits, providing health benefits [15].

The studies mentioned previously presented the beneficial effects of applying coatings and edible films made from cactus mucilage and its bioactive compounds for maintaining the postharvest quality of fruits and vegetables, such as weight loss, firmness, color, and biochemical parameters. They minimized the microbial contamination of several fruits and vegetables. Their combination with other ingredients improved the quality and efficacy of cactus mucilage, and different applicational methods also extended their shelf life. It is essential to highlight that variations in the mucilage extraction procedures presented in Table 1, including factors such as time, temperature, type, and volume of solvent, can influence the physicochemical characteristics of mucilage, which play a crucial role in the techno-functional properties of films and coatings that act in the preservation of fruits and vegetables [15,16].

6. Trends, Limitations, and Market Challenges of the Use of Films and Coatings Formulated with Cactus Mucilages and Their Bioactive Compounds in the Food Industry

Cactus mucilage has been used in the formulation of coatings and films to extend the shelf life of fresh and minimally processed fruits and vegetables. The results are promising, with significant improvements in quality indicator parameters after its application alone or combined with other components [13,21,26]. Thus, applying coatings and films formulated with cactus mucilage can reduce waste and the amount of fruits and vegetables discarded throughout the production chain. Additionally, cactus mucilage, as a polymer matrix, is inexpensive and widely available in semiarid environments, favoring income generation through a circular economy approach with the sustainable use of natural resources [19,22,23]. By incorporating additives and other polymeric materials, it is possible to develop formulations that can be used as a viable alternative to traditionally applied methods [23].

In addition, cactus mucilage-based films and coatings can also serve as carriers for other bioactive substances, such as plant extracts with high concentrations of phenolic or probiotic microorganisms. This combination can enhance their functional properties, including antimicrobial activity against spoiling microorganisms and food contaminants, and benefit human health. A pioneering study on the incorporation of probiotics in cactus mucilage-based coatings showed promising results, highlighting the formation of thicker films that effectively preserved the quality attributes and shelf life of minimally processed apples [15]. However, a reduction in probiotic viability was observed during storage at 4 °C for 7 days, indicating the need to add bioactive compounds, such as phenolic compounds, to stimulate bacterial metabolism and promote greater viability of microorganisms during storage [15]. The incorporation of plant extracts and prebiotics in cactus mucilage-based coatings and films has proven effective in reducing weight loss and maintaining quality parameters such as pH, titratable acidity, total soluble solids, and antimicrobial and antioxidant activities in fruits and vegetables, thereby extending their shelf life [13,33].

More research is needed to identify biopolymer film-forming mechanisms for industrial use and optimize their physicochemical, morphological, mechanical tensile, barrier, and transparency properties. This information will support further studies, leading to better biofilm formulations for industrial applications [23]. Moreover, developing commercialization studies, such as evaluating new processes, safety, and toxicity, is essential for market applications [16].

Despite significant progress in the research on cactus mucilage, several limitations still need to be addressed. Studies often utilize a limited number of cactus species and need more crucial information regarding the genotype of the plants, maturation stage, and harvesting period for mucilage extraction. Additionally, the absence of standardization in extraction methods and the few available data on the physicochemical and technological characterization of the mucilages used in formulating films and coatings to be applied on fruits and vegetables hinder an integrated analysis of the variables influencing product quality and the success of their application. Therefore, it is recommended that future research focus on the bioprospecting of new cactus species, optimizing extraction processes, exploring the physicochemical characteristics of mucilages, and developing films and coatings that ensure effective incorporation into fruits and vegetables.

7. Conclusions

This review considered studies that tested the production methods, chemical composition, and technological and functional properties of edible films and coatings formulated with mucilage and bioactive compounds from different cactus species and their effects on the quality parameters of fruits and vegetables. Cactus mucilage presents a sustainable alternative to synthetic coatings. It is also a viable option for large-scale applications in the food industry, besides adding value to regional products represented by cactus species, which are currently used mainly as forage. The cactus species most used for obtaining mucilage are concentrated in the genus *Opuntia*, and the cladode is the most widely used part in the extraction process and subsequent application as an edible coating/film in fruits and vegetables. Cactus mucilage exhibits important physicochemical properties for formulating edible films and coatings, which can be used alone or combined with other polymers and bioactive compounds, often incorporated through immersion. The main findings from the analyzed studies show that the formulated edible films and coatings exhibit ideal mechanical properties for prolonging the shelf life of fruits and vegetables.

These properties are achieved through effective barriers against gases, moisture, and ultraviolet light; the efficient control of respiration rates; reduced senescence; and decreased water permeability. These factors contribute to the maintenance of firmness, improvement of appearance, preservation of nutritional composition, and enhancements in antioxidant and antimicrobial activities, resulting in an extended shelf life for fruits and vegetables. It is important to emphasize that the techno-functional properties of the coatings are closely linked to the physicochemical characteristics of the extracted mucilages, including their chemical composition, polymer structure, viscosity, particle size, and affinity with the substances added to the formulation. This relationship highlights the importance of carefully selecting mucilages to optimize and ensure a better performance of the developed films and coatings. Despite the promising results presented in this review, some scientific gaps still need to be addressed for a practical application of cactus mucilage in fruits and vegetables, such as the influence of genotypic variability and the harvest time of plant matrices on the techno-functional properties on the formulated films and coatings.

Author Contributions: Conceptualization, V.P.B.d.M. and E.L.d.S.; methodology, V.P.B.d.M., K.Á.R.d.O., T.S.Q. and E.L.d.S.; validation: V.P.B.d.M., K.Á.R.d.O. and T.S.Q.; formal analysis, V.P.B.d.M. and E.L.d.S.; investigation, V.P.B.d.M., K.Á.R.d.O., T.S.Q. and E.L.d.S.; resources, E.L.d.S.; data curation: V.P.B.d.M.; writing—original draft preparation, V.P.B.d.M., K.Á.R.d.O., T.S.Q. and E.L.d.S.; writing—review and editing, V.P.B.d.M. and E.L.d.S.; visualization, V.P.B.d.M. and E.L.d.S.; supervision, E.L.d.S.; project administration, E.L.d.S.; funding acquisition, E.L.d.S. All authors have read and agreed to the published version of the manuscript.

Funding: This research was funded by Coordination for the Improvement of Higher Education Personnel—CAPES (Brazil, finance code 001). The financial support from CAPES was provided through a postdoctoral fellowship [88887.917913/2023-00].

Institutional Review Board Statement: Not applicable.

Informed Consent Statement: Not applicable.

Data Availability Statement: No new data were created or analyzed in this study. Data sharing is not applicable to this article.

Conflicts of Interest: The authors declare no conflicts of interest.

References

1. Abera, N.G.; Kebede, W.; Wassu, M. Effect of Aloe gel and cactus mucilage coating on chemical quality and sensory attributes of mango (*Mangifera indica* L.). *J. Postharvest Technol.* **2019**, *7*, 31–43. [CrossRef]
2. Olicón-Hernández, D.R.; Acosta-Sánchez, Á.; Monterrubio-López, R.; Guerra-Sánchez, G. Quitosano y mucílago de *Opuntia ficus-indica* (nopal) como base de una película polimérica comestible para la protección de tomates contra *Rhizopus stolonifer*. *TIP Rev. Espec. Cienc. Quím. Biol.* **2019**, *22*, e194. [CrossRef]
3. Daraghmah, F.S.; Qubbaj, T. Impact of gum arabic and cactus mucilage as potential coating substances combined with calcium chloride treatment on tomato (*Solanum lycopersicum* L.) fruit quality attributes under ambient storage conditions. *Can. J. Plant Sci.* **2022**, *102*, 375–384. [CrossRef]
4. Aloo, M.A.; Opiyo, A.M.; Saidi, M. Maintaining postharvest quality of bell pepper (*Capsicum annuum* L. cv. California Wonder) using cactus (*Opuntia stricta* L.) mucilage coating. *Adv. Hortic. Sci.* **2022**, *36*, 193–200. [CrossRef]
5. Treviño-Garza, M.Z.; García, S.; Heredia, N.; Alanís-Guzmán, M.G.; Arévalo-Niño, K. Layer-by-layer edible coatings based on mucilages, pullulan and chitosan and its effect on quality and preservation of fresh-cut pineapple (*Ananas comosus*). *Postharvest Biol. Technol.* **2017**, *128*, 63–75. [CrossRef]
6. Brindis-Trujillo, R.A.; Salinas-Hernández, R.M.; MxBrito-Vega, H.; Gómez-Méndez, E.; Salaya-Domínguez, J.M.; Mercado-Ruíz, J.N.; Ulín-Montejo, F. Effect on the preservation of minimally processed Rambutan (*Nephelium Lappaceum* L.) by cactus (*Opuntia ficus-indica*) mucilage coating. *Trends Hortic.* **2021**, *4*, 112. [CrossRef]
7. Cheng, D.; Ma, Q.; Zhang, J.; Jiang, K.; Cai, S.; Wang, W.; Sun, J. Cactus polysaccharides enhance preservative effects of ultrasound treatment on fresh-cut potatoes. *Ultrason. Sonochem.* **2022**, *90*, 106205. [CrossRef]
8. Sortino, G.; Inglese, P.; Farina, V.; Gullo, G.; Allegra, A. The role of mucilage of *Opuntia ficus-indica* Mill. on sweet cherry fruit during storage. *Acta Hortic.* **2022**, *1343*, 323–330. [CrossRef]

9. Sortino, G.; Guccione, E.; Casales, F.G.; de Chiara, M.L.V.; Passafiume, R.; Gallotta, A.; Allegra, A. Application of *Opuntia ficus-indica* Mucilage and Aloe Gel-Based Edible Coating to Enhance Postharvest Quality and Microbiological Aspects of Fresh Figs (*Ficus carica* L.). *Horticulturae* **2024**, *10*, 482. [CrossRef]
10. Greco, G.; Gargano, F.; Motta, M.L.; Gugino, I.M.; Liguori, G. Effect of Application of a Cactus Pear Mucilage-Based Edible Coating Enriched with Glycerol and L-Glutamine on Minimally Processed White-Flesh Loquats. *Agronomy* **2024**, *14*, 1246. [CrossRef]
11. Kumar, L.; Deshmukh, R.K.; Gaikwad, K.K. Antimicrobial packaging film from cactus (*Cylindropuntia fulgida*) mucilage and gelatine. *Int. J. Biol. Macromol.* **2022**, *215*, 596–605. [CrossRef] [PubMed]
12. Mukwevho, P.L.; Kaseke, T.; Fawole, O.A. Ascorbic acid-enriched cactus mucilage coatings maintained quality attributes of roasted 'Wichita' pecan nuts under accelerated storage conditions. *J. Agric. Food Res.* **2024**, *18*, 101301. [CrossRef]
13. Carrillo-Lomelí, D.A.; Cerqueira, M.A.; Moo-Huchin, V.; Bourbon, A.I.; Souza, V.G.L.; Lestido-Cardama, A.; de Rodríguez, D.J. Influence of edible multilayer coatings with *Opuntia stenopetala* polysaccharides and *Flourensia microphylla* extract on the shelf-life of cherry tomato (*Solanum lycopersicum* L.). *Sci. Hortic.* **2024**, *332*, 113224. [CrossRef]
14. Liguori, G.; Greco, G.; Gaglio, R.; Settanni, L.; Inglese, P.; Allegra, A. Influence of Cactus Pear Mucilage-Based Edible Coating on Marketability and Edibility Parameters of Minimally Processed Loquat Fruits. *Agronomy* **2022**, *12*, 2120. [CrossRef]
15. Todhanakasem, T.; Boonchuai, P.; Itsarangkoon Na Ayutthaya, P.; Suwapanich, R.; Hararak, B.; Wu, B.; Young, B.M. Development of Bioactive *Opuntia ficus-indica* Edible Films Containing Probiotics as a Coating for Fresh-Cut Fruit. *Polymers* **2022**, *14*, 5018. [CrossRef]
16. López-Díaz, A.S.; Barriada-Bernal, L.G.; Rodríguez-Ramírez, J.; Méndez-Lagunas, L.L. Characterization of pitahaya (*Hylocereus undatus*) mucilage-based films. *Appl. Food Res.* **2023**, *3*, 100266. [CrossRef]
17. da Costa, P.R.; da Silva Barbosa, E.; de Brito, F.A.L.; Santos e Silva, V.N.; de Araújo Nogueira Marcelino, A.S.; Vieira, L.J.; Melo, A.A.M.; da Mota Silveira, F.P.; da Silveira, L.M.; da Silva Ribeiro, J.E.; et al. Forage-cactus based edible coating, packaging, and refrigeration improve the visual and nutraceutical qualities of sweet potatoes. *Sci. Hortic.* **2024**, *332*, 113205. [CrossRef]
18. Beikzadeh, S.; Khezerlou, A.; Jafari, S.M.; Pilevar, Z.; Mortazavian, A.M. Seed mucilages as the functional ingredients for biodegradable films and edible coatings in the food industry. *Adv. Colloid Interface Sci.* **2020**, *280*, 102164. [CrossRef] [PubMed]
19. Gheribi, R.; Gharbi, M.A.; Ouni, M.E.; Khwaldia, K. Enhancement of the physical, mechanical and thermal properties of cactus mucilage films by blending with polyvinyl alcohol. *Food Packag. Shelf Life* **2019**, *22*, 100386. [CrossRef]
20. Christopoulos, M.V.; Gkatzos, D.; Kafkaletou, M.; Bai, J.; Fanourakis, D.; Tsaniklidis, G.; Tsantili, E. Edible Coatings from *Opuntia ficus-indica* Cladodes Alongside Chitosan on Quality and Antioxidants in Cherries during Storage. *Foods* **2022**, *11*, 699. [CrossRef]
21. Bernardino-Nicanor, A.; Montañez-Soto, J.L.; Conde-Barajas, E.; Negrete-Rodríguez, M.d.l.L.X.; Teniente-Martínez, G.; Vargas-León, E.A.; Juárez-Goiz, J.M.S.; Acosta-García, G.; González-Cruz, L. Spectroscopic and structural analyses of *Opuntia Robusta* mucilage and its potential as an edible coating. *Coatings* **2018**, *8*, 466. [CrossRef]
22. de Andrade Vieira, E.; Alcântara, M.A.; dos Santos, N.A.; Gondim, A.D.; Iacomini, M.; Mellinger, C.; de Magalhães Cordeiro, A.M.T. Mucilages of cacti from Brazilian biodiversity: Extraction, physicochemical and technological properties. *Food Chem.* **2021**, *346*, 128892. [CrossRef] [PubMed]
23. Pinheiro, J.C.; Silva, L.J.V.; Lopes, B.K.A.; Ferreira, N.L.; Fonseca, K.S.; de Brito, F.A.L.; do Nascimento Simões, A. Effects of cactus pear clone harvest seasons and times on the physicochemical and technological properties of resulting mucilage and biopolymeric films. *Int. J. Biol. Macromol.* **2024**, *257*, 128374. [CrossRef] [PubMed]
24. Allegra, A.; Inglese, P.; Sortino, G.; Settanni, L.; Todaro, A.; Liguori, G. The influence of *Opuntia ficus-indica* mucilage edible coating on the quality of "Hayward" kiwifruit slices. *Postharvest Biol. Technol.* **2016**, *120*, 45–51. [CrossRef]
25. Allegra, A.; Sortino, G.; Inglese, P.; Settanni, L.; Todaro, A.; Gallotta, A. The effectiveness of *Opuntia ficus-indica* mucilage edible coating on post-harvest maintenance of 'Dottato' fig (*Ficus carica* L.) fruit. *Food Packag. Shelf Life* **2017**, *12*, 135–141. [CrossRef]
26. Ventura-Aguilar, R.I.; Bosquez-Molina, E.; Bautista-Baños, S.; Rivera-Cabrera, F. Cactus stem (*Opuntia ficus-indica* Mill): Anatomy, physiology and chemical composition with emphasis on its biofunctional properties. *J. Sci. Food Agric.* **2017**, *97*, 5065–5073. [CrossRef] [PubMed]
27. Liguori, G.; Gaglio, R.; Settanni, L.; Inglese, P.; D'Anna, F.; Miceli, A. Effect of *Opuntia ficus-indica* Mucilage Edible Coating in Combination with Ascorbic Acid, on Strawberry Fruit Quality during Cold Storage. *J. Food Qual.* **2021**, *2021*, 976052. [CrossRef]
28. Liguori, G.; Gaglio, R.; Greco, G.; Gentile, C.; Settanni, L.; Inglese, P. Effect of *Opuntia ficus-indica* Mucilage Edible Coating on Quality, Nutraceutical, and Sensorial Parameters of Minimally Processed Cactus Pear Fruits. *Agronomy* **2021**, *11*, 1963. [CrossRef]
29. Shinga, M.H.; Fawole, O.A. *Opuntia ficus indica* mucilage coatings regulate cell wall softening enzymes and delay the ripening of banana fruit stored at retail conditions. *Int. J. Biol. Macromol.* **2023**, *245*, 125550. [CrossRef]
30. dos Santos Morais, M.A.; Fonseca, K.S.; Viégas, E.K.D.; de Almeida, S.L.; Maia, R.K.M.; Silva, V.N.S.; do Nascimento Simões, A. Mucilage of spineless cactus in the composition of an edible coating for minimally processed yam (*Dioscorea* spp.). *J. Food Meas. Charact.* **2019**, *13*, 2000–2008. [CrossRef]
31. Wu, S. Extending shelf-life of fresh-cut potato with cactus *Opuntia dillenii* polysaccharide-based edible coatings. *Int. J. Biol. Macromol.* **2019**, *130*, 640–644. [CrossRef]
32. Wu, S.; Zhang, Y.; Li, S.; Li, C. Effects of cactus *Opuntia dillenii* polysaccharide-based coatings loaded with glutathione on the preservation of freshly cut Chinese water chestnut. *Food Chem.* **2023**, *401*, 134187. [CrossRef] [PubMed]

33. Pérez-Soto, E.; Badillo-Solis, K.I.; Cenobio-Galindo, A.d.J.; Ocampo-López, J.; Ludeña-Urquizo, F.E.; Reyes-Munguía, A.; Pérez-Ríos, S.R.; Campos-Montiel, R. Coating of tomatoes (*Solanum lycopersicum* L.) employing nanoemulsions containing the bioactive compounds of cactus acid fruits: Quality and shelf life. *Processes* **2021**, *9*, 2173. [CrossRef]
34. Razali, Z.; Somasundram, C.; Nurulain, S.Z.; Kunasekaran, W.; Alias, M.R. Postharvest quality of cherry tomatoes coated with mucilage from dragon fruit and irradiated with UV-C. *Polymers* **2021**, *13*, 2919. [CrossRef]
35. Adetunji, C.O.; Omojowo, F.; Ajayi, E.S. Effects of *Opuntia* Cactus Mucilage Extract and Storage under Evaporative Coolant System on the Shelf Life of Carica papaya Fruits. *J. Agrobiotechnology* **2014**, *5*, 49–66.
36. Riaz, S.; Sultan, M.T.; Sibt-e-Abass, M.; Imran, M.; Ahmad, R.S.; Hussain, M.B.; Egorova, G.N. Extraction of polysaccharides from *Opuntia* cactus for its potential application in edible coating to improve the shelf life of citrus (*Kinnow Mandarin*) fruit. *J. Microbiol. Biotechnol. Food Sci.* **2018**, *8*, 745–750. [CrossRef]
37. Maiuolo, J.; Nucera, S.; Serra, M.; Caminiti, R.; Oppedisano, F.; Macrì, R.; Scarano, F.; Ragusa, S.; Muscoli, C.; Palma, E.; et al. Cladodes of *Opuntia ficus-indica* (L.) Mill. possess important beneficial properties dependent on their different stages of maturity. *Plants.* **2024**, *13*, 1365. [CrossRef]
38. Mayer, J.A.; Wone, B.W.M.; Alexande, D.C.; Guo, L.; Ryals, A.J.; Cushman, J.C. Metabolic profiling of epidermal and mesophyll tissues under water-deficit stress in *Opuntia ficus-indica* reveals stress-adaptive metabolic responses. *Funct. Plant Biol.* **2021**, *48*, 717–731. [CrossRef]
39. do Nascimento Souza, J.F.; de Andrada, L.V.P.; de Sousa Martins, L.D.C.; Brito, A.M.S.S.; Silva, L.J.V.; de Lima Silva, I.D.; Vinhas, G.M.; da Silva, T.G.F.; do Nascimento Simões, A. Storage potential of cactus mucilage powder for incorporation into foods and production of biopolymeric films. *ACS Omega* **2024**, *9*, 43624–43634. [CrossRef]
40. Messina, C.M.; Arena, R.; Morghese, M.; Santulli, A.; Liguori, G.; Inglese, P. Seasonal characterization of nutritional and antioxidant properties of *Opuntia ficus-indica* [(L.) Mill.] mucilage. *Food Hydrocoll.* **2021**, *111*, 106398. [CrossRef]
41. Mannai, F.; Mechi, L.; Alimi, F.; Alsukaibi, A.K.D.; Belgacem, M.N.; Moussaoui, Y. Biodegradable composite films based on mucilage from *Opuntia ficus-indica* (Cactaceae): Microstructural, functional and thermal properties. *Int. J. Biol. Macromol.* **2023**, *252*, 126456. [CrossRef] [PubMed]
42. Junior, O.V.; Costa, L.D.; Cuello, R.E.G.; Ramos, A.Q.; Otero, D.M. Innovation in cacti extraction: Evaluating green methods for bioactive compounds. *Food Res. Int.* **2024**, *196*, 115046. [CrossRef]
43. Quintero-García, M.; Gutiérrez-Cortez, E.; Bah, M.; Rojas-Molina, A.; Cornejo-Villegas, M.d.l.A.; Del Real, A.; Rojas-Molina, I. Comparative analysis of the chemical composition and physicochemical properties of the mucilage extracted from fresh and dehydrated *Opuntia ficus indica* cladodes. *Foods* **2021**, *10*, 2137. [CrossRef] [PubMed]
44. Mannai, F.; Elhleli, H.; Mosbah, M.B.; Khiari, R.; Nacer, S.N.; Belgacem, M.N.; Moussaoui, Y. Comparative study of conventional and combined ultrasound-assisted methods on the quality of mucilage extracted from *Opuntia ficus-indica* cladodes. *Ind. Crops Prod.* **2024**, *214*, 118566. [CrossRef]
45. Felkai-Haddache, L.; Remini, H.; Dulong, V.; Mamou-Belhabib, K.; Picton, L.; Madani, K.; Rihouey, C. Conventional and microwave-assisted extraction of mucilage from *Opuntia ficus-indica* cladodes: Physico-chemical and rheological properties. *Food Bioprocess Technol.* **2016**, *9*, 481–492. [CrossRef]
46. Liguori, G.; Greco, G.; Gaglio, R.; Settanni, L.; Gentile, C.; Inglese, P. Mucilage-based and calcium ascorbate edible coatings improve postharvest quality and storability of minimally processed cactus pear fruit stored under passive atmosphere. *Horticulturae* **2023**, *9*, 15. [CrossRef]
47. Medina, O.J.; Patarroyo, W.; Moreno, L.M. Current trends in cacti drying processes and their effects on cellulose and mucilage from two Colombian cactus species. *Heliyon* **2022**, *8*, e12618. [CrossRef]
48. Nabil, B.; Ouaabou, R.; Ouhammou, M.; Saadouni, L.; Mahrouz, M. Impact of particle size on functional, physicochemical properties and antioxidant activity of cladode powder (*Opuntia ficus-indica*). *J. Food Sci. Technol.* **2020**, *57*, 943–954. [CrossRef]
49. Mannai, F.; Elhleli, H.; Yılmaz, M.; Khiari, R.; Belgacem, M.N.; Moussaoui, Y. Precipitation solvents effect on the extraction of mucilaginous polysaccharides from *Opuntia ficus-indica* (Cactaceae): Structural, functional and rheological properties. *Ind. Crops Prod.* **2023**, *202*, 117072. [CrossRef]
50. Elshewy, A.; Blando, F.; Bahlol, H.; El-Desouky, A.; De Bellis, P.; Khalifa, I. Egyptian *Opuntia ficus-indica* (OFI) residues: Recovery and Characterization of fresh mucilage from cladodes. *Horticulturae* **2023**, *9*, 736. [CrossRef]
51. Mora-Palma, R.M.; Martinez-Munoz, P.E.; Contreras-Padilla, M.; Feregrino-Perez, A.; Rodriguez-Garcia, M.E. Evaluation of water diffusion, water vapor permeability coefficients, physicochemical and antimicrobial properties of thin films of nopal mucilage, orange essential oil, and orange pectin. *J. Food Eng.* **2024**, *366*, 111865. [CrossRef]
52. De Farias, P.M.; de Vasconcelos, L.B.; Ferreira, M.E.S.; Pascall, M.; Tapia-Blácido, D.R. Nopal cladode (*Opuntia ficus-indica*) flour: Production, characterization, and evaluation for producing bioactive film. *Food Packag. Shelf Life* **2021**, *29*, 100703. [CrossRef]
53. Du Toit, A.; De Wit, M.; Fouché, H.J.; Taljaard, M.; Venter, S.L.; Hugo, A. Mucilage powder from cactus pears as functional ingredient: Influence of cultivar and harvest month on the physicochemical and technological properties. *J. Food Sci. Technol.* **2019**, *56*, 2404–2416. [CrossRef] [PubMed]
54. Dick, M.; Dal Magro, L.; Rodrigues, R.C.; de Oliveira Rios, A.; Flôres, S.H. Valorization of *Opuntia monacantha* (Willd.) Haw. cladodes to obtain a mucilage with hydrocolloid features: Physicochemical and functional performance. *Int. J. Biol. Macromol.* **2019**, *123*, 900–909. [CrossRef] [PubMed]

55. González Sandoval, D.C.; Luna Sosa, B.; Cristian, G.; Martínez-Ávila, G.; Rodríguez, H.; Avendaño Abarca, V.H.; Rojas, R. Formulation and Characterization of edible films based on organic mucilage from Mexican *Opuntia ficus-indica*. *Coatings* **2019**, *9*, 506. [CrossRef]
56. Blancas-Benitez, F.J.; Montaño-Leyva, B.; Aguirre-Güitrón, L.; Moreno-Hernández, C.L.; Fonseca-Cantabrana, Á.; del Carmen Romero-Islas, L.; González-Estrada, R.R. Impact of edible coatings on quality of fruits: A review. *Food Control* **2022**, *139*, 109063. [CrossRef]
57. Olawuyi, I.F.; Kim, S.R.; Lee, W.Y. Application of plant mucilage polysaccharides and their techno-functional properties' modification for fresh produce preservation. *Carbohydr. Polym.* **2021**, *272*, 118371. [CrossRef] [PubMed]
58. Kumar, N.; Neeraj. Polysaccharide-based component and their relevance in edible film/coating: A review. *Nutr. Food Sci.* **2019**, *49*, 793–823. [CrossRef]
59. Jiang, H.; Zhang, W.; Xu, Y.; Cao, J.; Jiang, W. Properties of pectin-based films from white-fleshed pitaya (*Hylocereus undatus*) peel waste as affected by montmorillonite. *Food Packag. Shelf Life* **2022**, *34*, 100952. [CrossRef]
60. Sharma, C.; Pathak, P.; Yadav, S.P.; Gautam, S. Potential of emerging "all-natural" edible coatings to prevent post-harvest losses of vegetables and fruits for sustainable agriculture. *Prog. Org. Coat.* **2024**, *193*, 108537. [CrossRef]
61. Gheribi, R.; Puchot, L.; Verge, P.; Jaoued-Grayaa, N.; Mezni, M.; Habibi, Y.; Khwaldia, K. Development of plasticized edible films from *Opuntia ficus-indica* mucilage: A comparative study of various polyol plasticizers. *Carbohydr. Polym.* **2018**, *190*, 204–211. [CrossRef] [PubMed]
62. Makhloufi, N.; Chougui, N.; Rezgui, F.; Benramdane, E.; Freire, C.S.R.; Vilela, C.; Silvestre, A.J.D. Bio-based sustainable films from the Algerian *Opuntia ficus-indica* cladodes powder: Effect of plasticizer content. *J. Appl. Polym. Sci.* **2021**, *138*, 50450. [CrossRef]
63. Gupta, D.; Lall, A.; Kumar, S.; Patil, T.D.; Gaikwad, K.K. Plant-based edible films and coatings for food-packaging applications: Recent advances, applications, and trends. *Sustain. Food Technol.* **2024**, *2*, 1428–1455. [CrossRef]
64. Kumar, L.; Tripathi, S.; Gaikwad, K.K. Valorization of cactus biomass to manufacture sustainable packaging films: Moisture sorption behavior and influence of citric acid as crosslinking agent. *Biomass Convers. Biorefinery* **2024**, *14*, 23031–23043. [CrossRef]
65. Zhang, Y.; Kong, Q.; Niu, B.; Liu, R.; Chen, H.; Xiao, S.; Gao, H. The dual function of calcium ion in fruit edible coating: Regulating polymer internal crosslinking state and improving fruit postharvest quality. *Food Chem.* **2024**, *447*, 138952. [CrossRef]
66. Gao, Q.; Xiong, T.; Li, X.; Chen, W.; Zhu, X. Calcium and calcium sensors in fruit development and ripening. *Sci. Hortic.* **2019**, *253*, 412–421. [CrossRef]
67. Liu, S.; Yang, J.; Yu, Y.; Liang, D.; Li, Y.; Si, X.; Zhang, Y. Multifunctional flexible carbon fiber felt@nickel composite films with core–shell heterostructure: Exceptional Joule heating capability, thermal management, and electromagnetic interference shielding. *Chem. Eng. J.* **2024**, *494*, 153221. [CrossRef]
68. Jiang, H.; Zhang, W.; Pu, Y.; Chen, L.; Cao, J.; Jiang, W. Development and characterization of a novel active and intelligent film based on pectin and *undatus*). *Food Chem.* **2023**, *404*, 134444. [CrossRef]

MDPI AG
Grosspeteranlage 5
4052 Basel
Switzerland
Tel.: +41 61 683 77 34

Foods Editorial Office
E-mail: foods@mdpi.com
www.mdpi.com/journal/foods

www.ingramcontent.com/pod-product-compliance
Lightning Source LLC
Chambersburg PA
CBHW061455210926
51850CB00008B/1725